SOLAR
ASTROPHYSICS

SOLAR ASTROPHYSICS

PETER FOUKAL
Cambridge Research and Instrumentation, Inc.
Cambridge, Massachusetts

A WILEY-INTERSCIENCE PUBLICATION
John Wiley & Sons, Inc.
NEW YORK / CHICHESTER / BRISBANE / TORONTO / SINGAPORE

Copyright © 1990 by John Wiley & Sons, Inc.

All rights reserved. Published simultaneously in Canada.

Library of Congress Cataloging-in-Publication Data:
Foukal, Peter.
  Solar astrophysics/Peter Foukal.
    p. cm.

  "A Wiley-Interscience publication."
  Includes index.
  1. Sun. 2. Sun—Research. 3. Astrophysics. I. Title.

QB521.F68 1989
523.7—dc20   89-32525
ISBN 0-471-83935-3   CIP

Printed in the United States of America

10 9 8 7 6 5 4 3 2 1

To Lizzie

Preface

The aim of this book is to provide a balanced and quantitative description of the sun and of techniques used in solar research for readers with a good physics background. The level of presentation is intended for undergraduate physical science majors, graduate students in astrophysics and space physics, and scientists working in related fields, such as atmospheric physics, where the influence of the sun's variability is of increasing interest.

Several good monographs and compendia of review papers have appeared in the past few years, and these provide up-to-date access to the areas of solar research where efforts are presently most concentrated. But many of the most obvious questions lie outside these specialized topics, and there, quantitative discussions can be difficult to find. I hope that this book will provide a comprehensive and unified account of what is presently known about our star, including much of interest that has been understood for many decades, but seems to require periodic rediscovery by more elaborate means.

Chapter 1 presents a historical introduction to certain key instruments, ideas, and people that have substantially increased our understanding of the sun. For the most part, solar telescopes and spectrometers operate on familiar principles. But the design of a few specialized and important solar instruments merits attention, and these are described in the context of their development.

Chapters 2, 3, and 4 lay out the basic concepts of radiative transfer, atomic spectroscopy, and plasma dynamics that are necessary for analysis of solar phenomena. Most of this material is derived here from elementary principles of photometry, of modern physics, and of electromagnetism covered in undergraduate curricula. More complete discussions are contained in the standard texts listed as additional reading.

Chapters 5 through 13 describe the sun, beginning with observations of the photosphere, which provide most of our knowledge about the main bulk of our star. The discussion then takes up the generation of solar energy in the deepest interior and in subsequent chapters moves back outward through the sun's convection zone to the nonthermally heated layers of the solar atmosphere and into the solar wind. Chapter 13 provides a discussion of the sun in the context of other stars, including information on its variable outputs of light, charged particles, and fields.

Lists of additional reading are provided at the end of each chapter to guide the reader to the most recent reviews and monographs or to important older material. The exercises range from the very easy to some that may require consultation with the author. The units used in this book are cgs because solar astronomers still favor them. Electromagnetic units (emu) are used because they seem most convenient in MHD expressions, which constitute the main application of electrodynamics in this text.

I thank the following colleagues for their careful reading and commenting of material as follows: J. Eddy (Chapter 1), E. Avrett (Chapter 2), B. Lites, G. Victor (Chapter 3), G. Van Hoven (Chapter 4), S. Keil (Chapter 5), R. Rood (Chapter 6), J. Leibacher (Chapter 7), R. Moore (Chapter 8), D. Wentzel (Chapter 9), D. Rust (Chapter 10), A. Van Ballegoijen (Chapter 11), R. Kopp (Chapter 12), and R. Gilliland (Chapter 13). Any remaining errors or omissions are of course my responsibility. I also thank the numerous colleagues who have sent me preprints, data, and illustrations. I have done my best to assign appropriate credit in captions and apologize for any oversights.

Several people have made indirect but important contributions to the writing of this book. I am grateful to my parents, Jaroslav and Jarmila Foukal, for their support and encouragement in my pursuit of an unusual profession. Mr. I. Kudrnac provided much appreciated early guidance and hardware to a young amateur astronomer not yet aware of potential NSF support. Finally, I owe a great debt of gratitude to my wife, Elisabeth, whose inspiration and understanding have made this book possible.

PETER FOUKAL

Nahant, Massachusetts

Contents

SOLAR
ASTROPHYSICS

1

Development of the Ideas and Instruments of Modern Solar Research

1.1 EARLY TELESCOPIC DISCOVERIES ON THE SUN

Scientific inquiry into the nature of the sun began around 1610 with the first telescopic observations of sunspots (called maculae or blemishes) by J. Fabricius, Galileo Galilei, C. Scheiner, and others in Western Europe. Naked-eye sightings of dark markings on the sun's disk had been reported in China and elsewhere at least 15 centuries earlier, and this evidence was almost certainly familiar to Galileo and his contemporaries. But the telescope showed that the westward motion of the spots was fastest near disk center and relatively slower near the east and west limbs. Using this observation, Galileo was able to argue from the projection effects expected in circular motion that the spots must be dark markings on the sun's surface rather than planets or other bodies in distant orbit and transiting its disk.

By 1630, the Jesuit astronomer Scheiner had used the spots' daily motion to accurately measure the sun's equatorial rotation period of 27 days as seen from the orbiting earth and to determine the 7° tilt of its equator relative to the ecliptic plane. Scheiner also detected the longer rotation period of spots at high solar latitudes and inferred that the sun's angular rotation rate decreases toward the poles. The well-known confrontation between Galileo's presentation of these early discoveries on the sun and the Church's doctrine of immaculate and stationary celestial bodies made a direct impact on the cosmology of the day that remains unique in the history of solar research.

To understand the fate of solar research in the two centuries following these promising beginnings, we should recall that the sun's nature was about as dimly perceived in Galileo's time as is the nature of quasars today. A useful estimate of its distance and size became available only after 1673, when the distance of Mars from Earth was measured using that planet's parallax observed between Paris and French Guiana. The Earth–sun distance then followed from J. Kepler's third law (of 1619) to within 10% of the correct value. The sun's mass relative to that of the Earth (determined from the orbital periods and distances of the Earth and moon) was not published by Isaac Newton until some 50 years after Galileo's first observations. The Earth itself was properly "weighed" by H. Cavendish's experiment only in 1798. Even the order of magnitude of the sun's surface temperature remained uncertain until after the 1880s, when the fourth-power relationship between the temperature

Fig. 1-1 Eyepiece projection scheme used by J. Hevelius to observe sunspots and faculae. From his *Selenografia*. By permission of the Houghton Library, Harvard University.

of an opaque body and its emittance came to be accepted through the experimental work of J. Stefan and the thermodynamic arguments set forth by L. Boltzmann.

Given the slow development of the physics and chemistry required to interpret solar observations, it is not surprising that for roughly two centuries after the first telescopic discoveries solar research consisted mainly of sunspot observations. Observers in the early 17th century used nonachromatic refractors with spatial resolution of 10 to 15 arc sec (a modern pair of binoculars will do better), and for detailed work the solar image was usually projected onto a screen (Fig. 1-1). Drawings by J. Hevelius (Fig. 1-2) and Scheiner show

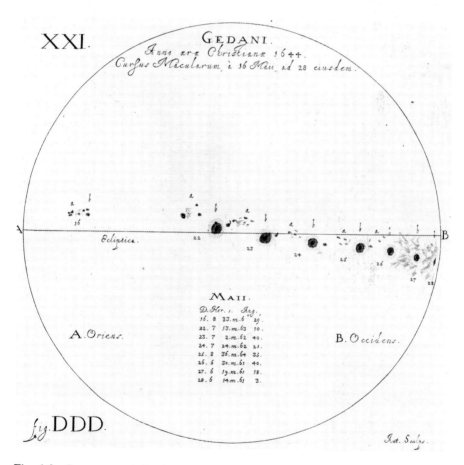

Fig. 1-2 Sunspots and faculae observed by J. Hevelius on May 16–28, 1644. The position of the large spot is shown on successive days as solar rotation moves it from east to west across the disk. Faculae (hatched) are seen only near the limb. From the *Selenografia*. By permission of the Houghton Library, Harvard University.

Fig. 1-3 Drawings of sunspots made by S. P. Langley on September 21, 1870 (top) and March 5, 1873 (bottom), using the 13-inch refractor at Allegheny Observatory. Note the light "bridge" across the umbra and the fine penumbral detail. Granulation can be seen in the photosphere outside the spots. From S. P. Langley, *The New Astronomy*, 1888.

that the darkest, roundish central umbra of the spot is often surrounded by a roughly annular lighter region, the penumbra. The relative positions and areas of individual spots within a group were observed to change from hour to hour.

These observers also noted bright irregular patches comparable in size to spots, that could often be observed in association with sunspots near the limb. Scheiner called them faculae or little torches. The contrast of faculae is much lower than that of spots, and they are not visible at all in white light near disk center. Both spots and faculae can often be observed to recur on subsequent solar rotations, implying lifetimes of up to several months. Figure 1-2 shows the march of faculae and a sunspot across the disk from day to day.

The development of more compact and achromatic refractors in the 19th century eventually made possible visual observations of greatly increased resolution, such as the superb sunspot drawings by S. Langley shown in Fig. 1-3. But the records compiled by 17th-century observers continue even now to provide new insights into the sun's behavior. For instance, we find that spots became rare after about 1645 and remained so until around 1715. This broad minimum of sunspot occurrence was rediscovered by E. W. Maunder in the late 19th century and has since been named after him. Since the processes responsible for sunspot generation and for the solar differential rotation may be related, efforts have also been made to reconstruct the solar rotation rate and its latitude profile from the rotation of spots observed immediately before 1645 and after 1715, to check for a possible change during the Maunder Minimum. The data used included Hevelius' drawings, such as Fig. 1-2. More than 300 years after their discovery, the mechanisms responsible for the generation of sunspots, the relation of spots to faculae, and the differential rotation of the sun's surface continue to present problems at the forefront of solar research.

Toward the end of the 18th century, advances in stellar astronomy associated in large part with the work of William Herschel gradually led to the view that the sun is a star similar to the thousands of bright points seen in the night sky, only located much closer to the earth. How much closer became clear in 1837 with the reliable measurement of the stellar parallax of 61 Cygni by F. Bessel. Study of the sun's physical structure and chemical composition as a valuable key to understanding the stars provided a powerful impetus to solar research in the 19th century that continues to the present day.

1.2 THE SPECTROSCOPE AND PHOTOGRAPHY

The spectroscope was the new instrument that enabled astronomers to carry the idea of the sun as a star to useful conclusions. After initial studies on the dispersion of solar light by Newton and others in the 17th and 18th centuries, J. Fraunhofer built the first spectroscope useful for quantitative analysis in 1814 and used it to measure the positions of over 500 of the remarkable dark lines seen in the solar spectrum. Fraunhofer's original spectrum is illustrated

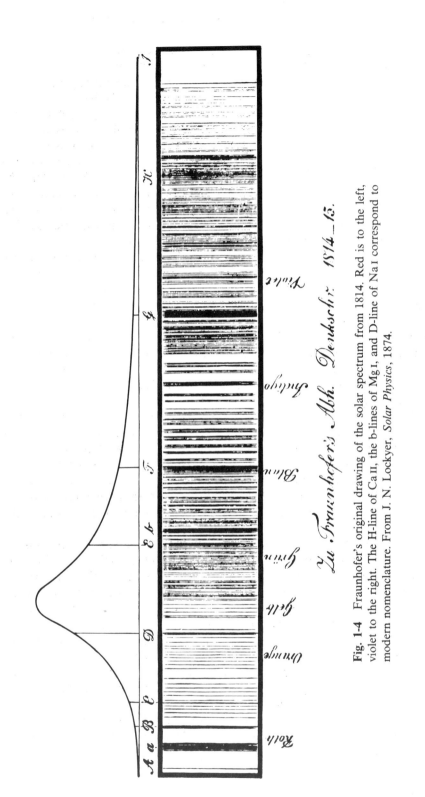

Fig. 1-4 Fraunhofer's original drawing of the solar spectrum from 1814. Red is to the left, violet to the right. The H-line of CaII, the b-lines of MgI, and D-line of NaI correspond to modern nomenclature. From J. N. Lockyer, *Solar Physics*, 1874.

in Fig. 1-4. A certain amount of experimental work on the spectra of the sun, stars, and incandescent laboratory sources ensued. But it required the insight of G. Kirchhoff almost 50 years later to discover the simple laws governing the emission and absorption of light from solid and gaseous bodies.

The most important impact of Kirchhoff's work on solar research lay in his demonstration that the thousands of dark absorption lines observed in the spectrum of the solar disk have, in general, a one-to-one correspondence to bright emission lines in incandescent vapors observed spectroscopically in the laboratory. From these foundations, Kirchhoff and others showed that the sun's surface consisted of hot gases made up of elements found on Earth. This discovery surely ranks with the most far-reaching findings of natural science, since it gave valuable support to the rather daring cosmological principle that all corners of the universe obey the same laws of nature.

The development of photography in the 1840s made possible more objective recording of structure seen on the sun's disk in white light. One of the first results was to demonstrate that the darkening of the solar disk toward the limb claimed from visual observations was real. This limb darkening of the white-light-emitting layer in the sun's atmosphere (later called the photosphere) was eventually explained by K. Schwarzschild, in a classic paper in 1906. Schwarzschild used newly developed ideas of radiative transfer to show that the observed limb darkening by a factor of roughly 2.5 can be understood in terms of the decreasing temperature outward through the photosphere if energy transport through this layer is by radiation.

By the 1870s, photographs such as Fig. 1-5 were obtained of the pattern of small mottles, called granulation, that cover the photosphere. The bright roundish granules (also seen in the drawings of Fig. 1-3) are typically a few arc seconds in diameter and are separated by narrower dark intergranular lanes. It was later recognized that these granules resemble the pattern of convective cells found in H. Bénard's experiments of 1901 on fluids heated from below, so that they probably represent hot rising gas elements convecting heat to the sun's surface.

This association of granules with convection is generally accepted on the basis of modern observations and simulations, but the validity of the early evidence discussed above has been questioned. The remarkable regularity of the granulation seen in Fig. 1-5 seems to arise at least in part from a recently discovered pattern of fine cracks in the emulsion. The cells seen in Benard's experiments are considered to have been caused by surface tension effects rather than by convection. This ironic episode shows how compensating errors in observations and interpretation can underlie a correct result.

It appears that, although most of the energy is carried by radiation through the photosphere, the granulation seems to represent the "overshooting" of convective elements from the deeper, opaque layers of the sun where convection carries most of the heat flux. The structure of a deep convection zone, extending roughly 30% of the sun's radius below the photosphere, was first worked out in 1930 by A. Unsöld.

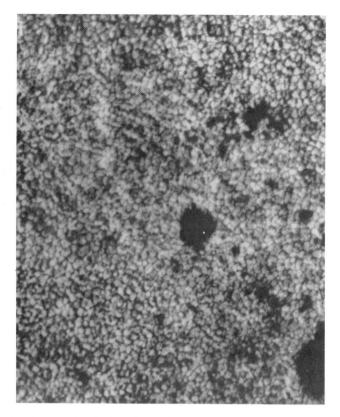

Fig. 1-5 A photograph of the photosphere, showing some spots and pores and perhaps granulation, taken in Paris by J. Janssen on July 5, 1885.

1.3 SOLAR-TERRESTRIAL RESEARCH AND THE NEW ASTRONOMY

A different line of inquiry that had great impact on 19th-century solar research, and continues to provide important motivation today, was initiated by the discovery of a significant 11-year periodic variation in the number of sunspots by H. Schwabe in 1843. Soon afterward, R. Wolf and E. Sabine connected these results to the modulation in geomagnetic storm occurrence discovered earlier by J. Lamont and by Sabine in research on magnetic disturbances in Germany and in Canada. This was the first good evidence for a variable solar influence on Earth, and the beginning of solar-terrestrial research.

More insight into the specific aspect of solar activity responsible for geomagnetic disturbances came after the first observations of great solar eruptions within spot groups, reported in white-light observations of the photosphere in 1859 by R. Carrington and R. Hodgson. The occurrence of a great geomagnetic storm within less than a day suggested that these "flares" on the sun, rather than the sunspots themselves, were directly responsible for the terrestrial disruptions. The impressive correlations found thereafter between geomagnetic storms and disk passage of large spot groups gradually led to our relatively recent acceptance of the view that corpuscular or electromagnetic wave emissions travel from the sun to Earth.

Observations during total solar eclipses provided the only method to study the solar atmosphere outside the bright photospheric disk until the 1860s. The ability to predict eclipses dates back to Babylonian times, and improvements in technique introduced by the great astronomer Hipparchus of Rhodes in the second century B.C. increased accuracy to a few hours. The extended white-light corona had been noted in the early 18th century, along with naked-eye sightings of the pinkish coronal condensations that we now call prominences. Eclipse watchers of about that time had also remarked on the red light of the narrow layer called the chromosphere, visible briefly just as the moon covers and uncovers the photosphere at beginning and end of totality.

During the American Revolutionary War hostilities were suspended in a part of the state of Maine for one day to permit the Reverend Prof. Williams and his collaborators from Harvard College to observe the eclipse of 1780. This expedition was reportedly the first to note the phenomenon of Bailey's beads, which was rediscovered at the 1836 eclipse by the English amateur astronomer F. Bailey, and is caused by solar light shining through the lunar valleys at the edge of the moon's disk.

But until the event of 1836, few physical observations of the sun were made, although the times of eclipse contacts were systematically recorded. At eclipses beginning in 1842, some of the foremost astronomers of the day set up experiments intended to determine whether the corona, prominences, and chromosphere were solar or whether they originated in the Earth's atmosphere. Good photographs obtained at two sites 250 miles apart during the 1860 eclipse established that prominences were indeed some kind of cloud or condensation in the sun's atmosphere.

Kirchhoff's work on the interpretation of spectra encouraged spectroscopy at the 1868 eclipse. Here J. Janssen observed the yellow (D_3) line λ 5876 in prominences, later found to be radiated by neutral helium, an element not discovered on Earth until 1895 by the chemist W. Ramsay. The green coronal emission line at λ 5303 was also first observed during the eclipse of 1869. This was the first of 24 coronal emissions to be discovered in the visible spectrum which could not be identified with any known element, and for many decades posed a vexing problem for spectroscopy. The explanation came in 1941, when the spectroscopist B. Edlèn, proceeding from earlier evidence put forward by W. Grotrian, showed how these lines ascribed to the mystery element "coronium" actually arose from forbidden transitions between low-lying fine structure levels of highly ionized heavy atoms such as Fe and Ca. In 1871, Janssen identified greatly broadened Fraunhofer absorption lines in the coronal continuum spectrum and showed that they originated in scattering of the photospheric Fraunhofer lines by coronal particles. He thus established that the white-light corona was solar in nature. Advances in understanding such as these, achieved through the application of spectroscopy to the sun and stars, came to be recognized as the New Astronomy, or astrophysics as we now call it.

The 1868 eclipse also led to the first use of monochromatic imaging of solar features, through the invention of the "spectrohelioscope" to observe prominences. Most of the emission from prominences occurs in a few narrow spectral lines, so the principle of this instrument was to use a prism spectroscope to form an image in a narrow spectral passband including one of these lines. Most of the continuum light of the solar disk scattered in the Earth's atmosphere was thus excluded. Using this technique, J. Janssen in France, and soon thereafter, J. Lockyer in England, were the first to observe prominences and also the chromosphere, outside of eclipse. Drawings of prominences and of the chromosphere made by W. Huggins and A. Secchi using spectrohelioscopes in the 1870s revealed structural details such as the chromospheric jets (seen at the limb in Fig. 1-6) that were rediscovered after 1945 and named spicules. These visual observers also documented the intricate shapes of prominences, such as the loops illustrated in Fig. 1-6, and their often rapid motions. Around this time, comparative study of the eclipsed corona near sunspot cycle maximum and near the minimum (see Fig. 1-7) first revealed the differences in morphology characteristic of these periods of high and low activity. The intricate loops and plumes visible near maximum activity led to the first suspicion that electromagnetic forces play a role in the dynamics of the solar atmosphere.

At the eclipse of 1870, C. Young observed the "flash" spectrum of the chromosphere, which blazes out for a few seconds immediately before and after totality. In 1913, S. Mitchell showed that this spectrum could not originate in scattering of the Fraunhofer lines since it contains high excitation lines weak or absent in the photospheric spectrum. His reasoning also led to the first suggestion that the chromospheric temperature might exceed that of the photosphere. The identification in 1941 of the "coronium" lines with forbidden transitions in Fe x, Fe xiv, Ca xv, and other highly stripped ions proved that the temperature of coronal gases lying above the chromosphere was very high, over one million degrees. These high temperatures continue to stimulate research into the processes of nonthermal heating in the chromosphere and corona. Dissipation of waves or electric currents seems to be required since thermal heating of these layers by conduction, convection, or radiation from the cooler photosphere would violate the second law of thermodynamics.

The structure of the chromosphere on the disk was first revealed by observations with the spectroheliograph, developed independently in 1891 by H. Deslandres and by G. Hale. This instrument combined the narrow passband of the spectrohelioscope with the advantages of permanent and accurate recording made possible by the photographic plate. Its principle of operation

Fig. 1-6 Coronal loops drawn by A. Secchi from spectrohelioscope observations in the Balmer alpha line on October 5, 1871. The thin radial structures at the limb are chromospheric spicules. From C. Young, *The Sun*, 1895.

11

Fig. 1-7 A drawing of the corona of 1868 observed at Mantawalok-Kekee, Malaysia. This was the eclipse at which helium was first detected spectroscopically. Drawings of the same eclipse by different observers usually showed considerable variation. From J. N. Lockyer, *Solar Physics*, 1874.

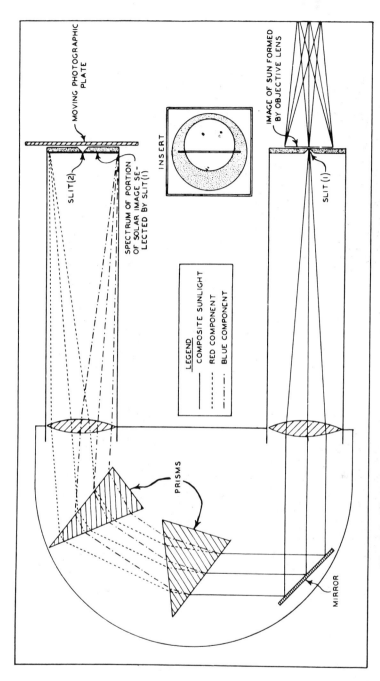

Fig. 1-8 The principle of the spectroheliograph. Slit 2 is adjusted in position and width to select a wavelength and passband width of interest. The photographic plate is then translated across slit 2 at exactly the same rate as the solar image is translated across slit 1. By permission of Mt. Wilson and Las Campanas Observatories, Carnegie Institution of Washington.

Fig. 1-9 The presence of intense magnetic fields in sunspots was suggested to Hale by vortex-like shapes in the chromospheric structures shown in this Hα spectroheliogram taken September 9, 1908. By permission of Mt. Wilson and Las Campanas Observatories, Carnegie Institution of Washington.

(Fig. 1-8) relies on stepping the entrance slit of a spectrograph acrc disk at exactly the same rate that the exit slit is moved across the pł plate. A full image of the solar disk can then be produced in any s_l or continuum region by setting the grating angle so that the sp_l forms an image of the entrance slit at the exit slit in the desired wavelength.

This instrument made it possible to photograph the spatial structure of the chromosphere on the disk in the resonance line of singly ionized calcium at 3934 Å, known as the Ca II K-line. Later, observations were obtained in the strongest Balmer absorption line Hα at 6563 Å when red-sensitized emulsions became available in 1908. Hale's spectroheliograms in these lines, one of which is illustrated in Fig. 1-9, revealed the intricate dark structures of chromospheric fibrils and mottles, which later work related to the spicules that are seen in emission at the limb in Fig. 1-6. Hale's spectroheliograms also showed the bright chromospheric plages seen when faculae are observed on the disk in the cores of the Hα and Ca K-lines. Hale and Deslandres were nominated jointly for the Nobel prize for their invention of this remarkable instrument.

1.4 SOLAR CHEMICAL COMPOSITION AND ENERGY GENERATION

Kirchhoff made the first solar identifications of absorption lines of sodium, iron, magnesium, and other heavy elements around 1860. The presence of hydrogen in the sun's composition was revealed about ten years later by A. Ångström and others. H. Rowland's development of excellent diffraction gratings led to his publication in 1897 of a superb atlas of the solar spectrum between atmospheric cutoff around λ 2975 and the visual limit near λ 7331. From these spectroscopic data, he was able to increase the number of elements identified in the sun to 36 by the end of the century.

The major advance to quantitative analysis of the solar line spectrum came in the 1920s after the development of N. Bohr's atomic theory and of M. Saha's ionization equation. This progress in atomic and statistical physics enabled H. Russell in 1928 to use Rowland's eye-estimates of line intensities to establish the rough relative abundances of elements in the solar atmosphere. The surprising result that the sun consisted mainly of hydrogen was by no means immediately accepted.

Around 1920, A. Eddington explained how hydrogen burning to helium might provide the energy to fuel the sun's luminosity. Building on earlier ideas, he suggested that the hydrogen fuel might be sufficient to account for the Earth's age implied by fossils. This age had been recognized by then to greatly exceed the 20×10^6 year time scale that H. van Helmholtz had estimated in 1854 for conversion of the sun's gravitational potential energy to radiation. Helmholtz's theory that a slow contraction of the sun provided the energy required to fuel its heat and light output remained the only reasonable explanation of the source of the sun's power output until about the turn of the

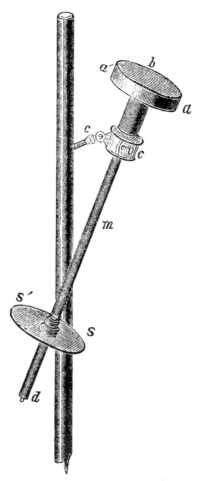

Fig. 1-10 The radiometer used by S. Pouillet in 1837 for measurement of the total solar irradiance. The hollow metal box marked *a-a'* was filled with a known volume of water. The upper surface *b* was blackened to absorb sunlight. The shaft *m* contained a thermometer-agitator to read the water temperature increase caused by exposure of the box to sunlight for a known interval of time. The measurement was absolute since the volume and heat capacity of the water, the blackened area, and the time interval were known.

century. The rate of steady contraction implied in the present epoch would be well below the detection limit for measurements of change in solar diameter.

Two sets of nuclear fusion reactions, called the carbon-nitrogen cycle and the proton-proton chain, were put forward in 1938 by H. Bethe, C. Critchfield, and C. von Weiszacker to specify how the nuclear burning might proceed. The first direct test of these mechanisms in the sun is provided by the neutrino experiment of R. Davis, located deep in the Homestake gold mine in South Dakota. This experiment has been running since 1969 and detects a significantly lower neutrino flux than predicted by standard solar models. The discrepancy might arise for a number of reasons discussed in Chapter 6.

The success of nuclear fusion in explaining the luminosity of stars now rests largely on the good agreement between the observed and calculated distribution of stars in the Hertzsprung–Russell plot of stellar luminosity against color. Impressive regularities found in the relative solar abundances of elements heavier than helium are also well explained through the theory of stellar evolution put forward by M. Burbidge, G. Burbidge, W. Fowler, and F. Hoyle in 1957. Their explanation is based on successive fusion episodes, which build up the elements lighter than iron.

The sun's enormous power output was demonstrated by the first good calorimeter measurements of the total solar irradiance or "solar constant" obtained in 1837 by S. Pouillet in France and by J. Herschel working in South Africa. Pouillet's radiometer is shown in Fig. 1-10. Measurements of the spectral distribution of solar power were advanced by S. Langley's invention of the thermoconducting bolometer around 1881. With this sensitive detector, Langley mapped the sun's spectrum to 5.3 μm. He also showed the importance of molecules in the Earth's atmosphere, such as H_2O, CO_2, and CO, in the seasonally variable terrestrial absorption of solar infrared light.

C. Abbot, working with Langley at the Smithsonian Institution, put into use better radiometers for the solar constant measurements. Together with his collaborators, he then carried out an epic program of daily solar constant observations between 1923 and 1952 at a worldwide chain of Smithsonian mountain stations. Abbot's measurements demonstrated that the solar radiation transmitted by the Earth's atmosphere was constant to better than 1% over several decades. But the absolute value of the solar constant remained uncertain at the 5% level until the sun's ultraviolet flux below 3000 Å was properly measured by rockets in the 1950s.

1.5 THE MT. WILSON ERA OF LARGE TELESCOPES

Using first the horizontal Snow telescope, and then the 60-ft and 150-ft tower telescopes built at Mt. Wilson near Los Angeles in 1907 and 1912 (Fig. 1-11), Hale and his collaborators made the first solar observations using high-dispersion spectrographs and large image scale to achieve good spectral and spatial resolution on the sun. Following the example of J. Lockyer, E. Frankland, and

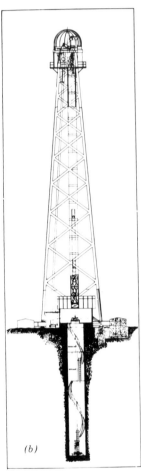

Fig. 1-11 (Left) The 150-ft tower telescope still in active use on Mt. Wilson and (right), cutaway view of the 150-ft solar tower. The deep pit below ground level contains the spectrograph, with its grating at the bottom. By permission of Mt. Wilson and Las Campanas Observatories, Carnegie Institution of Washington.

other chemists, they compared their solar spectra with those they obtained of laboratory plasmas at various temperatures and pressures. The laboratory at Mt. Wilson, where these comparisons were carried out, is shown in Fig. 1-12. The most important results had to do with the physics of sunspots and of the solar magnetic field. Hale's work confirmed (by comparison of the umbral spectrum with arc and spark spectra in the laboratory) that the dark umbra is cooler than the brighter photosphere. This fact was less obvious than it seems now, since the theory of radiative transfer in gaseous atmospheres was only beginning to emerge in the work of A. Schuster, K. Schwarzschild, and others.

Fig. 1-12   Interior of early physical laboratory on Mt. Wilson, showing spectrograph and magnet used in studies of the Zeeman effect. By permission of Mt. Wilson and Las Campanas Observatories.

High-dispersion sunspot spectra also showed a puzzling widening and sometimes splitting of the lines in umbrae. Hale's Hα spectroheliograms in 1908 had shown vortex-like chromospheric dark filaments wound around the spots (see Fig. 1-9) resembling the configuration of iron filings around a magnet, which led him to suspect strong umbral magnetic fields. Proceeding from P. Zeeman's (1896) analysis of the magnetic field splitting of spectral lines, Hale obtained umbral spectra through a Nicol prism and found the expected zigzag pattern caused by the opposite circular polarization of the Zeeman-split line-wing components in sunspot fields of up to 3,500 G.

This first discovery of extraterrestrial magnetic fields set the stage for the systematic study by Hale and S. Nicholson of the laws of sunspot polarity and their changing behavior over the spot cycle. Their work showed that the oscillation of sunspot number recognized since 1843 constituted half of a 22-year cycle of sunspot magnetic polarities. The most obvious aspect of this 22-year cycle is the reversal of polarity of the east and west spots in active regions that accompanies the onset of a new 11-year cycle in spot number.

Hale's further evidence for a general polar magnetic field of intensity roughly 50 G with opposite sign at the north and south poles, was never fully

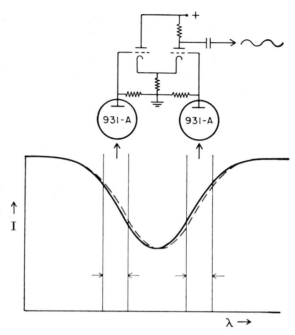

Fig. 1-13 Schematic diagram showing the principle of operation of the original Babcock magnetograph. Profiles of an absorption line for the two states of opposite circular polarization are shown solid and dashed. The symbols I and λ refer to intensity and wavelength respectively. The two detectors are type 931-A photomultipliers. By permission of the University of Chicago Press.

accepted. Only after the development of the photoelectric magnetograph by H. D. Babcock and H. W. Babcock in 1951 was a weak 1-5 G polar field reproducibly detected in spatial averages over the polar regions and observed to change in polarity every 11 years along with the spot fields.

The principle behind this very important instrument can be sketched as follows. Two exit slits of a high-dispersion spectrograph are placed in the red and blue wings of a magnetically sensitive photospheric absorption line, whose intensity profile is shown in Fig. 1-13. The solid and dashed $I(\lambda)$ curves denote the profiles from a magnetic region of the sun when left- or right-hand circularly polarized light is alternately admitted to the spectrograph. Since the two wings of a normal Zeeman triplet broadened by a magnetic field are oppositely circularly polarized, admitting alternate senses of polarization by rotating a quarter-wave plate placed in series with a Nicol prism, moves the profile back and forth. The difference in intensity measured between the two photomultipliers placed in the wings changes from one profile to the other. This variation of the difference signal is amplified and recorded. Its amplitude is roughly proportional to the *net* magnetic flux (the effect of opposite

Fig. 1-14 An early photoelectric magnetogram obtained with the Babcock magnetograph. Each trace was made by a separate scan across the sun. A deflection equal to the distance between traces is produced by a field (averaged over the spectrograph entrance slit) of about 10 G. By permission of the University of Chicago Press.

magnetic polarities will tend to cancel) in the solar region whose light is admitted through the entrance slit of the spectrograph.

The magnetograph revealed for the first time, the distribution of photospheric magnetic fields outside of spots and faculae (Fig. 1-14). Observations with improved versions of this instrument by the 1960s showed a network pattern of magnetic fields of characteristic cell dimensions 20,000–40,000 km covering the quiet photosphere. This magnetic network was found to coincide well with the chromospheric network defined by the bright emission in Ca K spectroheliograms outside of active regions.

The suggestion of a velocity shift of spectral lines by C. Doppler in 1842 eventually led to the first measurements of the relative blue- and redshifts of Fraunhofer absorption lines at the sun's east and west limbs and thus to spectroscopic measurement of the photospheric gas rotation rate. These measurements (by N. Dunér, H. Vogel, and C. Hastings between 1870 and 1890) showed a similar rotation rate and equatorial acceleration to the accurate sunspot rotation rates obtained by Carrington in 1865. But the plasma rates could be extended to latitudes $\pm 75°$, whereas spots could be used as tracers only over the active latitudes to roughly $\pm 40°$. Motions around sunspots also proved interesting. J. Evershed's 1909 observations in Kodaikanal, India, revealed a strong horizontal outflow extending several spot diameters from the umbral edge. This Evershed effect has yet to be incorporated into a satisfactory dynamical theory of spot coolness and stability.

The improved accuracy of line wavelength determinations achieved by upgrading the Rowland atlas in 1928 to use the cadmium red line interferometric wavelength standard led to increased awareness of the limb redshift of Fraunhofer lines relative to their disk center position. The nature of this shift, noted earlier by Hale and his collaborators at Mt. Wilson, generated lively debate. A small redshift relative to the laboratory wavelength, of constant magnitude across the disk, was expected from general relativity. But the conspicuous center-to-limb redshifts of up to 0.5 km s^{-1} required a different explanation. Only in the 1950s was a viable mechanism put forward in terms of the net Doppler blueshift near disk center caused by the bright upflowing granules, relative to the smaller contribution of dark downflowing material of intergranule lanes.

In the late 1950s, R. Leighton devised a technique for studying periodic velocity signals on the sun, using the spectroheliograph at the 60-ft Mt. Wilson tower. The slit was placed in the wing of a line, and the spectroheliograph was scanned across the sun and back taking P minutes in each direction. A negative of the scan in one direction was then overlaid in perfect spatial registration on a transparent positive of the scan in the opposite direction. Points on the sun that happened to be Doppler shifted in the same sense on both the negative and positive were registered strongly as dark or bright on the resultant print, whereas points whose Doppler shift changed sign during the time, cancelled to a neutral gray in the Doppler summation print. The variable time delay across the summed print, was ideally suited to detecting oscillations of unknown period in the solar velocity field over the range of periods shorter than the maximum time delay $2P$.

Using this technique, Leighton and his graduate students, R. Noyes and G. Simon, detected and studied the 5-min oscillation of the photosphere. The fruitful interpretation of the oscillation as standing acoustic waves trapped in resonant cavities below the photosphere was made in the early 1970s, when the observations of F. Deubner and the calculations of R. Ulrich independently showed the rich mode structure of these oscillations when their oscillatory power is plotted in the plane of spatial wavenumber k versus temporal frequency ω. The position of the modes in the k-ω plane, their width, and their splitting yield valuable information on the structure of the sun's interior, on the dynamics of wave damping by turbulent solar convection, and on the sun's internal rotation profile.

The Doppler cancellation techniques developed by Leighton also showed that the network magnetic fields coincided roughly with the edges of an outflowing horizontal velocity field of roughly 0.5 km s^{-1} centered within each of the cells. These 20,000–40,000 km diameter velocity cells were called supergranular convection. The strong magnetic fields measured at their boundaries in magnetograms were thought to be intensified from weaker fields pushed to the edges by the ram pressure of the convergent supergranule flows. More recent magnetograms obtained with much higher spatial resolution have shown that the network magnetic fields consist of individual vertical flux tubes in which the magnetic field intensity is of order 10^3 G, thus similar to that of a spot umbra.

1.6 ADVANCES IN CORONAL PHYSICS AND IN THE THEORY OF SOLAR ACTIVITY

Instruments used for coronal studies did not progress much for 50 years after the introduction of the spectrohelioscope at the 1868 eclipse. Only a few minutes per year were available to study the corona outside of the brightest prominences (which were visible outside of eclipse through a spectrohelioscope) using portable equipment set up at remote sites. The idea of blocking out the photospheric disk at the image plane of a telescope had occurred to many, but the difficulties of overcoming scattered light were not solved until B. Lyot implemented some ingenious precautions and built the first working coronagraph at the Pic du Midi in 1930.

The principle of the coronagraph is illustrated in Figure 1-15. The photospheric disk is imaged on a convex occulting disk at the prime focus of a refracting telescope and reflected out of the beam. One of Lyot's important advances was in placing a field lens (D) behind the occulting disk (C) to image the primary objective lens (B) onto a circular diaphragm (E). This diaphragm was made somewhat smaller than the objective lens image, so that rays originating at the edge (A) of the lens were blocked. Lyot had found that diffraction from the edge of the lens contributed a large fraction of the total scattered light.

Through this arrangement, only coronal light gathered by the central part of the objective was used to form the final image of the corona. To reduce scattering in the lens glass itself to a minimum, Lyot used a single lens of bubble-free glass instead of an achromatic doublet, which causes multiple internal reflections from the glass-air surfaces (unless modern antireflection coatings unavailable in 1930 are used). Finally, he is reported to have taken

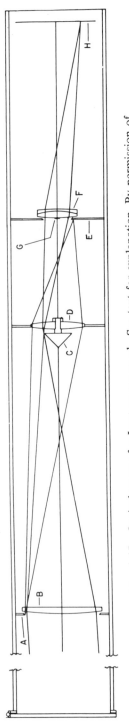

Fig. 1-15 Optical parts of a Lyot coronagraph. See text for explanation. By permission of the University of Chicago Press.

Fig. 1-16 The 16-inch aperture coronagraph at Sacramento Peak Observatory in New Mexico. This instrument is in active use for coronal studies and other observations requiring low scattered light. National Solar Observatories photograph.

care to clean the objective with oil from the tip of his nose, applied with a well-laundered handkerchief. A large coronagraph is shown in Fig. 1-16.

Lyot also co-invented the birefringent filter (with Y. Öhman of Sweden), which enabled him to take monochromatic pictures of coronal and chromospheric phenomena far more rapidly than could be done with a spectroheliograph. The principle of the filter relies on the interference produced between the ordinary and extraordinary rays passing through a birefringent crystal such as calcite. Since the refractive index of the crystal differs for the two rays, they emerge with a phase difference that depends (for a given calcite thickness) on the wavelength of the light. Transmission is minimal for wavelengths at which the block of calcite yields a phase difference of π. A practical filter consists of several calcite elements, the thinner elements remove unwanted transmission sidebands at wavelengths other than that selected. The birefringent filter can be built to produce narrow passbands of 0.25 Å or less, with a few percent transmission, uniform over a field of view of roughly 1°. Almost all high-resolution photographs and time-lapse films of the sun in narrow passbands are made using variants on this type of filter.

The results of these instrumental advances were rewarding; Lyot, R. McMath, and others made time-lapse movies of coronal structures at the limb in the green and red coronal lines λ 5303, λ 6374, or in Hα. These films followed the slow changes in the intricate shapes of quiescent prominences, and the rapid changes in active prominences near spots, including the predominant downflow of cool material from the corona called coronal rain. They also captured the occasional magnificent brightening of the chromosphere in an active region during the eruption of a solar flare. Sometimes the films showed the outward acceleration of material in flare-related surges (where the ascending material is later seen to return downwards along the same path) and sprays (where its return is not observed). Then the flare brightening expands upward, eventually forming intricate arches of postflare loops similar to those seen in Fig. 1-6. Very rarely, a flare is so bright as to be seen in white light, as was the case of the first flare ever reported in 1859.

The energy released in flares, and the intricate structure of prominences, represent two aspects of the control that electromagnetic forces exert over the solar plasma. Important advances in plasma electrodynamics were achieved through the work of H. Alfvén, T. Cowling, L. Biermann, T. Gold, and others in the 1940s and 1950s. Alfvén showed in 1942 that, under the conditions of reasonably good electrical conductivity and enormous scale commonly encountered in cosmic plasmas, their self-induction is exceedingly high and the motions of magnetic field lines and plasma particles can be usefully visualized as "frozen" together. That is, the plasma can move along the field lines but not perpendicular to them. This makes it easier to visualize, for example, how horizontal magnetic fields might support the cool dense material of prominences against gravity.

The energy density of the field measured in spots and plages was known to be amply sufficient to provide the energy for flares. But here the self-induction of the plasma intrudes, since it limits the rate of electric current discharge. The mechanism of energy release in flares is still not clear, but some lines of evidence indicate that the resistive dissipation of electric currents into heat takes place in very thin layers, perhaps a kilometer or less. The self-induction is limited here by the diminished thickness, and intense compression may increase the plasma resistivity through creation of turbulence. Alfvén also showed in 1942 that plasma motions perpendicular to the field could give rise to transverse oscillations whose restoring force is the tension in the magnetic field lines. The propagation and dissipation of Alfvén waves combined in various ways with longitudinal pressure waves have been widely studied in attempts to explain how the high temperature of the chromosphere and corona are maintained.

Using these new concepts of magnetohydrodynamics, and regularities he observed in solar magnetic field behavior, H. W. Babcock put forward in 1961 a highly successful phenomenological model of the sunspot cycle. In this model the differential rotation (in both latitude and depth) of the sun's convection zone provides the power to wind up (and thus intensify) a weak, initially poloidal field. This intensification leads to the 11-year oscillation in sunspot number and also produces the global change in direction of the field.

The physical processes behind Babcock's model were based on new ideas in dynamo theory suggested by Cowling and others. E. Parker's calculations in 1955 on magnetic buoyancy, which tends to make fields appear at the photosphere, and field helicity, which is necessary to explain the field reversal process, put these ideas on a firmer basis. Leighton added to the model in 1964 by showing how the evolution of the supergranular velocity cells he discovered could cause the magnetic fields to carry out a random walk at the photosphere, thus leading to a general dispersion of the field producing a relaxation of the oscillator.

A regenerative dynamo mechanism is not required to explain the existence of a solar magnetic field. Unlike the Earth, the sun is a large enough conducting sphere that its self-induction would prevent the ohmic decay of any primordial field that might have been present at its birth 4.5 billion years ago. But some kind of oscillator is required to understand the regular 11-year and 22-year variability of the solar field. To produce more detailed models, a better understanding of solar convection and its interaction with magnetic fields is required, and numerical simulations on large computers are beginning to provide useful insights.

1.7 OBSERVATIONS AT RADIO, ULTRAVIOLET, AND X-RAY WAVELENGTHS

The sun's emission at radio frequencies was first identified in 1942 by J. Hey in the United Kingdom and by J. Southworth in the United States using World War II radar receivers. For reasons of military secrecy, their results were not published until after 1945. Rapid progress was then made in categorizing solar radio signals both spatially and spectrally in Australia, the United Kingdom, and the United States. The thermal continuum background at centimeter and decimeter wavelengths was found to be slowly varying with the rotation and evolution of plages, and various types of bursts were classified, together with their relation to flares and other disturbances of the active sun. Meter wave emissions from nonthermal sources were studied using rapid sweeping in frequency. Observations in meter waves could track plasma ejections from flares and eruptive prominences as they moved outward through the coronal plasma. Centimeter wave observations provided some of the first data used to estimate the thickness of the transition region, the height interval above the ten-thousand-degree chromosphere in which plasma temperatures rise to the million-degree values of the corona.

The low spatial resolution achievable with single radio dishes because of diffraction, and the relatively wide range of plasma temperatures contributing to the radio emission at a given frequency, led solar astronomers to seek other techniques in the ultraviolet and X-ray spectral regions for more detailed and easily interpreted data on the corona and transition region. Recent advances in the use of radio interferometry, and in aperture synthesis (utilizing many dishes electronically locked together), have changed the situation dramatically. The highest spatial resolution imaging (approximately 0.1 arc sec or 70 km on the sun) could, in principle, be achieved in centimeter waves using aperture synthesis at the Very Large Array (VLA). Unfortunately, it is difficult to achieve coverage of extended solar features in reasonable times at these spatial resolutions.

The sun's ultraviolet radiation down to roughly 3000 Å was photographed in 1840 by A. Becquerel. A rough idea of the sun's emissions at ultraviolet wavelengths below the atmospheric cut-off imposed by ozone (O_3) and molecular oxygen in the Earth's atmosphere was inferred from the results of ionospheric studies in the 1920s and 1930s. Some early detections of solar UV light were reported using balloons flown to roughly 20 km by K. Kiepenheuer and his colleagues in the 1930s. But the first solar UV spectra were obtained by workers at the Naval Research Laboratory (NRL) starting in 1946 using spectrometers flown on captured German V2 rockets. These spectra down to λ 2100 showed that the solar UV flux over this spectral range was significantly lower than would be expected from black-body extrapolation of the visible-light radiation curve to wavelengths below 3000 Å.

Subsequent rocket- and satellite-borne spectrometers showed that at even lower UV wavelengths the flux observed far exceeded that expected from such a black-body curve and was increasingly variable over all time scales at wavelengths below approximately 2000 Å. It is now well established that some of this variation is caused by bright ultraviolet plage regions (also bright in the visible light of Ca K and Hα) moving across the disk as the sun rotates. But despite their great importance to stratospheric chemistry and dynamics, measurements have not yet been made with errors small enough ($\leq 10\%$) to clearly reveal the 11-year ultraviolet flux variations that must exist at these wavelengths.

The ultraviolet region below 2000 Å contains the strongest lines of most of the abundant ions formed in the temperature transition between the chromosphere and corona. Since these lines are each formed in a restricted temperature regime within this 10^4–10^6 K temperature interval, they offer a unique opportunity to "unwrap" the physical structure of this interesting layer of the solar atmosphere. Images in strong ultraviolet emission lines at $\lambda \leq 1500$ Å show the intricate magnetic structures of the chromosphere, transition regions, and corona on the solar disk since the continuum emission of the photosphere is very weak at these wavelengths. The diagnostic techniques of ultraviolet spectroscopy were applied to studies of the plasma densities and temperatures using spectrometers flown on the *OSO 3* and *OSO 6* satellites, on *Skylab*, and most recently on *OSO 7*, *OSO 8*, and the Solar Maximum Mission.

The first rockets bearing X-ray detectors were flown by NRL in 1949, and solar radiations were detected in the spectral region below 100 Å. Below roughly 300 Å optical imaging using normal incidence mirrors fails because of excessive absorption from the metal surfaces. X-ray telescopes using, instead, grazing incidence optics (Fig. 1-17) were first flown on rockets by a team at

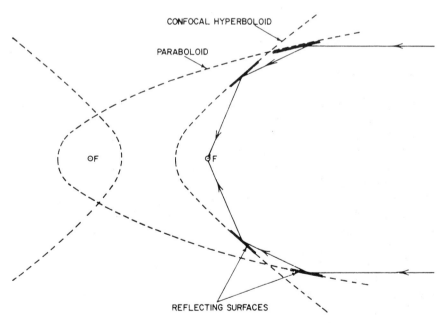

Fig. 1-17 The principle of a soft X-ray imaging telescope. Light entering from the right is reflected at grazing incidence from surfaces of confocal hyperboloid and paraboloid mirrors and is brought to the focus at F. Film is usually used as a detector. By permission of L. Golub.

American Science and Engineering in the late 1960s. As in the ultraviolet, the photosphere and chromosphere are too cool to emit a significant X-ray flux, so such telescopes image the million-degree structures of the corona on the disk. This advance made possible study of the hottest coronal and flare emissions of $T \gtrsim 5 \times 10^6$ K on the disk.

1.8   THE SOLAR WIND AND HELIOSPHERE

The X-ray and ultraviolet techniques for observing the solar corona and transition region on the disk led to the discovery, in the late 1960s, of extended, sharply demarcated areas of very low coronal emission, called coronal holes. For decades, correlations between geomagnetic disturbances and passage of active regions across the disk had indicated that seemingly quiet areas of the sun outside of active regions seemed to be responsible for many 27-day recurrent geomagnetic storms. These mysterious quiet areas were called M-regions. The good correlation found between disk passage of long-lived coronal holes and increased solar particle fluxes near earth showed that these areas of low coronal plasma density are the long-sought M-regions responsible for a substantial fraction of the sun's output of particles and fields into the interplanetary medium.

The idea of intermittent particle flows from the sun was widely accepted by the turn of the century from the evidence of a several-day time lag between solar flares and their geomagnetic disturbances. In the 1950s, the work of L. Biermann also showed that deflections of comet plasma tails could not be explained by solar radiation pressure, but required charged particles. The concept of a steady, rather than intermittent, outflow of coronal plasma was put forward already in the 1930s by J. Bartels and others, but it was predicted from a hydrodynamical model by Parker in 1958 on the theoretical grounds that solar gravity could not entirely contain the gases of the million-degree corona. This solar wind of roughly 400 km s^{-1} speed and 1–10 particles cm^{-3} density, whose existence was highly controversial for several years, was confirmed by *in situ* measurements from the *Mariner 2* space probe to Mars in 1962,

Sporadic (not recurrent) geomagnetic storms can often be linked to flares or to the eruption of filaments from the chromosphere. The disturbance travels as a shock wave at ~ 500 km s^{-1} and produces particle and field enhancements at 1 AU within 2–3 days. The solar eruption is much more promptly heralded by ultraviolet and X-ray enhancements that affect the ionosphere within about 8 min of the flare occurrence, and within roughly an hour by the arrival (for the largest flares) of relativistic particles in the energy range 0.1–1 GeV. The production of these relativistic particles in some flares is not well understood. Recent observations of flare-produced γ-rays, and also primary neutrons, indicate nuclear processes occurring within the acceleration region. The first such observations were made in 1972 from the *OSO 7* spacecraft, using a sodium iodide scintillation counter.

This new evidence may also shed light on the variable isotope composition of helium nuclei detected at Earth from flares. Important information on how flare- or filament-produced disturbances move through the corona and interplanetary medium is being gathered from white-light coronagraphs flown in spacecraft. In these coronagraphs an additional external occulting disc is mounted well in front of the objective. Scattered light levels below 10^{-7} of disc brightness can be achieved in the absence of atmospheric scattering, using such an arrangement, although the external occultation limits the field of view to the outer corona.

Further from the sun than a few solar radii, the progress of interplanetary disturbances is monitored by spacecraft such as *Helios* placed at about 1/3 AU, and from the Earth's orbit by means of very low frequency radio waves emitted by plasma oscillations in these disturbances. These radio waves, in the kilometer wavelength range, can only be observed by antennas on spacecraft, since their frequencies lie below the cut-off imposed by the Earth's ionosphere.

The solar wind punches out a volume in the interstellar material through which it passes, called the heliosphere. Its dimensions are still quite uncertain. In the upstream direction, where the cavity is compressed by the sun's proper motion through the gases of the galactic plane at roughly 20 km s^{-1}, its boundary, the heliopause, might be as close as 50 times the sun–Earth distance. The *Pioneer* and *Voyager* space probes have detected no sign so far of the heliopause on their voyages toward the orbit of Pluto and beyond. If their plasma sensors and telemetry are still functioning when these probes cross the heliopause, they will inaugurate a new era, of interstellar space exploration.

1.9 FUTURE DIRECTIONS IN SOLAR INSTRUMENTATION

The sun has been observed in emissions ranging from neutrinos to high-energy charged particles, and over essentially the full spectrum of photon energies between γ-rays and kilometric radio waves. Some of these spectral ranges, most notably the visible region, have been exploited to a far greater degree than others. For instance, infrared spectra of the sun, unencumbered by absorption lines of the Earth's atmosphere, have only been obtained from a spacecraft in the past few years. But in contrast to the situation even 30 years ago, it can no longer be said that entirely untouched wavelength frontiers exist for future observation.

At the same time, the gradual development of basic detection techniques over this wide range of solar emissions affords opportunities for observations that yield new insights into the workings of the sun. For instance, over a dozen experiments are either underway or in planning to discriminate between neutrinos in various energy ranges. Neutrino "spectrometers" and "imaging telescopes" are quite possible. The first proper look at the structure of the solar interior is also emerging from helioseismological observations begun in the early 1970s. A global network of telescopes optimized for such solar oscillation measurements is being constructed to provide essentially uninterrupted 24-h coverage and improve resolution of the oscillation modes.

The achievement of higher spatial resolution has, in the past, provided one of the main sources of new insight into solar phenomena. Plans for a large diffraction-limited solar telescope in orbit were put in motion in the mid-1960s. A 1-m class instrument now envisioned by NASA would provide better spatial resolution than achievable from the best ground-based solar telescopes. But perhaps more important, the absence of atmospheric "seeing" effects would enable a resolution of about 0.1 arc sec to be achieved for extended periods, thus making possible the long exposures necessary to achieve adequate signals from small areas of the solar surface, with high-dispersion spectrographs.

The construction of an even larger aperture ground-based solar telescope is being explored, to gather enough light for high spatial and spectral resolution polarimetric studies of the vector field in the magnetic flux tubes that cover the solar surface. A design for such a 2.5-m telescope, perhaps to be sited on Mauna Kea in Hawaii, is shown in Fig. 1-18. Its light-gathering power would exceed even that of the giant McMath telescope at Kitt Peak, whose 60-inch diameter primary mirror makes it the largest aperture solar telescope in the world. Achievement of the diffraction-limited angular resolution of such a large aperture depends on advances in rapid servo systems designed to sense and remove deformations in the wavefront entering the telescope. Such adaptive optics may be able to reduce the image blurring caused by the Earth's atmosphere.

The ultimate in "spatial resolution" would be achieved by a spacecraft called the Solar Probe, discussed by NASA for a mission into the solar atmosphere. This specially heat-shielded craft might be placed in an eccentric solar orbit with perihelion near 3 solar radii above the photosphere. A wide range of interesting observations have been envisioned for such a probe, including study of its orbital behavior for information on the sun's gravitational quadruple moment, for in-situ plasma and field measurements of the sun's outer corona, and for high spatial resolution imaging of the sun's blazing surface with telescopes operating in the UV, X-ray, and visible regions. Unfortunately, the time available for data taking near perihelion would be very short, considering the expense and complexity of such a mission.

As regards the sun's variable effects on the Earth, three observations of great practical importance urgently need to be made. For one, the 11-year variation of the sun's spectral irradiance in the ultraviolet region responsible for ozone formation and destruction (i.e., roughly between 1600 Å and 3100 Å) has not yet been successfully measured. The difficulty lies in constructing UV spectrometers whose calibration is stable (or can be corrected) to 1% or better over many years of exposure to the space environment.

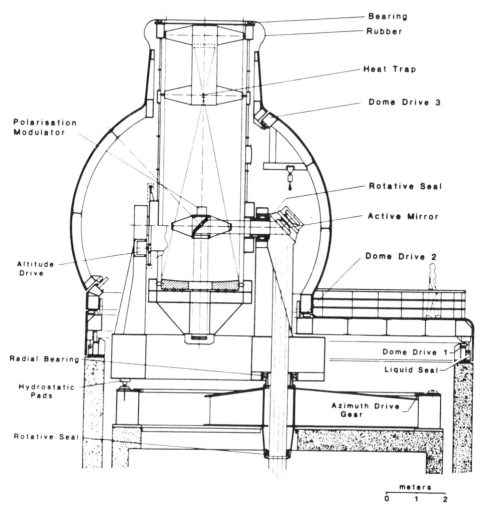

Bearing

Rubber

Heat Trap

Dome Drive 3

**Polarisation
Modulator**

Rotative Seal

Active Mirror

Dome Drive 2

**Altitude
Drive**

Dome Drive 1

Liquid Seal

Radial Bearing

**Hydrostatic
Pads**

**Azimuth Drive
Gear**

Rotative Seal

meters
0   1   2

Fig. 1-18 The design of a large solar telescope under study by an international consortium for the 1990s. Note that the dome is sealed with an optical window to minimize turbulence caused by airflow through an open slit. By courtesy of the High Altitude Observatory.

A related issue is measurement of the 11-year variation in the sun's EUV flux between about 100 Å and the Lyman α-line at 1216 Å. These are the radiations that mainly account for the heating of the Earth's ionosphere and thus determine the density of the Earth's upper atmosphere above about 100 km. This is the orbital environment of most large spacecraft, and its density determines the drag, and thus their lifetimes.

Finally, the great importance of possible climate perturbations caused by slow variations in solar luminosity makes it interesting to fly pyrheliometers whose calibration can be maintained to the level of 0.01% over many decades. An uninterrupted series of measurements spanning several solar cycles is required to inform us whether important climatic perturbations of the past may have been caused by variation of solar luminosity.

Many of the most important applications of solar research require stable and continuous observational programs of key solar phenomena. Our understanding of the sun's magnetic variability rests in large part on such "synoptic" observations of indices such as the sunspot number and the microwave flux at 10.7 cm. Recent advances in understanding of solar and stellar variability place increasing demands on the individuals and institutions with the dedication to maintain synoptic programs of the behavior of the sun's atmosphere, its effects on the Earth, and of the behavior of other sunlike stars.

Some of the new observations described above are expensive, and their viability depends as much on budgetary and political issues in the 1990s as on their scientific merits. Fortunately, there are still relatively unexplored areas of solar behavior that require only imagination, some experimental skill, and perseverance. These opportunities for relatively small-scale fundamental research complement the grander programs that require highly developed organizational skills, to make solar research such a stimulating subject.

ADDITIONAL READING

Historical Development of Solar Research

A. Berry, *A Short History of Astronomy*, Dover, 1961.

L. Goldberg, "Introduction," in *The Sun*, G. Kuiper (Ed.) University of Chicago Press, p. 1, 1953.

A. Meadows, *Early Solar Physics*, Pergamon, 1970.

Solar Instrumentation

H. Zirin, "Solar Observations," in *Solar Astrophysics*, Cambridge University Press (1988).

M. Kundu, "Techniques of Solar Observations," in *Solar Radio Astronomy*, Wiley-Interscience (1965).

J. Brandt, "Plasma Probes and Magnetometers," in *Introduction to the Solar Wind*, W. H. Freeman (1970).

J. Evans et al., "Empirical Problems and Equipment," in *The Sun*, G. Kuiper (Ed.), University of Chicago Press, 1953.

R. Dunn, "High Resolution Solar Telescopes," *Solar Phys.*, **100**, No. 1 (1985).

R. Dunn (Ed.), "Solar Instrumentation: What's Next?" Sacramento Peak Observatory Publication (1981).

EXERCISES

1. Explain how Galileo was able to distinguish between sunspots and planets transiting the solar disc using his observations of their foreshortening, and of their decreasing rate of motion, near the limb.

2. Describe the triangulation method used by Aristarchus to estimate the sun's distance (A. Berry, *A Short History of Astronomy*, Dover, 1961, p. 34). Compare the precision of his technique to that used in the 17th and 18th centuries, based on the transits of Venus (op. cit., p. 284), and to modern methods based on radar echoes from Venus, Mercury, and Mars and particularly from the asteroid Eros [see, e.g., D. Muhleman, *Monthly Notices Roy. Astron. Soc.*, **144**, 151 (1969)].

3. Describe the principle behind determinations of the sun's mass (see, e.g., K. Lang, *Astrophysical Formulae*, Springer-Verlag, 1980, p. 544) and give the main sources of error and the accuracy of modern values [see, e.g., L. Goldberg, in *The Sun*, G. Kuiper (Ed.), University of Chicago Press, 1953, p. 17].

4. Compare the principle of operation of Pouillet's calorimeter (Fig. 1-10) and of modern pyrheliometers [R. Willson, *Appl. Opt.*, **18**, 179 (1979)]. Explain why such relatively insensitive thermal detectors rather than more common astronomical detectors, such as photomultipliers or photodiodes, must be used to measure the total solar irradiance.

5. Explain the basic operating principles of a) a spectroheliograph, b) a magnetograph, c) a coronagraph, using a diagram in each case. State the main advance in solar observations achieved with each instrument.

2

Radiative Transfer in the Sun's Atmosphere

Most of our information about conditions on the sun has been derived from analysis of solar electromagnetic radiation intercepted by detectors near the Earth. Charged particles and fields blown out in the solar wind have been collected for about 25 years from satellites and probes outside the magnetosphere, and primary cosmic rays accelerated in solar flares have been captured at balloon altitudes for somewhat longer. The flux of uncharged, massless neutrinos from the sun's interior has also been investigated, and plans have even been put forward to drop a solar probe into the solar atmosphere to conduct "in situ" measurements of charged particles and fields. But we continue to rely on the analysis of photons for the bulk of our information on the solar atmosphere and interior.

Although 19th-century solar astronomers had excellent telescopes at their disposal, their physical interpretations of the "clouds" and "incandescent vapors" they described in the sun's atmosphere seem naive to us, mainly because ideas on how light is emitted and absorbed in a semi-transparent fluid remained rudimentary until after the turn of the century. A theory of radiative transfer then emerged through the work of K. Schwarzschild, A. Schuster, A. Eddington, E. Milne, S. Chandrasekhar, and others who put at our disposal techniques to determine the physical structure of the sun's atmosphere to an accuracy presently limited mainly by the angular resolution of our telescopes.

The purpose of this chapter is to provide some idea of the power of radiative transfer techniques and of their limitations when applied to modeling the inhomogeneous solar atmosphere. The tenuous coronal plasmas are quite transparent, so radiative transfer plays little role in their analysis. In the photosphere and solar interior, the high density makes important simplifications possible, so insights into the physical state of these layers are accessible using the relatively simple techniques we shall study here. More complex computations are required to obtain a quantitative understanding of the chromosphere, where the plasma is not transparent but also not of sufficient density to justify the simplifications allowed when thermodynamic equilibrium holds locally. This chapter should at least give the reader some appreciation of the physical principles behind the results of these more elaborate computations.

2.1 PHOTOMETRIC PRINCIPLES

To relate measurements of solar radiative power at the Earth to the rate of energy flow through the sun's atmosphere, we define two useful quantities: the radiative intensity and the net outward flux. These describe the energy passing per unit time through an elemental area da located at a point $\mathbf{r}$ and oriented so its normal vector is $\mathbf{n}$.

2.1.1 The Radiative Intensity

The intensity $I_\lambda(\mathbf{r}, \mathbf{s})$ is defined as the radiative energy per second or power dP_λ in the wavelength interval between λ and $\lambda + d\lambda$ flowing through da in the direction of the vector $\mathbf{s}$, within a bundle of rays whose solid angle is $d\omega$. Figure 2-1 illustrates this definition. In the cases of interest here, the intensity is assumed to be symmetrical about $\mathbf{n}$, so its value in the direction $\mathbf{s}$ is fully specified by the polar angle θ. We can then write

$$dP_\lambda = I_\lambda(\mathbf{r}, \theta)\, da \cos\theta\, d\omega\, d\lambda. \tag{2-1}$$

The cgs units of I_λ are ergs cm^{-2} s^{-1} sr^{-1}. The intensity measures the rate of flow of electromagnetic energy along a pencil of rays whose base is the projected area $da \cos\theta$ and within the cone or solid angle defined by $d\omega$. Astronomers often use the intensity to represent the photometric brightness or radiance of an extended surface, although da can be made arbitrarily small, so I_λ can also represent the radiative power passing through, or originating at, a point at $\mathbf{r}$. The definition of I_λ ensures its invariance with distance from da,

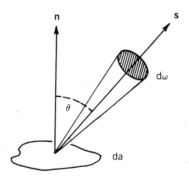

Fig. 2-1   Illustration of the concept of intensity.

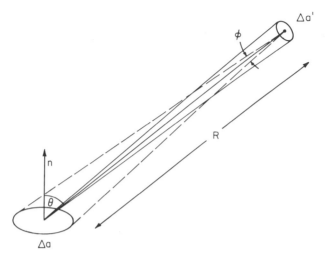

Fig. 2-2   Illustration of a measurement of intensity.

provided the rays intersect no absorbing material. The intensity also remains unchanged when the rays are refracted or reflected in passing through an optical system. These important consequences are left to be demonstrated in the exercises.

To determine the intensity of an area Δa in the solar atmosphere as viewed at an angle θ to its normal, we measure the radiative power falling on an absolutely calibrated detector of collecting area $\Delta a'$ located at a distance R (Fig. 2-2). The radiative power passing through Δa in the solid angle of rays also passing through $\Delta a'$ is

$$\Delta P_\lambda = I_\lambda(\mathbf{r}, \theta)\,\Delta a \cos \theta \left(\Delta a'/R^2\right) d\lambda. \tag{2-2}$$

Provided the projected area subtends a large enough solid angle $\Delta \omega' = (\Delta a \cos \theta)/R^2$ to be spatially resolved by an observing instrument at the distance $R = 1$ AU, its intensity then follows from measured quantities as

$$I_\lambda\, d\lambda = \Delta P_\lambda/(\Delta \omega'\, \Delta a'). \tag{2-3}$$

To the precision usually of interest in solar photometry, $\Delta \omega' \sim \phi^2$, where ϕ is the small angle subtended by a fraction of the solar disk.

The radiation pattern $I_\lambda(\theta)$ emitted by a surface element of the photosphere can be measured by observing how the intensity of the photospheric disk varies along a solar diameter. Figure 2-3 illustrates how observations at increasing heliocentric angle θ from disk center reveal the photospheric intensity at increasing values of the angle θ from the normal. The limb darkening of the photosphere (Fig. 2-4) therefore implies that this layer emits

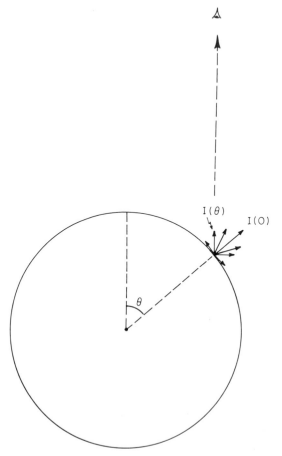

Fig. 2-3 Illustration of how the angular distribution of photospheric intensity is observed with measurements obtained at various heliocentric angles. The vector length is roughly proportional to the intensity.

a pronouncedly anisotropic intensity pattern when viewed from outside the sun's atmosphere at different angles from the local vertical.

2.1.2 The Net Outward Flux and the Solar Constant

The monochromatic power passing through da into the total solid angle of 2π sr above the photospheric surface is obtained by summing the power in all outward directions passing through da. This yields the net outward flux F_λ given by

$$F_\lambda = \int^{2\pi} I_\lambda(\theta)\cos\theta \, d\omega. \qquad (2\text{-}4)$$

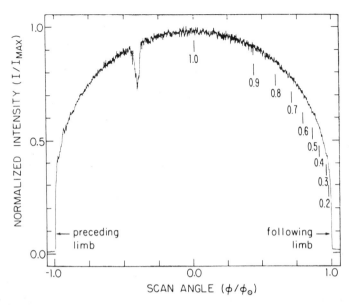

Fig. 2-4 A scan of the photospheric intensity along a diameter, showing limb-darkening at a wavelength of $\lambda = 4451$ Å. Numbers along the curve indicate values of $\mu = \cos \theta$. The large dip is caused by a spot. Other fluctuations are mainly caused by granulation.

The element of solid angle can be expressed (see Fig. 2-5) as

$$d\omega = dA/R^2 = 2\pi \sin \theta \, d\theta, \tag{2-5}$$

and we have

$$F_\lambda = 2\pi \int_0^{\pi/2} I_\lambda(\theta) \cos \theta \sin \theta \, d\theta. \tag{2-6}$$

It is common to simplify this integral by adopting the variable $\mu = \cos \theta$, so that equation (2-6) becomes

$$F_\lambda = 2\pi \int_0^1 I_\lambda(\mu)\mu \, d\mu. \tag{2-7}$$

For a surface of uniform intensity I_λ, integration of the relation (2-7) yields $F_\lambda = \pi I_\lambda$. The integral (2-7) represents the net flux passing outward through any horizontal layer of the solar atmosphere. If we imagine the photosphere as

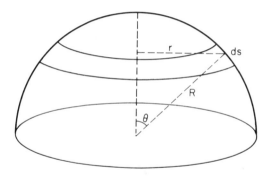

$$r = R \sin\theta, \quad ds = Rd\theta$$

Fig. 2-5 Illustration of the calculation of net outward flux through a surface.

a surface, it represents the monochromatic emittance of this surface into space, which can be calculated once we have measured the limb-darkening distribution $I_\lambda(\mu)$. Calculations in astrophysical radiative transfer are often carried out using a different definition of the flux as $\pi F' = F$, where F' is the so-called astrophysical flux. We shall use the more physically clear flux definition of F given above throughout this book.

To understand how the photospheric emittance can be measured from the Earth, we consider the relation of F_λ to the flux f_λ received from the whole solar disk, falling on a detector of 1-cm^2 area at the Earth–sun distance R. From the geometry of Fig. 2-6 we see that, provided $R \gg R_\odot$, rays reaching the detector from all points on the solar disk are effectively parallel. Then the

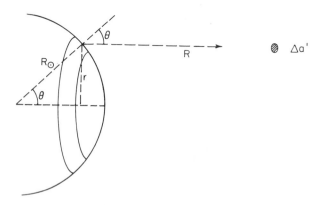

Fig. 2-6 Illustration of the relation between solar flux received at a detector placed at the Earth and photospheric emittance.

contribution df_λ to the flux falling on the detector from a photospheric annulus of projected area $2\pi r\,dr$ is

$$df_\lambda = I_\lambda(\theta)\,d\omega = \frac{2\pi r\,dr}{R^2}I_\lambda(\theta). \tag{2-8}$$

Expressing the solid-angle element in equation (2-8) in terms of the heliocentric angle θ, we can see from Fig. 2-6 that

$$f_\lambda = \frac{2\pi R_\odot^2}{R^2}\int_0^{\pi/2}I_\lambda(\theta)\sin\theta\cos\theta\,d\theta. \tag{2-9}$$

From the definition of F_λ in equation (2-6), we thus see that the detected flux f_λ and the photospheric emittance are related by

$$F_\lambda = \pi f_\lambda/\omega_s = 4.62\times10^4 f_\lambda, \tag{2-10}$$

where $\omega_s = 6.8\times10^{-5}$ sr is the solid angle subtended by the solar disk. Note that since $\omega_s = A_s/R^2$, the flux of solar radiation, unlike its intensity, falls off as $1/R^2$.

The values of f_λ corrected to the mean Earth–sun distance of 1 AU (either measured from satellites or rockets or corrected for the Earth's atmospheric transmission) define the curve of solar spectral irradiance. Its integral over wavelength, given by

$$S = \int_0^\infty f_\lambda\,d\lambda, \tag{2-11}$$

is called the total solar irradiance or solar constant. The values of f_λ for 0.2 μm $\le\lambda\le5$ μm are given in Table 3-2. The total irradiance around solar activity minimum has been determined from satellite and rocket radiometry to be 1.367×10^6 ergs cm^{-2} s^{-1} to an absolute accuracy of roughly 0.2%. From equation (2-10) this value of S implies that the emittance of the photosphere is 6.31×10^{10} ergs cm^{-2} s^{-1} or about 6.3 kW cm^{-2}. Assuming that the photospheric emittance is independent of solar latitude, we then have for the solar luminosity the enormous power output

$$L_\odot = 4\pi R_\odot^2 F = 3.84\times10^{33}\text{ ergs s}^{-1}. \tag{2-12}$$

2.2 THE RADIATIVE TRANSFER EQUATION

2.2.1 The Optical Depth and Source Function

To deal with radiation in a medium where both absorption and emission occur along a pencil of rays, we consider the transfer of a beam of intensity $I_\lambda(\theta, z)$,

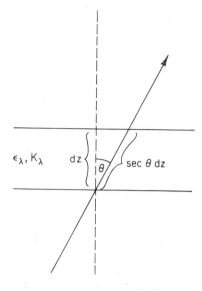

Fig. 2-7 Slab geometry used in deriving the radiative transfer equation.

making an angle θ to the normal, through a plane-parallel element of atmosphere of thickness dz (Fig. 2-7). The monochromatic extinction and emission coefficients per unit volume in the element are respectively $\kappa_\lambda(z)$ and $\epsilon_\lambda(z)$. The change in beam intensity on passing through the element is

$$I_\lambda(\theta, z + dz) - I_\lambda(\theta, z) = [\epsilon_\lambda(z) - \kappa_\lambda(z)I_\lambda(\theta, z)]\sec\theta\, dz, \quad (2\text{-}13)$$

or

$$\mu\frac{dI_\lambda(\theta, z)}{dz} = \epsilon_\lambda(z) - \kappa_\lambda(z)I_\lambda(\theta, z). \quad (2\text{-}14)$$

We next introduce the optical depth

$$d\tau_\lambda = -\kappa_\lambda\, dz, \quad (2\text{-}15)$$

as a measure of the opacity of the element. Here κ_λ is the linear absorption coefficient (cm^{-1}), which is proportional to ρ, the gas density. We have $\kappa_\lambda = k_\lambda\rho$, where k_λ is the mass absorption coefficient (g^{-1}), which is calculated from the composition and physical properties of the medium, as discussed further in Chapter 3. Note that a negative sign has been introduced in equation (2-15) so that the optical depth increases looking into the solar atmosphere in which z is taken to increase outward. In the absence of any emission, equation (2-14) reduces to

$$dI_\lambda/I_\lambda = -\kappa_\lambda\, dz, \quad (2\text{-}16)$$

so that the emergent and incident intensities I'_λ and I_λ are related by

$$I'_\lambda / I_\lambda = \exp(-\tau_\lambda). \tag{2-17}$$

A layer of optical depth $\tau_\lambda = 1$ thus reduces the intensity of a transmitted beam to $e^{-1} = 0.368$ of its incident intensity. Media with optical depths less than unity and exceeding unity, are called optically "thin" and optically "thick" respectively.

In the more general situation where the gas also emits ($\epsilon_\lambda \neq 0$), we have from (2-14)

$$\mu \frac{dI_\lambda}{d\tau_\lambda} = I_\lambda - S_\lambda, \tag{2-18}$$

where we have substituted the optical depth τ_λ as the new depth coordinate. The function $S_\lambda = \epsilon_\lambda / \kappa_\lambda$ is called the source function. This standard form of the radiative transfer equation shows that it is the behavior of S_λ, rather than of ϵ_λ or κ_λ individually, that determines the emergent intensity from a medium where emission and absorption occur together.

2.2.2   Solution for Constant Source Function

The general solution of this linear first-order equation can be obtained by use of integrating factors when S_λ is given throughout a plane-parallel slab bounded by the surfaces $\tau_1 < \tau_2$. The geometry is illustrated in Fig. 2-8. Dropping the λ subscript from the monochromatic optical depths for clarity

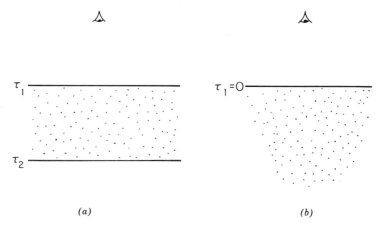

(a)   (b)

Fig. 2-8 Boundary conditions appropriate for integrating the radiative transfer equation over (a) a slab and (b) a semi-infinite atmosphere.

in this expression, we have

$$I_\lambda(\tau_1, \mu) = I_\lambda(\tau_2, \mu)\exp\left(\frac{-(\tau_2 - \tau_1)}{\mu}\right) + \frac{1}{\mu}\int_{\tau_1}^{\tau_2} S_\lambda \exp\left(\frac{-(\tau - \tau_1)}{\mu}\right) d\tau.$$

(2-19)

To obtain the emergent intensity $I_\lambda(0, \mu)$ emitted at the top of a semi-infinite atmosphere such as the photosphere, we have

$$I_\lambda(0, \mu) = \frac{1}{\mu}\int_0^\infty S_\lambda(\tau_\lambda)\exp\left(\frac{-\tau_\lambda}{\mu}\right) d\tau_\lambda,$$

(2-20)

since the first term of equation (2-19) vanishes and $\tau_1 = 0$, $\tau_2 = \infty$ (see Fig. 2-8). That is, the intensity emergent from the slab at an angle $\theta = \cos^{-1}\mu$ is the sum along that slant path of all the contributions from the local source functions at each optical depth, each weighted exponentially by the optical depth between that point and the top of the slab.

We note that if the source function is constant with depth, the intensity emergent from a slab of optical depth τ_2 is just

$$I_\lambda(0, \mu) = S_\lambda \int_0^{\tau_2} \exp\left(\frac{-\tau_\lambda}{\mu}\right) d\tau_\lambda,$$

(2-21)

which yields for the intensity in the normal ($\mu = 1$) direction

$$I_\lambda(0) = S_\lambda(1 - e^{-\tau_2}).$$

(2-22)

If the slab is optically thick ($\tau_2 \gg 1$), then

$$I_\lambda(0) = S_\lambda,$$

(2-23)

whereas if the slab is optically thin ($\tau_2 \ll 1$),

$$I_\lambda(0) = \tau_\lambda S_\lambda.$$

(2-24)

These relations are useful in developing some intuition for the behavior of the emergent intensity. It is seen to increase with the source function, and through τ, with the density and the mass absorption coefficient, which are intrinsic properties of the gas. It also scales as the line-of-sight dimension of the atmosphere. Furthermore, these relations tell us that if the optical thickness of the atmosphere is increased, the emergent intensity will increase, as we might expect, but only up to a maximum value equal to the source function itself. These relations will become more useful when we point out below the simple connection that exists between S_λ and the temperature of a gas in thermodynamic equilibrium.

2.2.3   Solution for a Linear Source Function: The Eddington-Barbier Relation

The behavior of the emergent intensity when the source function varies linearly with depth is also instructive. We represent the source function as

$$S_\lambda(\tau_\lambda) = S_\lambda(0) + b\tau_\lambda, \qquad (2\text{-}25)$$

where b is a constant.

Equation (2-21) then becomes

$$I_\lambda(0, \mu) = \int_0^{\tau_2} \frac{1}{\mu} S_\lambda(0)\exp\left(\frac{-\tau_\lambda}{\mu}\right) d\tau_\lambda + b\int_0^{\tau_2} \frac{1}{\mu}\tau_\lambda \exp\left(\frac{-\tau_\lambda}{\mu}\right) d\tau_\lambda. \quad (2\text{-}26)$$

Setting $\tau_2 = \infty$ for a semi-infinite atmosphere, we have

$$I_\lambda(0, \mu) = S_\lambda(0) + b\mu. \qquad (2\text{-}27)$$

Comparing equations (2-25) and (2-27), we see that the emergent intensity observed at the slant angle $\theta = \cos^{-1}\mu$ is equal to the source function at the optical depth $\tau_\lambda = \mu$. That is

$$I_\lambda(0, \mu) = S_\lambda(\tau_\lambda) \qquad \text{at} \quad \tau_\lambda = \mu. \qquad (2\text{-}28a)$$

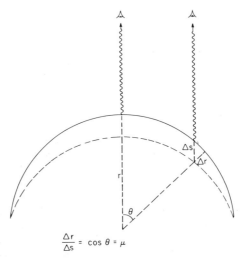

Fig. 2-9 Schematic illustration of the varying depth of penetration, Δr, into the solar photosphere implied by the Eddington-Barbier relation.

This result, known as the Eddington-Barbier relation, implies that if $S_\lambda(\tau_\lambda)$ is a linear function, the emergent monochromatic radiation viewed at angle μ can be interpreted to be formed at a surface whose depth is given by $\tau_\lambda = \mu$ (Fig. 2-9). As illustrated in Fig. 2-10, the photospheric radiation at any wavelength is formed over a range of geometrical depth described by a smooth contribution function. It will be seen in Chapter 5 that linearity of the source

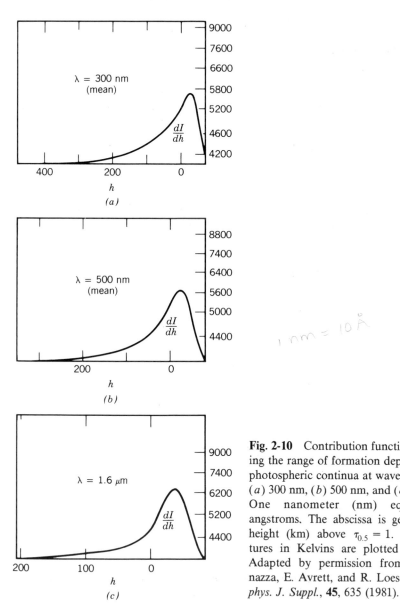

$1\ nm = 10\ \overset{\circ}{A}$

Fig. 2-10 Contribution functions showing the range of formation depths of the photospheric continua at wavelengths of (*a*) 300 nm, (*b*) 500 nm, and (*c*) 1.6 μm. One nanometer (nm) equals 10 angstroms. The abscissa is geometrical height (km) above $\tau_{0.5} = 1$. Temperatures in Kelvins are plotted at right. Adapted by permission from J. Vernazza, E. Avrett, and R. Loeser, *Astrophys. J. Suppl.*, **45**, 635 (1981).

function with τ_λ can be a good enough approximation in the photosphere to make the Eddington-Barbier relation very useful.

2.3   THERMODYNAMIC EQUILIBRIUM

Calculation of the source function usually requires that we know the chemical constitution and ionization of the gas and the atomic structure of the radiating species, as well as the temperature and density. Examples of such calculations will be discussed in Chapter 3, but for the relatively dense plasma found throughout the solar interior up to the highest photospheric layers, great simplification is achieved by taking into account the properties of matter and radiation in thermodynamic equilibrium (TE).

2.3.1   The Planck Function

The concept of TE is most easily visualized by considering the radiation field inside an adiabatic cavity (Fig. 2-11) whose interior walls are at a uniform temperature T that is constant in time, i.e., in steady equilibrium. This equilibrium requires that each surface element emit and absorb radiative power at the same rate, independent of the material or geometry of the cavity walls. If this did not hold, we could divide the cavity so that the outward and inward intensities passing through a suitably chosen aperture were unequal. A mechanical device such as a light-mill could then be placed in this aperture to extract energy from the difference in fluxes. The experimental evidence supporting the second law of thermodynamics indicates that this cannot be achieved, so the intensity must be identical at all surfaces within the cavity. The intensity of cavity radiation is thus uniquely a function of the temperature given by

$$I_\lambda = B_\lambda(T), \qquad (2\text{-}28b)$$

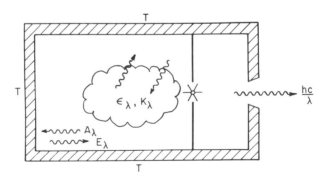

Fig. 2-11   Schematic illustration of a black-body cavity.

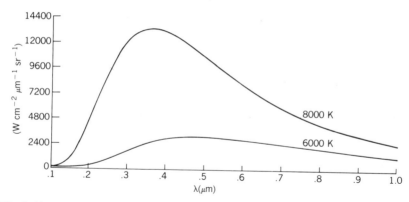

Fig. 2-12 Behavior of the Planck function at temperatures of 6000 K and 8000 K.

where $B_\lambda(T)$ is the Planck function

$$B_\lambda(T) = \frac{2\pi hc^2}{\lambda^5} \frac{1}{\exp(hc/k\lambda T) - 1}. \qquad (2\text{-}29)$$

Figure 2-12 illustrates the shape of the Planck function emitted by a 6000 K black body of temperature similar to the photospheric value. Integration of this function over wavelength yields the total intensity of cavity radiation

$$\int_0^\infty B_\lambda(T) \, d\lambda = \frac{\sigma}{\pi} T^4, \qquad (2\text{-}30)$$

where $\sigma = 5.67 \times 10^{-5}$ erg cm^{-3} s^{-1} deg^{-4} is the Stefan-Boltzmann constant. The factor π arises from the relation $F_\lambda = \pi I_\lambda$, found from integration of equation (2-7), for a surface of uniform intensity. For small cavity temperature changes ΔT, the change in shape of the Planck function illustrated in Fig. 2-12 can be neglected, and we can derive a useful estimate of the resultant small changes in monochromatic intensity from

$$\frac{\Delta B_\lambda}{B_\lambda} \simeq \frac{4 \Delta T}{T}. \qquad (2\text{-}31)$$

2.3.2 Kirchhoff's Law

Since $B_\lambda(T)$ is the intensity of the universal radiation field within an adiabatic cavity, each of its surface elements da emits and absorbs radiation (in a steady state) at the equal rate

$$E_\lambda \, d\lambda \, da \cos\theta \, d\omega = A_\lambda B_\lambda(T) \, d\lambda \, da \cos\theta \, d\omega, \qquad (2\text{-}32)$$

where E_λ and A_λ are respectively the emissivity and absorptivity of the cavity surface. Here $A_\lambda = 1 - R_\lambda$, where R_λ is the reflectivity of the surface. It follows that

$$E_\lambda/A_\lambda = B_\lambda(T), \qquad (2\text{-}33)$$

which is the form of Kirchhoff's law applied to surfaces. By blackening the interior so that $A_\lambda \sim 1$, one can construct a cavity that absorbs essentially all radiation and has the appearance of a "black body." In this case, (2-33) reduces to $E_\lambda = B_\lambda(T)$, and we have a Planckian radiator. For this reason, Planckian emission is often referred to as black-body radiation.

Suppose we now introduce a radiating gas into the adiabatic cavity. The above reasoning leads us to write for the emission and absorption from each volume element of the gas in steady equilibrium

$$\epsilon_\lambda \, d\lambda \, dV \, d\omega = \kappa_\lambda B_\lambda(T) \, d\lambda \, dV \, d\omega, \qquad (2\text{-}34)$$

and thus

$$\epsilon_\lambda/\kappa_\lambda = B_\lambda(T), \qquad (2\text{-}35)$$

which is the form of Kirchhoff's law appropriate to volume emission and absorption in a cavity.

2.3.3   Local Thermodynamic Equilibrium (LTE)

We now return to the evaluation of the source function in TE. Kirchhoff's law in the form of relation (2-35) states that the source function $S_\lambda = \epsilon_\lambda/\kappa_\lambda$ within a cavity in TE is equal to the value of the Planck function at the cavity temperature. This suggests a great simplification of the radiative transfer problem. Over the deeper layers of the photosphere (that contribute most of the sun's radiation), the gas is sufficiently dense that most photons are absorbed and thermalized (see Chapter 3) before they travel a distance over which the plasma temperature changes appreciably. The photospheric radiation from these layers can then be approximated as arising within an isothermal cavity whose adiabatic walls consist of opaque gas, and whose radiation field is in local thermodynamic equilibrium (LTE) with the local gas temperature. The radiation field at any depth is of course determined to some extent by contributions from deeper, hotter layers, and thus cannot be entirely in equilibrium with the local gas kinetic temperature. Nevertheless, this LTE approximation turns out to be highly useful.

The LTE approximation is especially powerful in the interpretation of photospheric continuum radiation and in understanding many aspects of Fraunhofer line formation, as will be seen below. But it does require that the temperatures of radiation and of the plasma be closely coupled by frequent atomic absorptions, over distances that are small compared to the scale of temperature change in the atmosphere. This requirement already becomes impossible to defend at the higher photospheric layers, where the far infrared continuum is formed and where the cores of many of the stronger Fraunhofer lines originate. In the tenuous plasmas of the chromosphere and corona, the plasma temperature becomes completely decoupled from the radiation field and the radiation intensity falls many orders of magnitude below the intensity expected of a Planckian radiator at the high plasma temperatures in these layers.

2.3.4   The Brightness- and Effective Temperatures

Whether or not radiation is Planckian, it is often useful to characterize its intensity I_λ in terms of a brightness temperature T_b defined as the temperature of a black body whose intensity B_λ at wavelength λ equals that of the radiating atmosphere. That is,

$$I_\lambda = B_\lambda(T_b). \tag{2-36}$$

Extending this concept to the wavelength-integrated intensity, we introduce the effective temperature, defined by

$$\int_0^\infty I_\lambda \, d\lambda = \frac{\sigma T_{eff}^4}{\pi}, \tag{2-37}$$

where T_{eff} is the temperature of a black body whose total radiation equals that measured. Most often, the effective temperature is used as a parameter of the whole atmosphere, so that

$$F = \frac{L_\odot}{4\pi R_\odot^2} = \sigma T_{eff}^4, \tag{2-38}$$

where $L_\odot$ and $R_\odot$ are the solar luminosity and radius.

Note that in strict TE, $T_b = T_{eff} = T$, where T is the plasma kinetic temperature.

2.4   THE GRAY ATMOSPHERE

2.4.1   Formulation of the Problem

Some analytical solutions to the radiative transfer equation, of use in probing photospheric structure, can be obtained without any prior assumptions regarding the form of $S(\tau)$ if we make two approximations beyond that of LTE. The first is to take κ as a constant independent of wavelength, that is, "gray." We shall see in Chapter 5 that grayness might be a reasonable approximation in the solar photosphere since the opacity is found to vary by less than a factor of 2 over the wavelength range 0.4–2.6 μm that includes about 90% of the sun's radiative flux. We also assume purely radiative equilibrium as a working hypothesis. This requires that energy be carried through the atmosphere by radiation alone, so competing thermal processes (most significantly convection) and also nonthermal processes, are neglected in our discussion.

In the gray approximation and in LTE, the radiative transfer equation becomes

$$\mu \frac{dI_\lambda}{d\tau} = I_\lambda - B_\lambda(T),\qquad(2\text{-}39)$$

where τ is now calculated from some appropriately defined average opacity (see below) over the wavelength range of interest. The condition of radiative equilibrium requires that the total (wavelength-integrated) energy emitted and absorbed must balance for any given volume. In LTE, the total emitted energy integrated over solid angle is

$$4\pi\epsilon = 4\pi\kappa B(T),\qquad(2\text{-}40)$$

where ϵ and $B(T)$ represent the total values

$$B(T) = \int_0^\infty B_\lambda(T)\,d\lambda\qquad(2\text{-}41)$$

and

$$\epsilon = \int_0^\infty \epsilon_\lambda\,d\lambda.\qquad(2\text{-}42)$$

From the balance between total emission and total energy absorbed per unit volume, we have

$$4\pi B(T) = \int_0^{4\pi} I\,d\omega = 2\pi\int_{-1}^{+1} I\,d\mu.\qquad(2\text{-}43)$$

With this expression, we can rewrite (2-39) as

$$\mu\frac{dI}{d\tau} = I - \frac{1}{2}\int_{-1}^{+1} I\,d\mu,\qquad(2\text{-}44)$$

where I is now also an integral over wavelength. The integral

$$J = \frac{1}{4\pi}\int_0^{4\pi} I\,d\omega = \frac{1}{2}\int_{-1}^{+1} I\,d\mu\qquad(2\text{-}45)$$

encountered in (2-43) expresses the mean intensity of the radiation field. It can be shown that the mean intensity and source function in an atmosphere are related by a solution of the radiative transfer equation of the form

$$J_\lambda(T_\lambda) = \frac{1}{2} \int_0^\infty S_\lambda(t_\lambda) E_1(t_\lambda - \tau_\lambda)\, dt_\lambda, \tag{2-46}$$

where

$$E_1(X) = \int_1^\infty \frac{1}{w} e^{-wx}\, dw \tag{2-47}$$

is called an exponential integral of the first kind.

Now, since from (2-43) and (2-45) we have

$$B \equiv J, \tag{2-48}$$

it follows from (2-46) that solution of the gray radiative transfer equation (2-44) will take the form

$$B(\tau) = \frac{1}{2} \int_0^\infty B(t) E_1(t - \tau)\, dt. \tag{2-49}$$

From considerations beyond the scope of our discussion, it can be shown further that the general solution of (2-49) is of the form

$$B(\tau) = \frac{3}{4}(\tau + q(\tau))\frac{F}{\pi}, \tag{2-50}$$

where the function $q(\tau)$ has the value $\frac{2}{3}$ if we make the further approximation that

$$\int I\, d\mu = 3 \int I \mu^2\, d\mu. \tag{2-51}$$

This relation holds exactly for isotropic radiation and is valid for $\tau \gg 1$. Following an approximation made by Eddington, we assume that (2-51) holds throughout the atmosphere. Substituting this source function defined by equation (2-50) into expression (2-20) for the emergent intensity, we have

$$I(0, \mu) = \frac{3}{4}\frac{F}{\pi} \int_0^\infty \left(\tau + \frac{2}{3}\right) e^{-\tau/\mu} \frac{d\tau}{\mu} = \frac{3}{4}\frac{F}{\pi}\left(\mu + \frac{2}{3}\right). \tag{2-52}$$

2.4.2 Gray Limb Darkening in the Eddington Approximation

It follows from expression (2-52) for the emergent intensity at angle $\theta = \cos^{-1}\mu$ that in the Eddington approximation, the limb darkening can be

TABLE 2-1   Comparison between Eddington Limb Darkening and Observations of $I(\mu)\,/\,I(\phi)$ at Three Wavelengths

μ	$\frac{3}{5}(\mu + \frac{2}{3})$	Observed		
		$\lambda\ 4000$	$\lambda\ 6000$	$\lambda\ 8000$
0.8	0.876	0.835	0.900	0.924
0.6	0.756	0.663	0.788	0.843
0.4	0.636	0.490	0.664	0.744
0.2	0.516	0.308	0.508	0.615
0.1	0.456	0.222	0.412	0.533

expressed as

$$\frac{I(0, \mu)}{I(0, 1)} = \frac{3}{5}\left(\mu + \frac{2}{3}\right). \tag{2-53}$$

Table 2-1 gives this function for comparison with observations. The agreement to a few percent with the limb darkening observed in white light was first used in 1906 by K. Schwarzschild to argue that radiative equilibrium describes energy flow through the photosphere. More elaborate radiative transfer models, which use detailed atomic and molecular capacities, do not assume either LTE or radiative equilibrium and do not rely on the Eddington approximation, can match the observed limb darkening throughout the violet, visible, and near infrared wavelength ranges to a few percent.

The major source of error in such models generally remains the assumption that the photosphere consists of a single "atmosphere." In fact, high-resolution photographs show the bright granules and dark intergranular lanes, and the observed limb-darkening curve is some average over these two intermingled atmospheres, as discussed in Chapter 5.

2.4.3   The Photospheric Level Identified with Radiation at T_{eff}

The gray approximation also provides an estimate of the mean optical depth from which the photospheric flux emerges. Recalling that

$$B(T) = \frac{\sigma}{\pi}T^4 \tag{2-30}$$

and

$$F = \sigma T_{eff}^4, \tag{2-38}$$

we have by substitution into (2-50)

$$\sigma T^4 = \tfrac{3}{4}\left(\tau + \tfrac{2}{3}\right)\sigma T_{eff}^4. \tag{2-54}$$

We deduce from this expression that the emergent flux can be characterized by the black-body temperature of the atmospheric layer at a depth $\tau = \frac{2}{3}$.

Testing this prediction against values observed for the photosphere, we find that $T_{\text{eff}} = 5776$ K as calculated from (2-38) using measurements of the total solar irradiance and of $R_{\odot}$. By comparison, photospheric models (Chapter 5) give $T(\tau = \frac{2}{3}) \simeq 6050$ K, so the agreement suggested by the gray model holds to within 4%. Thus, in a limited sense the $\tau = \frac{2}{3}$ level can be thought of as the "surface" whose black-body temperature characterizes the emittance of the photosphere. In fact, the contributions of various photospheric layers to the continuum radiation at a given wavelength are described by peaked curves, as illustrated in Fig. 2-10.

Optical and geometrical depth scales in a gray atmosphere can be related if we define a suitable wavelength-integrated opacity κ that can be calculated from atomic data given the chemical composition, electron pressure, and temperature. A useful estimate is provided by the harmonic mean opacity weighted by the radiative flux. This quantity is called the Rosseland opacity, defined as

$$\bar{\kappa}_R = \frac{\displaystyle\int_0^{\infty} F_\lambda \, d\lambda}{\displaystyle\int \frac{1}{\kappa_\lambda} F_\lambda \, d\lambda}. \tag{2-55}$$

Values of $\bar{\kappa}_R$ are tabulated as a function of electron temperature and pressure. For example, the Rosseland opacity in the photosphere around $\tau = 1$ is roughly 10^{-6} cm^{-1}. It follows that the geometrical scale over which the gray optical depth doubles in the deep photosphere is of order $l \sim \bar{\kappa}_R^{-1} \sim 10$ km. This relatively high opacity is partly responsible for the sharpness of the photospheric limb and certainly justifies the concept of the photosphere as a geometrical surface whose thickness is much less than $R_{\odot}$. It is worth noting also that $\bar{\kappa}_R$ is a very steep function of temperature (roughly as T^{10}!). This has important consequences for the interpretation of observations of photospheric structures like granulation.

2.4.4 Radiative Diffusion

Study of radiative transfer below the photosphere can be simplified by taking into account its diffusionlike properties. By expanding the Planckian source function as a power series in the optical depth $\Delta\tau_\lambda$ about a given point at depth $\tau_\lambda \gg 1$, we have

$$B_\lambda(\Delta\tau) = B_\lambda(\tau_\lambda) + \frac{dB_\lambda}{d\tau_\lambda}\Delta\tau_\lambda + \frac{1}{2}\frac{d^2 B_\lambda}{d\tau_\lambda^2}\Delta\tau_\lambda^2 + \cdots. \tag{2-56}$$

Substituting this expansion of the source function into the radiative transfer equation, we see that the intensity can be expressed as a series

$$I_\lambda(\tau_\lambda, \mu) = B_\lambda(\tau_\lambda) + \mu \frac{dB_\lambda}{d\tau_\lambda} + \cdots . \tag{2-57}$$

Using the definition of the flux [equation (2-4)], we then find that

$$F_\lambda(\tau_\lambda) = \frac{4\pi}{3} \left[\frac{dB_\lambda}{d\tau_\lambda} + \frac{d^3B_\lambda}{d\tau_\lambda^3} + \cdots \right], \tag{2-58}$$

and for large τ_λ, it can be shown that the higher-order terms can be neglected, so that

$$F_\lambda(\tau_\lambda) = \frac{4\pi}{3} \frac{\partial B_\lambda}{\partial \tau_\lambda} = \left(\frac{4\pi}{3\bar{\kappa}_R} \frac{\partial B_\lambda}{\partial T} \right) \frac{\partial T}{\partial z}, \tag{2-59}$$

where $\bar{\kappa}_R$ is the Rosseland mean opacity.

It can be seen that expression (2-59) has a form

$$F_{rad} = k' \nabla T \tag{2-60}$$

similar to the thermal conduction equation, where

$$k' = \frac{4\pi}{3\bar{\kappa}_R} \frac{\partial B_\lambda}{\partial T} = \frac{16\pi\sigma T^3}{3\bar{\kappa}_R} \tag{2-61}$$

can be considered the radiative "conductivity" of the medium. It follows that the radiative contribution to energy balance can be described by a diffusion equation of the form

$$\nabla \cdot F_{rad} = \nabla \cdot (k' \nabla T). \tag{2-63}$$

2.5 RADIATIVE TRANSFER IN THE FRAUNHOFER LINES

The techniques developed above apply to the transfer of both line and continuum radiation. We concern ourselves now with the Fraunhofer absorption lines, where the opacity peaks sharply around certain discrete wavelengths. The simple model we consider shows how the depths of different lines and their center-to-limb variation can depend on the kind of atomic transition responsible for the line and also on the temperature structure of the photosphere. More sophisticated models extend this theory to the interpretation of line-profile shapes in terms of the temperature, pressure, and velocity fields with height. Such models provide the main source of information on conditions in the highest photospheric layers (where visible continuum opacity is low) and in the chromosphere.

2.5.1   Formation of Fraunhofer Lines

The dark Fraunhofer lines observed in the solar disk spectrum are caused by a selective process of line scattering or line absorption, accompanied by continuum absorption. We may visualize an atom with discrete lower and upper levels 1 and 2, between which a transition can occur at wavelength λ_0. As a beam of photons of this wavelength passes through the atmosphere, it excites atoms to level 2, whose spontaneous decay back to level 1 produces photons of closely similar wavelength λ_0 emitted in all directions, thus reducing the intensity of the original beam. In addition, each such scattering increases the distance that a photon must travel to escape from the atmosphere.

A second ingredient is the presence of a continuum opacity source which in the photosphere of the sun is mainly due to the H^- ion. The increased path length of the scattered photon increases the probability that it will encounter an H^- ion and be absorbed. Each absorption produces a free electron whose kinetic energy is rapidly shared through collisions with the general electron pool. This rapid thermalization ensures that, by the time the free electron recombines with a hydrogen atom to form another H^- ion, it will have a substantially different kinetic energy and its recombination will lead to emission of a photon of wavelength significantly different from λ_0. The flux of photons of wavelength λ_0 is thus depleted, and the energy emerges at other wavelengths distributed smoothly throughout the photospheric continuum.

Of course, atoms consist of more than two discrete levels. The most important modification to the simple case described above takes into account that each atomic species has its own continuum c. Particularly in studying the formation of lines due to transitions between higher atomic levels (rather than the resonance transitions between the ground state and first excited level, for which the scattering model considered above is most appropriate), we must take into account that excitation from level 1 to 2 by the photon of wavelength λ_0 may be quickly followed by another photon capture of wavelength λ_1. This capture can cause the transition $2 \rightarrow c$, so the energy of the original (λ_0) photon ends up in the kinetic energy of a free electron, instead of merely being scattered. This is called selective absorption instead of selective scattering, and again the flux of photons at the discrete wavelength λ_0 is depleted, producing a dark line. We now consider a quantitative model of Fraunhofer line formation that extends these qualitative considerations.

2.5.2   The Transfer Equation for Lines

In the model considered here, we assume that a fraction η of the radiation removed from the beam is selectively absorbed into thermal motions of the atoms and electrons while the remaining fraction $(1 - \eta)$ is scattered isotropically and coherently. Coherent scattering requires that the scattered and incident photons have identical wavelength. This makes the mathematical treatment simpler than if we consider the more realistic situation allowing for inevitable small changes in wavelength between the incoming and outgoing light caused by fine structure of the atomic levels, Doppler shift, and other line-broadening mechanisms discussed in Chapter 3. The influence of such re-distribution of photons throughout the line profile is briefly considered later.

The coefficient for true absorption integrated over the whole line is then $\eta\kappa_l$, and in a steady state, the energy absorbed from the radiation field of intensity $B_\lambda(T)$ must be returned, so the volume coefficient of thermal emission is

$$\epsilon_\lambda^t = \eta\kappa_l B_\lambda(T), \tag{2-64}$$

while the scattering contribution to the emission coefficient is

$$\epsilon_\lambda^s = (1 - \eta)\kappa_l J_\lambda. \tag{2-65}$$

Additional contributions to the solar opacity and emissivity arise from continuum (e.g., H^- ion) absorption and emission with absorption coefficient κ_c and emission coefficient $\kappa_c B_\lambda(T)$. The transfer equation becomes

$$\mu\frac{dI_\lambda}{dz} = -(\kappa_l + \kappa_c)I_\lambda + \kappa_c B_\lambda(T) + \eta\kappa_l B_\lambda(T) + (1 - \eta)\kappa_l J_\lambda. \tag{2-66}$$

We next define an optical depth scale

$$d\tau_\lambda = -(\kappa_l + \kappa_c)\,dz \tag{2-67}$$

and set

$$\beta = \kappa_l/\kappa_c. \tag{2-68}$$

With these parameters, the simplified transfer equation becomes

$$\mu\frac{dI_\lambda}{d\tau} = I_\lambda - \frac{[1 + \eta\beta]B_\lambda(T) + [(1 - \eta)\beta]J_\lambda}{1 + \beta}, \tag{2-69}$$

which can be further reduced to the form

$$\mu\frac{dI_\lambda}{d\tau_\lambda} = I_\lambda - \alpha B_\lambda(T) - (1 - \alpha)J_\lambda, \tag{2-70}$$

known as the Milne-Eddington equation, where

$$\alpha = \frac{1 + \eta\beta}{1 + \beta}. \tag{2-71}$$

2.5.3 The Milne-Eddington Model

We consider the solution of equation (2-70) under the simplest assumption that the importance of line opacity relative to total continuum opacity (given by β) remains constant with depth, as does also the parameter α. This is more realistic than another simple picture of Fraunhofer line formation, the Schuster-Schwarzschild model, in which the lines are formed in a separate "reversing" layer imagined to overlie the layer of continuum formation. We assume further that the Planck function $B_\lambda(T)$ can be linearly related to the continuum optical depth by

$$B_\lambda(T) = a + p\tau_\lambda, \tag{2-72}$$

where

$$p = b/(1 + \beta). \tag{2-73}$$

If we assume the Eddington approximation (2-51) holds, a solution for the emergent flux can be given as

$$F_\lambda(0) = \frac{2\pi}{3} \frac{p_\lambda + (3\alpha)^{1/2}a}{1 + \alpha^{1/2}}. \tag{2-74}$$

This solution can be used to determine the residual flux in an absorption line. Since in the continuum $\beta = 0$ and $\alpha = 1$, we have

$$F_c(0) = \frac{2\pi}{3} \frac{b + a\sqrt{3}}{2}, \tag{2-75}$$

and the residual flux is given by

$$\frac{F_l}{F_c} = \frac{p + (3\alpha)^{1/2}a}{1 + \alpha^{1/2}} \frac{2}{b + a\sqrt{3}}. \tag{2-76}$$

Equation (2-76) predicts some easily observable differences in the central depth of Fraunhofer lines. In the case of a line formed by pure scattering [$\eta = 0$ and $\alpha = (1 + \beta)^{-1}$], we find that $F_l/F_c \to 0$ in the limit of $\beta \to \infty$ appropriate for a very strong line with a large absorption coefficient κ_l. This shows that a very strong line formed by scattering (in the presence of continuum absorption) can in principle be entirely black.

On the other hand, if the line is formed by thermal absorption, $\eta = 1$ and $\alpha = 1$, so we find that

$$\frac{F_l}{F_c} = \frac{a\sqrt{3} + b(1 + \beta)^{-1}}{a\sqrt{3} + b}, \qquad (2\text{-}77)$$

which approaches the asymptotic value

$$\frac{F_l}{F_c} = \frac{1}{1 + b/(a\sqrt{3})}, \qquad (2\text{-}78)$$

as $\beta \rightarrow \infty$.

2.5.4 Comparison with Observations of Line Depth near Disk Center

Values of a and b can easily be derived from continuum limb-darkening observations using the Eddington-Barbier approximation, and when they are substituted into equation (2-78), we find $F_l/F_c \sim 0.4$, in agreement with depths of moderately strong Fraunhofer lines. Figure 2-13 shows some examples of Fraunhofer line profiles traced with a high-dispersion spectrometer. Comparison of the maximum line depths expected for lines formed by pure scattering and by true absorption helps to explain why, for instance, the hydrogen Balmer lines (which originate in transitions between the $n = 2$ and higher states of H I) are not as deep as the resonance lines originating at the $n = 1$ level of certain much less abundant ions such as Ca II and Na I.

For a resonance line such as Ca II H and K or Na I D, excitation of the upper ($n = 2$) level is most probably followed by spontaneous radiative decay back to the $n = 1$ ground state, so such a line tends to be formed under conditions more closely approximated by scattering. In a transition between higher levels as in the Balmer lines, the upper state can be de-excited by many possible paths, so that the probability of a direct decay to the original state is much lower. The formation of such lines tends to be better described by the concept of selective absorption, and according to (2-78), we expect them to be less deep than resonance lines, regardless of how high the elemental abundance (and thus κ_l and β) may be.

The discussion of Fraunhofer line depth given here is only qualitatively useful. For instance, the observation that even strong resonance lines such as the Na I D- and Mg I b-lines have considerable ($\sim 10\%$) central intensity requires that we take into consideration the smearing influence of redistributed, scattered photons within the line profile. Also, departures from purely resonance scattering caused by further upward radiative excitations from the $n = 2$ level seem to contribute a certain absorption-type character to the formation of these lines. An accurate representation of all the factors determining the depth and profile of a line such as Balmer α has only recently

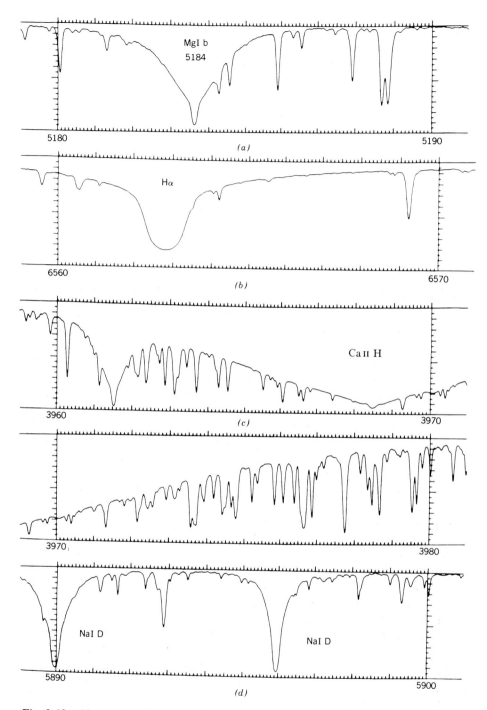

Fig. 2-13 Observed profiles of (*a*) the Mg I b-line at $\lambda = 5184$ Å, (*b*) the Balmer line Hα at $\lambda = 6563$ Å, (*c*) the Ca II H-line at $\lambda = 3968$ Å, and (*d*) the Na I D-lines at $\lambda\lambda = 5890, 5896$ Å. Adapted from the Sacramento Peak Observatory Atlas by courtesy of the Air Force Geophysical Laboratory.

become available through computer solutions of the equations of statistical equilibrium for a model atom consisting of many levels (see Chapter 3).

2.5.5 Comparison with Observed Center-to-Limb Behavior

The center-to-limb behavior of the line depth can be predicted in our model by inserting the line source function S_λ of the form used in equation (2-70) into equation (2-20) for the emergent intensity, which yields

$$I(0, \mu) = \int_0^\infty [B_\lambda(T) + (1 - \alpha)(J_\lambda - B_\lambda(T))] \frac{1}{\mu} \exp\left(\frac{-\tau_\lambda}{\mu}\right) d\tau_\lambda. \quad (2\text{-}79)$$

A solution can be obtained of the form

$$I_l(0, \mu) = a + p\mu + \frac{(p - a\sqrt{3})(1 - \alpha)}{\sqrt{3}(1 + \sqrt{\alpha})(1 + \mu\sqrt{3\alpha})}. \quad (2\text{-}80)$$

Again we have in the continuum $\alpha = 1$ and $\beta = 0$, so

$$I_c(0, \mu) = a + b\mu. \quad (2\text{-}81)$$

For the case of a pure scattering line $[\eta = 0, \alpha = (1 + \beta)^{-1}]$, and as $\beta \to \infty$, equation (2-80) reduces to

$$I_l(\mu) \equiv 0 \quad (2\text{-}82)$$

for all μ-values, which predicts that strong lines formed by scattering should remain dark to the limb. On the other hand, for a pure absorption line ($\eta = 1$, $\alpha = 1$), we have

$$\frac{I_l(\mu)}{I_c(\mu)} = \frac{1}{1 + (b/a)\mu}\left(1 + \frac{b}{a}\frac{\mu}{1 + \beta}\right) \quad (2\text{-}83)$$

which approaches unity as $\mu \to 0$ at the limb, so the line disappears.

Figure 2-14 shows the observed center to limb behavior of $I_l(\mu)/I_c(\mu)$ for some weak and moderately strong Fraunhofer lines. It can be seen that the

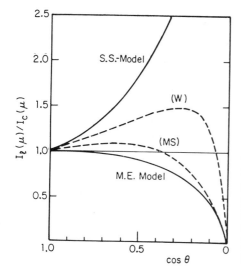

Fig. 2-14 Center-to-limb behavior of the line wing intensity (relative to continuum) for the Milne-Eddington model (M.E.) compared to observed behavior for some weak (W) and moderately strong (MS) lines. S.S. refers to the prediction for the Schuster-Schwarzschild model. Adapted from A. Unsöld, *Physik der Sternatmosphären*, 2nd ed., 1955, by permission of Springer-Verlag, Heidelberg.

curve for the stronger lines shows some evidence of the decrease with μ predicted by the Milne-Eddington model for lines whose formation is dominated by true absorption. However, these tend to be the same lines that require a strong component of resonance scattering to explain their great depth at disk center. Further information on the center-to-limb behavior of some strong lines is given in Table 2-2.

TABLE 2-2   Central Intensities and Equivalent Widths of Some Strong Fraunhofer Lines Measured at $\mu = 1$ and $\mu = 0.3$

Ion	Wavelength (Å)	Central Intensity (% of continuum)		Equivalent Width (Å)	
		$\mu = 1$	$\mu = 0.3$	$\mu = 1$	$\mu = 0.3$
Ca II	3934	6	16	19	16
	3968	6	11	14	12
Mg I	5167	13	18	0.88	0.74
	5177	10	14	1.3	1.3
	5184	9	14	1.6	1.5
Na I	5890	4	6	0.83	0.85
	5896	5	6	0.60	0.59
H I	6563	16	20	4.1	1.9
	4861	15	22	3.8	1.9

Adapted from C. Allen, *Astrophysical Quantities* 3rd ed., Athlone Press, (1976).

The comparison with center-to-limb data thus brings out an important contradiction which (along with the difficulties in providing a quantitative explanation of disk center line depths that were mentioned above) indicates the insufficiency of our simple model for obtaining insight into more than the most basic features of Fraunhofer line formation. However, we will see in Chapter 5 that this model is adequate to obtain an accurate picture of the sun's chemical composition. This capability is one of the most important results of Fraunhofer line analyses.

ADDITIONAL READING

D. Mihalas, *Stellar Atmospheres*, 2nd ed., W. H. Freeman, 1978.

M. Minnaert, "The Photosphere," in *The Sun*, G. Kuiper (Ed). University of Chicago Press, p. 88, 1953.

A. Unsöld, *Physik der Sternatmosphären* (in German), 2nd ed., Springer-Verlag, 1955.

D. Gray, *The Observation and Analysis of Stellar Photospheres*, Wiley, 1976.

EXERCISES

1. Show that the intensity of a source is independent of the detector distance. Also, using Snell's law ($n_1 \sin \theta_1 = n_2 \sin \theta_2$), show that the intensity of a pencil of rays is unaltered in passing from a medium of one refractive index n_1 to another of index n_2. Discuss the implications of these results for solar intensity measurements made through the Earth's atmosphere and through optical systems.

2. Derive the formulas (2-7) and (2-10) for the flux from a plane surface element and for the relation between photospheric emittance and flux detected at 1 AU.

3. Calculate the decrease of total solar irradiance caused by a dark sunspot of area A_S (millionths of the solar hemisphere) and photometric contrast (integrated over wavelength) given by $C_S = I_S/I_{phot}$. Assuming C_S is independent of the spot's heliocentric angle, θ, and taking Eddington limb darkening to hold for the photosphere, show that the relative decrease of irradiance depends on the spot's limb distance according to the formula

$$\frac{\Delta S}{S} = \mu A_S (C_S - 1) \left(\frac{3\mu + 2}{2} \right),$$

where $\mu = \cos \theta$.

4. Given the scattering coefficient per electron $\sigma_e = 0.66 \times 10^{-24}$ cm^2 and the particle density of order 10^7 cm^{-3} characteristic of coronal electrons en-

countered along the line of sight passing some $0.5R_\odot$ above the limb, estimate the optical depth of the white-light corona seen at eclipse. Calculate how much brighter the coronal radiations would be if their density were increased sufficiently to provide an optical depth of unity. Find the mass of a typical comet, and comment on the suggestion that disintegration of a comet in the corona could provide enough material to produce a fireball of sufficient brightness to cause eye damage if viewed from Earth.

5. Calculate a linear regression fit to the photospheric source function $B_\lambda(T)$ given the photospheric temperature profile $T(\tau)$ in Chapter 5 and so determine the coefficients a and b in equation (2-78). Use these values to estimate the maximum central depth of lines formed by thermal absorption in the Milne-Eddington model.

3

Solar Spectroscopy

3.1 A SURVEY OF THE SUN'S SPECTRUM

Newton's experiments in 1666 showed that a prism disperses the sun's white light into the rainbow of brilliant colors that constitute the solar visible spectrum. From measurements corrected for the Earth's atmospheric absorption, we now know that approximately 40% of the sun's radiative output lies in the range of wavelengths detectable to the human eye, between roughly 0.38 μm and 0.7 μm. Ultraviolet radiation of $\lambda \lesssim 0.38$ μm accounts for an additional 7% of the output, and more than half of the sun's power (measured above the Earth's atmosphere) lies in the infrared. The X-ray wavelengths (below ~ 100 Å) and the millimeter and radio waves make a negligible contribution.

The spectral distribution of the sun's radiation as observed from the Earth's orbit is shown in Fig. 3-1. This curve of the flux above the Earth's atmosphere, at the mean sun–Earth distance of 1 astronomical unit or AU, is referred to as the solar spectral irradiance. Its integral over wavelength constitutes the total solar irradiance or "solar constant" introduced in Chapter 2. We see that the maximum irradiance occurs in the blue-green near 0.45 μm. But most of the sun's visible light lies at longer wavelengths. This distribution of solar radiation, together with the eye's peaked response near 0.55 μm, explains the whiteness of sunlight.

The spectral irradiances for wavelengths between 0.1 μm and 50 μm, together with the percentage of the solar constant emitted at wavelengths shorter than a given value of λ, are given in Table 3-1. It is worth remembering that all except 0.1% of the sun's output lies between the wavelengths 0.18 μm and 10 μm. The disk-center intensities and their disk-averaged values for wavelengths in the range 0.2 μm–5 μm are given in Table 3-2. On account of limb darkening, the disk-averaged intensities are lower and peak slightly

66

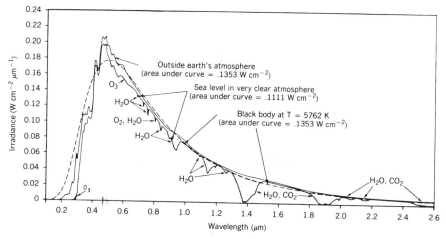

Fig. 3-1 Solar spectral irradiance at ground level and in space, shown together with a black body for $T = 5762$ K. From a curve given by A. Drummond and M. Thekaekara, in *The Extraterrestrial Solar Spectrum*, 1973, by permission of the Institute of Environmental Sciences.

further to the red (as do also the spectral irradiances) than do the disk-center intensities, which arise from a deeper and hotter level in the photosphere.

The solar radiation curve illustrated in Fig. 3-1 lies close to the curve calculated for a 5760 K black body, at wavelengths redward of about 0.5 μm. But in the blue and violet, at $\lambda \lesssim 0.4$ μm, the solar flux is significantly lower than this Planck curve would predict. The reduction in the sun's flux in the blue is caused by the "line blanketing" effect of increasing numbers of closely packed Fraunhofer absorption lines, whose effect is to increase the mean opacity of the atmosphere at those wavelengths. Further in the violet, the continuum opacity also increases, at wavelengths sufficient to ionize relatively abundant neutral metals such as magnesium and aluminum. Both effects increase the atmospheric opacity, so that the relatively hot layers visible near disk center at $\lambda \gtrsim 0.5$ μm can no longer radiate directly into space. Instead we see the cooler strata higher in the photosphere.

Figure 3-2 shows the main absorption lines of the visible spectrum photographed with a large spectrograph. In the red we easily locate the Hα line at 6563 Å, the strongest of the Balmer series of hydrogen. Progressively weaker and more closely spaced lines of the Balmer series are found in the blue-green at 4861 Å (Hβ) and down to the Balmer series limit at 3646 Å in the violet. The strongest photospheric lines in the yellow are the two closely spaced D-lines of neutral sodium at 5896 Å and 5890 Å, which figured prominently in early laboratory analyses of flame spectra by Kirchhoff and others. Moving toward the peak of continuum brightness, we identify the triplet of neutral magnesium, called the Mg I b-lines, at 5184 Å, 5173 Å, and 5167 Å, the

TABLE 3.1 The Solar Spectral Irradiance

λ	E_λ	$D_{0-\lambda}$	λ	E_λ	$D_{0-\lambda}$	λ	E_λ	$D_{0-\lambda}$
0.115	.007	1×10^{-4}	0.43	1639	12.47	0.90	891	63.37
0.14	.03	5×10^{-4}	0.44	1810	13.73	1.00	748	69.49
0.16	.23	6×10^{-4}	0.45	2006	15.14	1.2	485	78.40
0.18	1.25	1.6×10^{-3}	0.46	2066	16.65	1.4	337	84.33
0.20	10.7	8.1×10^{-3}	0.47	2033	18.17	1.6	245	88.61
0.22	57.5	0.05	0.48	2074	19.68	1.8	159	91.59
0.23	66.7	0.10	0.49	1950	21.15	2.0	103	93.49
0.24	63.0	0.14	0.50	1942	22.60	2.2	79	94.83
0.25	70.9	0.19	0.51	1882	24.01	2.4	62	95.86
0.26	130	0.27	0.52	1833	25.38	2.6	48	96.67
0.27	232	0.41	0.53	1842	26.74	2.8	39	97.31
0.28	222	0.56	0.54	1783	28.08	3.0	31	97.83
0.29	482	0.81	0.55	1725	29.38	3.2	22.6	98.22
0.30	514	1.21	0.56	1695	30.65	3.4	16.6	98.50
0.31	689	1.66	0.57	1712	31.91	3.6	13.5	98.72
0.32	830	2.22	0.58	1715	33.18	3.8	11.1	98.91
0.33	1059	2.93	0.59	1700	34.44	4.0	9.5	99.06
0.34	1074	3.72	0.60	1666	35.68	4.5	5.9	99.34
0.35	1093	4.52	0.62	1602	38.10	5.0	3.8	99.51
0.36	1068	5.32	0.64	1544	40.42	6.0	1.8	99.72
0.37	1181	6.15	0.66	1486	42.66	7.0	1.0	99.82
0.38	1120	7.00	0.68	1427	44.81	8.0	.59	99.88
0.39	1098	7.82	0.70	1369	46.88	10.0	.24	99.94
0.40	1429	8.73	0.72	1314	48.86	15.0	4.8×10^{-2}	99.98
0.41	1751	9.92	0.75	1235	51.69	20.0	1.5×10^{-2}	99.99
0.42	1747	11.22	0.80	1109	56.02	50.0	3.9×10^{-4}	100.00

Note: The wavelength λ is given in μm, E_λ is the irradiance in W m^{-2} μm^{-1} averaged over a small band centered at λ, and $D_{0-\lambda}$ is the percentage of the solar irradiance at wavelengths shorter than λ. Adapted from A. Drummond and M. Thekaekara, The Extraterrestrial Solar Spectrum, 1973, by permission of the Institute of Environmental Sciences.

TABLE 3-2   Solar Spectral Intensity and Flux Values

μm	I_λ [10^{10} ergs cm^{-2} (sr μm s)$^{-1}$]	$\bar{I}_\lambda$	f_λ [erg cm^{-2} (Å s)$^{-1}$]
0.20	0.03	0.02	1.3
0.30	1.21	0.76	52
0.40	2.9	2.05	140
0.45	3.86	2.90	198
0.50	3.63	2.83	193
0.60	3.16	2.58	175
0.70	2.50	2.10	144
0.8	1.96	1.69	115
1.0	1.21	1.08	73
1.6	0.403	0.375	25.5
2.0	0.183	0.171	11.6
3.0	0.041	0.0386	2.6
5.0	0.0057	0.0055	0.4

Note: I_λ is the disk-center intensity, $\bar{I}_\lambda$ is the mean intensity of the solar disk, and f_λ is the spectral irradiance at 1 AU. Adapted from C. Allen, *Astrophysical Quantities*, 3rd ed., Athlone Press, (1976).

strongest lines in the green. The increasing packing density of lines toward the shorter wavelengths makes it difficult to find the true continuum level below 0.45 μm. Most of these lines are transitions in the rich spectra of neutral iron, titanium, and other metals. The strongest atomic lines in the visible spectrum are the H- and K-lines of singly ionized calcium at 3968 Å and 3934 Å respectively. Although their depth relative to continuum is roughly 98%, they are only dimly perceived in the violet near the limit of the eye's sensitivity when we look into a spectrograph eyepiece to focus the spectrum. The absorption of molecules is most easily seen in Fig. 3-2 in the hydrocarbon (CH) band near 4300 Å.

The strong Fraunhofer lines have all been identified with atomic and molecular transitions whose intensities and wavelengths are known from laboratory measurements, but the task of identifying the thousands of weaker lines is much more difficult. Roughly 22,000 lines have been listed between 2935 Å and 8770 Å in the most recent revision of the Rowland Solar Spectrum Tables. Roughly 70% of these lines are well identified with some 68 elements.

Multiplet rules that govern the relative intensity of groups of lines of a given element greatly facilitate identifications. The Doppler shift of solar rotation at the east and west limbs is used to discriminate solar and terrestrial lines. Other factors such as a line's asymmetry, sharpness, and width are helpful, as well as its relative strength in the photosphere and in the spectrum of a (cooler) sunspot. The amount of its magnetic splitting in the sunspot magnetic field can also assist in identification. The most recent work on line

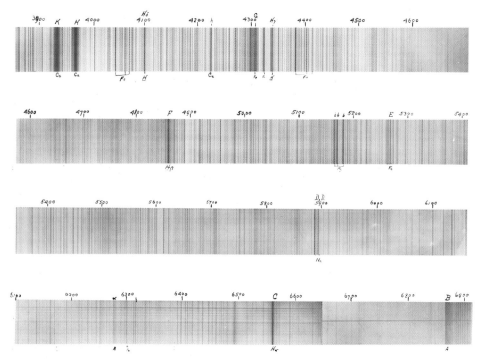

Fig. 3-2 The solar spectrum between 3900 Å and 6900 Å. By permission of Mt. Wilson and Las Campanas Observatories, Carnegie Institution of Washington.

identifications is based on comparisons between the observed spectrum and that calculated from detailed photospheric models which include the atomic parameters for millions of transitions.

In the ultraviolet, the character of the solar spectrum changes from absorption lines superposed on a continuum at $\lambda \gtrsim 2000$ Å to a spectrum dominated by strong emission lines at $\lambda \lesssim 1500$ Å, as illustrated in Fig. 3-3, The reason for this change is not difficult to understand. The continuum intensity expected from a black body at photospheric temperatures follows a rapid exponential decrease from the peak near 0.45 μm into the ultraviolet (see exercises). The lines go into emission relative to this weak continuum, because their opacity is much greater, so we see the more intense radiation of higher and hotter chromospheric layers above the temperature minimum. By far the strongest of these is the Lyman α resonance line of hydrogen at $\lambda = 1216$ Å;

Fig. 3-3 The solar ultraviolet spectrum between approximately 1400 Å and 280 Å obtained in 1969 by the Harvard spectrometer on the *OSO 6* satellite. The upper spectrum in each of the three panels is of an active region, the lower refers to the quiet sun. Solid calibration curves are in ergs cm^{-2} s^{-1} sr^{-1}, dashed are in photons cm^{-2} s^{-1} sr^{-1}. From A. Dupree et al. *Astrophys. J.* **182**, 321 (1973).

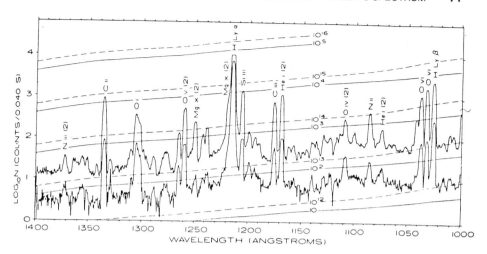

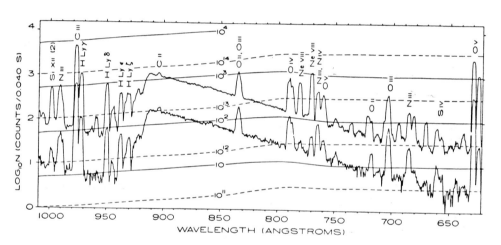

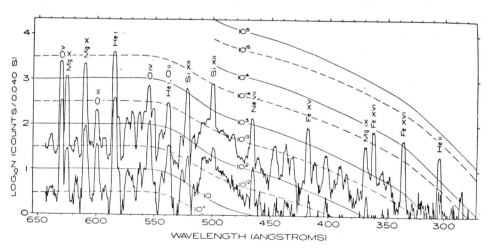

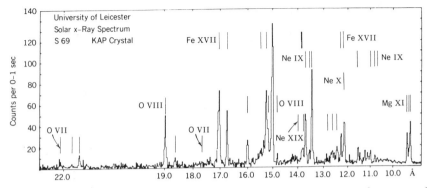

Fig. 3-4 The soft X-ray spectrum of an active region, between 22 Å and 10 Å, recorded by a spectrometer flown by J. Parkinson of the University of Leicester. Reproduced by permission from *Ann. Rev. Astron. Astrophys.*, **12** (1974). Copyright 1974 by Annual Reviews Inc.

its flux exceeds the combined emissions at all shorter wavelengths. The Lyman continuum of hydrogen is also prominent in emission at $\lambda < 912$ Å.

Most of the ultraviolet emission lines turn out to be optically thin on account of the relatively low densities encountered above the temperature minimum, so their intensity is governed by the relation $I_\lambda = \tau B_\lambda(T)$, where $\tau \ll 1$. But the intensities can still be high, because some, such as the strong resonance lines of nine times ionized magnesium (Mg x) at $\lambda = 610$ Å and 625 Å, are formed at temperatures as high as 10^6 K or more in the corona. Optically thin emission lines from highly ionized metals also dominate the solar X-ray spectrum, portions of which are illustrated in Figs. 3-4 and 3-5.

In the infrared, as illustrated in Fig. 3-6, the spectrum remains a continuum with dark absorption lines, but their number decreases rapidly beyond about 1 μm. Atlases of the solar spectrum show hundreds of lines throughout the infrared, but almost all of these are molecular absorptions by the Earth's atmosphere. Figure 3-6 shows a portion of the infrared spectrum around 12 μm, where the Earth's atmosphere is relatively transmissive. This spectrum shows some weak solar OH molecular lines. Only a few dozen solar atomic absorption lines (belonging to neutral metals) have been identified beyond 2 μm, and they disappear altogether at wavelengths beyond about 10 μm. The Paschen α- and Brackett α-lines of hydrogen are detected near 1.8 μm and 4 μm, and the fundamental and first overtone bands of the solar CO molecule are seen around 4.5 μm and 2.3 μm. No solar emission lines were known in the infrared until the recent discovery of some weak features near 12 μm (Fig. 3-6), identified as upper level transitions in Mg I.

The photospheric opacity in the infrared increases roughly as λ^2 from its minimum value near 1.65 μm (we see deepest into the sun near that wavelength), so that observing at longer wavelengths we see progressively higher layers in the photosphere. At wavelengths between 75 μm and 300 μm the

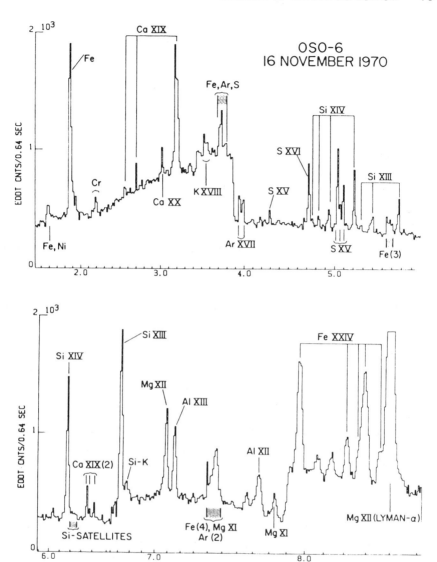

Fig. 3-5 The hard X-ray spectrum of a flare between about 1 Å and 9 Å recorded by a spectrometer flown by the Naval Research Laboratory on the *OSO 6* satellite. The spectrum is dominated by lines of hydrogenlike and heliumlike ions. The abrupt edges in the continuum are of instrumental origin. Reproduced by permission of Kluwer Academic Publishers, from G. Doschek in *International Astronomical Union Symposium*, No. 68, S. Kane, (Ed.), p. 67, 1975.

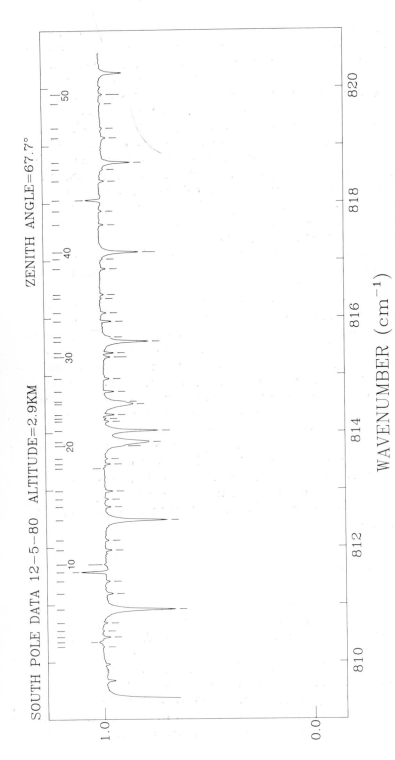

Fig. 3-6 Solar infrared spectrum between about 12.20 and 12.35 μm showing the recently discovered infrared emission lines and some weak solar OH lines. The strongest absorptions are all due to terrestrial CO_2 and H_2O. The data were obtained from a balloon-borne spectrometer flown by the University of Denver. By permission of D. Murcray.

radiation originates in the temperature-minimum region where temperatures below 4000 K have been reported. At even longer wavelengths, the radiation arises from the chromospheric layers above the temperature minimum, where temperatures rapidly rise to values around 7000 K for 3-mm radiation.

Given the minimum of temperature in the photosphere, and the increasing opacity through the infrared region, we can understand the transition between absorption and emission lines observed in the infrared. In the visible or near infrared, the continuum radiation originates at levels well below the height of T_{min}, so the increase of opacity encountered in a line results in less intense line radiation from the progressively cooler upper photospheric layers, and thus a dark absorption line. But further in the infrared, the continuum itself arises from higher photospheric layers close to the level of T_{min}, so that the additional opacity in a line results in more intense radiation from the hot chromosphere above the height of T_{min}, and thus an emission line. Models of the photosphere discussed in Chapter 5 confirm this basic picture and predict that the transition between absorption and emission lines should occur around $\lambda \sim 10$–15 μm, roughly as observed.

The general shape and intensity of the continuum, and the qualitative behavior of the Fraunhofer lines, can be understood from thermodynamic principles, as discussed above. But most of the fundamental questions arising from observations of the solar spectrum require a look at the mechanics of atoms and molecules. For instance, the observation that the H- and K-lines of singly ionized calcium are the strongest absorption lines in the solar visible spectrum is surprising considering that calcium is almost a million times less abundant at the photosphere than hydrogen. On the other hand, helium, the second most abundant element in the sun, emits strongly in the ultraviolet but hardly reveals its presence at all in the absorption spectrum, except for weak lines at 5876 Å (the D_3-line) and at 10,830 Å. To understand these and many other questions that arise in the quantitative study of the solar spectrum, we now assemble the atomic physics required to relate the measured properties of solar radiations to the conditions of chemical composition and excitation in the sun's atmosphere.

3.2 ATOMIC STRUCTURE

We begin our brief review of the aspects of atomic structure most relevant to our discussions of solar phenomena with a diagram illustrating the energy levels of the hydrogen atom, shown in Fig. 3-7. The discrete, bound levels of negative total atomic energy ($E < 0$), are defined by solutions of the Schrödinger equation. They correspond to standing waves of the electron probability distribution function. The total energy of the hydrogen atom in these levels is to a very good approximation the sum of the electron kinetic energy $\frac{1}{2}m_e v_e^2$, and its electrostatic interaction energy with the nucleus, $z_e^2/4\pi\epsilon_0 r$, where m_e, v_e, e, and r are respectively the electron mass, velocity,

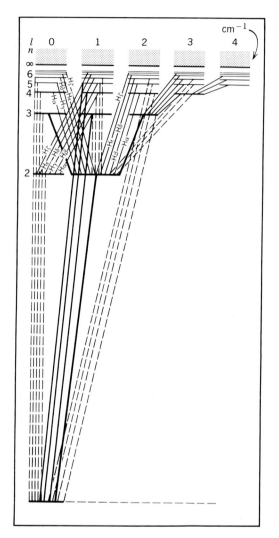

Fig. 3-7 Energy levels of the H I atom. Transitions giving rise to the strongest Balmer lines are indicated by letters. Reproduced by permission from G. Herzberg, *Atomic Spectra and Atomic Structure*, Dover, 1945.

charge, and radial distance from the nucleus of charge number $Z = 1$. The two particles constituting the hydrogen atom can also exist as an ion and an unbound electron of kinetic energy $\frac{1}{2}m_e v_e^2$. This ionized atom has unbound continuum levels of positive energy $E > 0$, as shown also in Fig. 3-7.

Each bound energy level of the atom is characterized by four quantum numbers. For hydrogen, the Bohr theory associates the principal quantum number n, with the semimajor axis $n^2 a_0 / Z$ of the electron orbit, where $a_0 = 0.529$ Å is the radius of the first Bohr orbit ($n = 1$) of hydrogen. These n energy levels are further divided into fine structure by the different states of orbital and spin angular momentum of the electron. The orbital angular momentum is measured in units of $l \times h/2\pi$, where l is allowed the values $0, 1, 2, 3, \ldots, n - 1$. An electron is designated by the letter s, p, d, f, according to whether $l = 0, 1, 2, 3$, for that individual electron. The spin angular momentum of the individual electron that codetermines the hydrogen atom's fine structure state takes the values $s \times h/2\pi$, where $s = \pm \frac{1}{2}$. The total angular momentum of the single hydrogen electron is then the vector sum $\mathbf{j} = \mathbf{l} + \mathbf{s}$. In the hydrogen ground state of $n = 1$, for instance, the electron's orbital angular momentum is $\mathbf{l} = 0$, and its total angular momentum is $\mathbf{j} = \frac{1}{2}$; it is referred to as a $1s_{1/2}$ electron.

The group of energy levels corresponding to all possible combinations of a given set of l and s values is called a term. The number of these possible combinations is called the multiplicity of the term and is given by $r = 2s + 1$, provided $l > s$. If $l < s$, it is equal to $2l + 1$. In hydrogen, all terms having $l > s$, i.e., $l > \frac{1}{2}$, yield $r = 2$ and thus are doublets. Even though the $l = 0$ terms are single, they are still referred to somewhat confusingly as doublets.

The differences in energy of the hydrogen fine structure levels are caused in part by the electromagnetic interaction between the orbital and spin magnetic moments of the electron which we describe later. In part, the energy differences are caused by differences in the electron's kinetic and potential energy for different angular momentum states (l) when relativistic effects are taken into account. To order of magnitude, the fine structure splitting of the nth hydrogen level is given by $\Delta E = E_n \alpha^2 / 4n^2$, where $\alpha = 1/137$ is the fine structure constant. ΔE is also a rapidly decreasing function of l. For $n = 1$, $\Delta E / E \sim 10^{-4}$, so the hydrogen fine structure energy splitting is too small compared to differences in energy of consecutive levels of increasing n, to be seen in Fig. 3-7. The splitting of the terms also decreases rapidly for the levels of increasing n.

In more complicated atoms or ions with several electrons, the fine structure splitting can be much larger, compared to energy differences caused by changes in the principal quantum number n. For these multielectron atoms or ions, it is most often found that the total angular momentum $\mathbf{J}$ of the atom can be represented as the vector sum $\mathbf{J} = \mathbf{L} + \mathbf{S}$ of the total orbital angular momenta $\mathbf{L} = \Sigma \mathbf{l}_i$ and of the total spin angular momenta $\mathbf{S} = \Sigma \mathbf{s}_i$. The quantum number J can have values $|L - S| \leq J \leq L + S$. This scheme of combining the individual angular momenta is called LS or Russell-Saunders coupling. As in hydrogen, a pair of L and S values again constitutes a term.

In analogy to the s, p, d, f designation of individual electrons, the total orbital angular momentum yields terms S, P, D, F corresponding to $L = 0, 1, 2, 3$. Again, provided $L > S$, the term splits into $r = 2S + 1$ energy levels with different J, and $r = 2S + 1$ denotes the multiplicity of the term even if

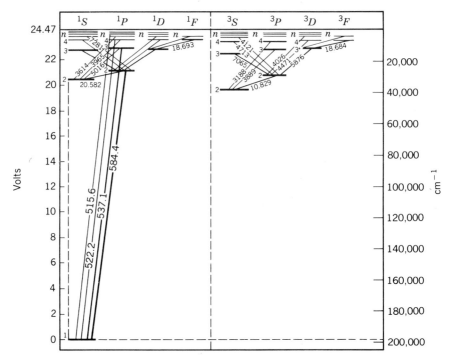

Fig. 3-8 Energy levels of the He I atom. Excitation energies are given in volts and in inverse wavelength units. Reproduced by permission from H. White, *Introduction to Atomic Spectra*, McGraw-Hill, 1934.

$L < S$ and the number of levels is given by $2L + 1$. The multiplicity is placed to the upper left of the uppercase letter denoting the orbital angular momentum of the term, e.g., $^1S_0$. The J-values of the levels that make up the term are placed to the lower right.

The energy levels of neutral helium are illustrated in Fig. 3-8. The two electrons orbiting the helium nucleus can line up so that their spins are either antiparallel ($S = 0$) or parallel ($S = 1$). In the $S = 0$ case, the multiplicity $r = 2S + 1 = 1$, and we have singlet terms ("parhelium") including the neutral helium ground state, which is $^1S_0$. When $S = 1$, we obtain a set of triplet terms ("orthohelium") each of which is split into three separate energy levels too close together to illustrate in Fig. 3-8. The He I ionization potential of 24.47 V is the energy required to remove the electron to the continuum from the $^1S_0$ ground state.

In multielectron atoms, the fine structure within a configuration of given principal quantum number n is caused by several effects. In order of importance in the two most common cases these are differences in spin-spin correlation energy of the multiple electrons (differences between, e.g., triplets and singlets) or equivalently in their mutual electrostatic energy (between, e.g.,

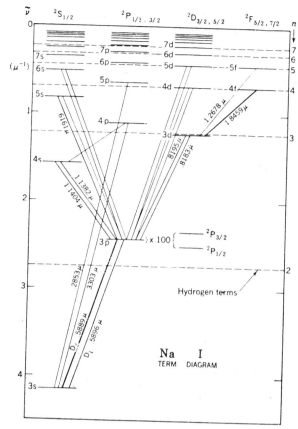

Fig. 3-9 Energy levels of the Na I atom. Reproduced by permission from H. White, *Introduction to Atomic Spectra*, McGraw-Hill, 1934.

^3P and ^3D terms) and in their spin-orbit energy. This ordering in importance of the effects gives rise to the so-called Russell-Saunders or *LS* coupling mentioned above. It can be recognized by much smaller energy-level differences within a term than between terms (i.e., weaker spin-orbit interaction than mutual electrostatic or spin correlation interactions). Less frequently, we find that the spin-orbit interaction is more important than electrostatic and spin effects, giving rise to so-called *j-j* coupling.

It can be seen from the energy-level diagrams of neutral helium and of neutral sodium (Fig. 3-9) that the terms of different orbital angular momentum, e.g., the S or P terms, can be as widely separated in these heavier atoms as terms of successive *n*-values. In solar spectroscopy, the fine structure of even the lowest hydrogen levels merely produces a small broadening of the levels (0.15 Å for the $n = 2$ level that gives rise to the Balmer series), but the fine structure levels of heavier atoms are sufficiently separated to give rise to widely separated lines. For instance, the sodium D-lines at 5889 Å and 5896 Å comprise a doublet arising from transitions between the $^2S_{1/2}$ ground term ($n = 1$) of Na I and the $^2P_{1/2,3/2}$ term of $n = 2$ (see Fig. 3-9).

3.3   SPACE QUANTIZATION AND THE ZEEMAN AND STARK EFFECTS

The fourth quantum number M represents the component of the total angular momentum $\mathbf{J}$ in the direction of an external magnetic or electric field. From experimental results on the deflection of neutral atoms in inhomogeneous fields, the allowed values of M are found to be limited to integral (when J is integral) or half-integral (J half-integral) multiples of $h/2\pi$. This space quantization of the $\mathbf{J}$ vector determines the allowed orientations of the atom's total magnetic moment relative to an external magnetic field and thus its magnetic energy. It also determines the values of electrostatic energy that the atom can assume in an electric field. This quantization in external magnetic and electric fields leads to the splitting of spectral lines referred to respectively as the Zeeman and Stark effects.

3.3.1   The Zeeman Effect

The magnetic moment of the atom is produced in part by the current loop formed by the electrons orbiting the nucleus. The magnetic moment for a single electron is calculated classically to be

$$\mu_e = -\frac{1}{2}\frac{e}{m}\mathbf{p}, \tag{3-1}$$

where e and m are the electron's charge and mass (assumed in this classical picture to be uniformly distributed around the Bohr orbit) and $\mathbf{p}$ is its angular momentum. In addition, the spin angular momentum σ of the electron is assumed in quantum mechanics to produce a magnetic moment given by

$$\mu_s = -\frac{e}{m}\sigma. \tag{3-2}$$

The interaction between the magnetic moments associated with the electrons' spin and orbital motions leads to precession of both the angular momentum vectors $\mathbf{L}$ and $\mathbf{S}$. When an external magnetic field is applied, the $\mathbf{J}$ vector processes about the field direction.

As shown in Fig. 3-10, the constraint that M assume integral or half-integral values means that $\mathbf{J}$ can only align itself so that its component M in the field direction assumes the $2J + 1$ discrete values $J, J - 1, \ldots, -J$. These $2J + 1$ levels are separated by a constant energy difference for the following

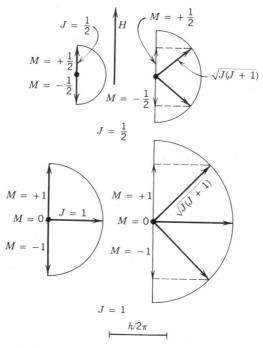

Fig. 3-10 Schematic diagram illustrating space quantization for $J = \frac{1}{2}$ (top) and $J = 1$ (bottom), To the left is the "naive" representation taking the magnitude of the total angular momentum equal to $J(h/2\pi)$. To the right it is given the (exact) magnitude $\sqrt{J(J+1)}\,h/2\pi$.

reason. The total energy W of an atom (or a compass needle) of magnetic moment μ in an external magnetic field **B** is

$$W = W_0 - (\mu \cdot \mathbf{B}), \qquad (3\text{-}3)$$

where W_0 is the total energy in the absence of an external field. Given the dependence of the orbital and spin magnetic moments of the electron [equations (3-1) and (3-2)] on the orbital and spin angular momenta, it is clear that the atom's total magnetic moment should depend on the vector sum of **L** and **S**. The way in which these are combined to form μ is beyond the scope of this discussion, but the magnitude of the result is

$$\mu = +\frac{1}{2}\left(\frac{e}{m}\frac{h}{2\pi}\right)gM. \qquad (3\text{-}4)$$

The dependence of μ on L and S is contained in the value of the Landé g-factor, whose value is given by

$$g = 1 + \frac{J(J + 1) + S(S + 1) - L(L + 1)}{2J(J + 1)}, \tag{3-5}$$

assuming the atom has a magnetic moment (i.e., $J \neq 0$).

Substituting equation (3-4) into (3-3), we have

$$W = W_0 - \frac{1}{2}\left(\frac{e}{m}\frac{h}{2\pi}\right)gM. \tag{3-6}$$

This result shows that the external magnetic perturbation $\Delta W = W - W_0$ to a level of given J, L, and S is proportional to M, so that the energy sublevels defined by the $2J + 1$ equally spaced values of M are also equally spaced in energy. The Zeeman structure of a transition between two levels of different J is determined by the g-value of each level (which determines the amount of splitting between its $2J + 1$ sublevels) and by the selection rule that governs transitions between the upper and lower levels of the line. This selection rule is $\Delta M = 0, \pm 1$, with the provision that $M = 0$ to $M = 0$ is forbidden if $\Delta J = 0$.

Figure 3-11 illustrates the simplest case: g is the same for the upper and lower levels, so that the spacing of all the levels is proportional to M alone and thus constant. In this relatively rare case of "normal" Zeeman effect, all transitions governed by $\Delta M = 0$ produce radiation of the same wavelength, as do those of $\Delta M = +1$ and $\Delta M = -1$. Figure 3-11 also illustrates how this leads to splitting of a singlet transition into three components: the $\Delta M = 0$ component lies at the unshifted wavelength λ_0, and the $\Delta M = +1, -1$ are shifted to the red and blue respectively by the wavelength difference

$$\Delta\lambda = \frac{\pi e}{m_e}\frac{\lambda^2}{c}gB = 4.7 \times 10^{-13}\lambda^2 gB. \tag{3-7}$$

Here, λ is in angstroms and B in gauss.

Typical values of g for Fraunhofer lines lie between 0 and 3. Lines of $g = 0$ are optimal for Doppler velocity measurements since line width variations due to the Zeeman effect are avoided. The splitting of a $g = 2.5$ line such as Fe I λ 6302.5 in a sunspot field of $B \sim 3000$ G, is roughly 0.15 Å, which can easily be detected and measured with a ruler on a spectrogram of a sunspot such as Fig. 3-12.

A solar magnetograph can detect much smaller magnetic fields of order 0.05 G, which broaden the line profile by much less than a milliangstrom. To achieve this sensitivity, the Babcock magnetograph (see Chapter 1) relies on the polarization structure of a Zeeman triplet illustrated in Fig. 3-13. In this classical, but still useful, model first put forward by Zeeman in 1896, the

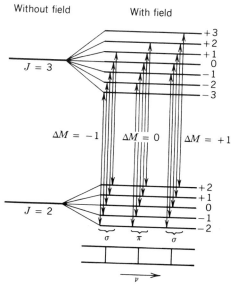

Fig. 3-11 Schematic diagram illustrating line splitting under normal Zeeman effect for a combination $J = 3$ to $J = 2$. The arrows representing transitions form three groups indicated by the braces. The arrows in each group are of equal length and thus give rise to only the three lines shown below. Reproduced by permission from G. Herzberg, *Atomic Spectra and Atomic Structure*, Dover, 1945.

unshifted ($\Delta M = 0$) component can be visualized as arising from a linear vibration along the magnetic vector, while the shifted ($\Delta M = \pm 1$) components arise from circular motions of the emitting charge (in opposite directions) around the field line.

Observing this emitter at right angles to the field direction, we see the unshifted component (called the π-component), polarized linearly along the field, and the two wing components (called the σ-components), polarized linearly transverse to the field. If, on the other hand, we observe along the field direction, then the π-component is extinguished (since vibrations along the line of sight do not produce a transverse electromagnetic wave), and we see only the two σ-components, polarized circularly in opposite senses. The polarization seen from an arbitrary angle to the field vector can then be visualized as the vector sum of the polarizations seen along these two orthogonal axes, e.g., the σ-components will in general be elliptically polarized. The relative fluxes in the individual σ- and π-components of a normal triplet are equal, but the relative intensities observed will vary according to viewing angle, most obviously for the π-component, which is extinguished when observed along the field.

The Babcock magnetograph makes use of this polarization structure by admitting into the spectrograph light that is alternately left- or right-circularly

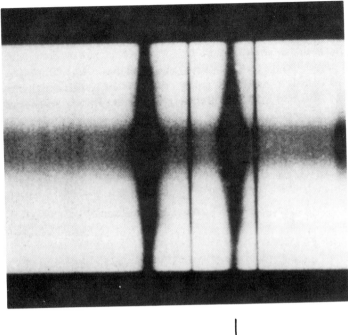

6303 Å

Fig. 3-12 Zeeman effect in a sunspot, observed in the λ 6303 line of Fe I. The two narrow lines arise from absorptions of terrestrial oxygen. From H. von Kluber, in *Solar Physics*, C. Xanthakis (Ed.), 1967, by permission of Wiley-Interscience Publishers.

polarized (see Chapter 1 and Fig. 1-13). But it will not detect even a strong field oriented transverse to the line of sight since it operates only on the magnetic **B**-field component along the line of sight, which gives rise to circular polarization in the σ-components. This is an important restriction, and vector polarimeters have been built to also measure the linearly polarized components of the Zeeman-affected lines. These are difficult measurements, since spurious linear polarization can arise, for instance, through oblique reflections from metallic surfaces on mirrors. Nevertheless, useful data on the complete magnetic vector (not only its longitudinal component) have been obtained, as will be seen in later chapters. It should be noted that any measurement of actual splitting of a Zeeman triplet (such as Hale and his collaborators first achieved in the very intense fields of spot umbrae—see Fig. 3-12) directly measures the magnitude of the magnetic vector.

Since in most cases the *g*-values of the upper and lower levels of a line are not equal, the occurrence of normal Zeeman effect is rare. More often the line splits into a large number of components. But even in the complex patterns of these so-called anomalous components, the σ-components lie in the wings (the

Fig. 3-13 Zeeman's (1896) model of the polarization structure of a Zeeman triplet. From H. von Kluber, in *Solar Physics*, C. Xanthakis, (Ed.), 1967, by permission of Wiley-Interscience Publishers.

two senses of opposite circular polarization occupy opposite wings) and the π-components lie near the center of the pattern. For this reason, successful operation of the Babcock magnetograph does not require that the line used exhibit normal Zeeman effect.

3.3.2   The Stark Effect

As mentioned previously, space quantization of J also takes place in an electric field, and M is again limited to the projections $+J, J - 1, \ldots, -J$ on the electric-field direction. An electric field interacts with the atom by displacing the centers of gravity of the positive nucleus and negative electrons within the atom, and thus inducing an electric dipole moment if there is not one already. This electric dipole attempts to set itself in the direction of the imposed field, and in general the energy perturbation ΔW is again given by equation (3-3), where μ is now the electric dipole moment and $\mathbf{B}$ is replaced by $\mathbf{E}$. In the case of the induced dipole moment, the moment is itself proportional to the intensity of the applied electric field, giving rise to so-called quadratic Stark effect.

In hydrogen, the linear effect predominates at the relatively low field strengths of interest in the solar atmosphere. An important difference between the electric and magnetic dipole moments is that the electric dipole moment does not change sign if the sign of J (i.e., the net sense of motion of the electrons) is flipped. This means that the dipole moment, and thus the sign of the energy perturbation ΔW in equation (3-3), is the same for the positive and negative values of M. The number of Stark term components is therefore just $J + 1$ or $J + 1/2$ depending on whether J is integral or half-integral.

Electric fields of the intensity ($E \gtrsim 10$ V cm^{-1}) required to produce an observable Stark effect in hydrogen (greater fields are required to produce the quadratic Stark effect observed in heavier atoms) exist on interatomic scales in the dense photospheric plasma, through electrostatic interactions between atoms and their immediate neighbors. This Stark effect is discussed later as pressure broadening of Fraunhofer lines. The high electrical conductivity of the solar plasma makes it difficult to sustain significant electrostatic fields over spatial scales much larger than the plasma Debye length (equation 4-92) and over time scales exceeding the plasma frequency (see Chapter 4).

This reasoning has so far inhibited exploitation of the Stark effect as a plasma diagnostic to anywhere near the level of the work on the Zeeman effect, which has been extremely fruitful in probing the magnetic fields of the sun's photosphere and, to a lesser extent, the chromosphere and corona. This situation is likely to change with increasing recognition that both the dc and wave-related electric fields that seem to be present in energy release and charged particle acceleration within flares and other transient solar phenomena, may be detectable through measurements of Stark effect in the neutral hydrogen component of these plasmas.

A limited number of solar measurements of Stark effect have been made from the Balmer emission line broadening in flares, prominences, and the limb chromosphere. These measurements are based on the fact that, in hydrogen, the shift ΔW of the Stark components from their field-free energies increases roughly as the square of the upper principal quantum number, through the formula

$$\Delta W = \frac{3h}{8\pi^2 mec} n(n_2 - n_1)E = 6.42 \times 10^{-5} n(n_2 - n_1)E, \qquad (3\text{-}8)$$

where E is the electric field in V cm^{-1}, and ΔW is the shift of the term values in wave numbers.

Here n is the principal quantum number of the term, and n_2, n_1 are the electric quantum numbers with allowed values

$$n_1 = 0, 1, 2, 3, \ldots, n - 1 \quad \text{and} \quad n_2 = 0, 1, 2, 3, \ldots, n - 1.$$

From the way in which the allowed values of n_1, n_2 can be combined it can be shown that the maximum values of the term shifts, and thus of the line broadening caused by transitions between such Stark-affected levels, increases roughly as the square of n. Formula (3-8) predicts wavelength shifts $\Delta\lambda = -\lambda^2 \Delta W / hc$, thus of order 0.1 Å for $E = 10^2$ V cm^{-1} at $n = 20$.

The Stark components of hydrogen lines are distributed symmetrically around the unshifted wavelength of the line (Fig. 3-14), although the asymmetric shifts seen in Stark effect of heavier elements gives rise to displacements of

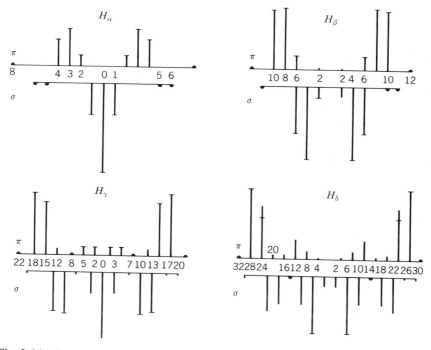

Fig. 3-14 Stark-effect splitting and polarization structure in the four strongest Balmer lines of hydrogen. The lengths of the lines show the theoretically calculated relative intensities of the components. From E. Condon and G. Shortley, *The Theory of Atomic Spectra*, 1959. Reprinted by permission of Cambridge University Press.

the line centers that can be of comparable size to the line broadening itself. The linearly and circularly polarized π- and σ-components shown in Fig. 3-14 can be understood from a model similar to that used for the Zeeman components and shown in Fig. 3-13. An important difference is that in the Stark effect the circularly polarized σ-components lie inside the π-components (Fig. 3-14). When observing transverse to **E**, one sees them linearly polarized transverse and parallel to the field vector, as for the Zeeman effect. Another difference is that when viewed along the field, the visible σ-components are unpolarized rather than oppositely circularly polarized as in Zeeman effect.

3.4 MULTIPLET RULES FOR TRANSITIONS

The energy level diagrams for H I, He I, and Na I in Figs. 3-7, 3-8, and 3-9 also show some of the principal transitions. A line arises from a transition between two energy levels. The set of possible transitions between all the levels in two terms produce a neighboring group of lines called a multiplet. The allowed electric dipole transitions are restricted in LS coupling by the following selection rules.

(a) Transitions occur only between terms of opposite parity, i.e., levels in a term for which the arithmetic sum Σl_i is odd or even. Odd terms are sometimes designated with a "0" to the upper right, e.g., $^2P^0_{3/2}$.

(b) For the atom as a whole, $\Delta S = 0$, $\Delta L = 0, \pm 1$, $\Delta J = 0, \pm 1$ (but $J = 0$ to $J = 0$ is forbidden), and $\Delta M = 0, \pm 1$ (but $M = 0$ to $M = 0$ is forbidden if $\Delta J = 0$).

Referring to the hydrogen energy-level diagram (Fig. 3-7), we see that allowed transitions in the Balmer series, for example, arise between $n = 2$ and any upper levels $n > 2$, and only transitions of $\Delta l = \pm 1$ are allowed. In principle, Balmer α contains fine structure components arising from transitions between the three $n = 2$ fine structure levels $^2S_{1/2}$, $^2P_{1/2,3/2}$, and the five $n = 3$ levels $^2S_{1/2}$, $^2P_{1/2,3/2}$, and $^2D_{3/2,5/2}$. In fact, the transitions that determine hydrogen fine structure do not follow simply from the selection rules for LS coupling since the multiplet splitting is larger than the term-to-term energy differences, which are due mainly to relativistic corrections in the orbital energy of the single electron. Five Balmer α components have been predicted, but only two have been resolved, even using special precautions to minimize line width and with very high dispersion spectrometers.

In neutral helium, the resonance line at 584 Å arises from a transition between the ^1S ground ($n = 1$) state and the 2^1P level; the selection rules ($\Delta S = 0$) prohibit intercombination of the singlet and triplet terms. The large energy difference between the singlet $n = 1$ and $n = 2$ states makes it difficult to populate these singlet upper levels by collision from the 1^1S state. On the other hand, since the triplets cannot decay to the singlet ground state, a considerable population of the 2^3S term can accumulate. Relatively small energies are required for upward excitation from this term to the ^3P terms of $n = 2, 3, 4$. The resulting downward transitions produce some of the strongest neutral helium radiations of the solar chromosphere at $\lambda\lambda$ 10,830, 3889, and 3188 respectively.

Transitions that do not obey the electric dipole selection rules are called forbidden. Although these transitions do not correspond to changes in atomic electric dipole moment which are responsible for the energy emitted or absorbed in the allowed transitions, they do often produce much weaker lines through changes in the atom's electric quadrupole or magnetic dipole moments. The broken lines in the energy-level diagrams for H I and He I show some forbidden transitions. Some of the best-known examples in solar spectroscopy are the "coronium" lines in the green (λ 5303) and red (λ 6374), which arise from transitions between the fine structure levels within the ground terms of highly ionized iron. The red line λ 6374 arises from a transition between the $^2P_{1/2}$ and $^2P_{3/2}$ fine structure levels of the ground state

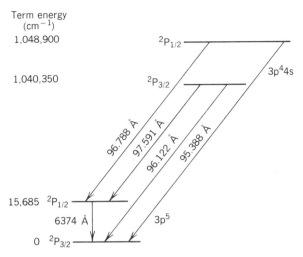

Fig. 3-15 Term splitting in the ground state of Fe x giving rise to the λ 6374 forbidden coronal line.

of Fex (see Fig. 3-15), while the green line is emitted by a transition between $^2P_{3/2}$ and $^2P_{1/2}$ levels of the Fe xiv ground term. The resonance lines of these highly stripped ions lie in the soft X-ray region around 100 Å, so the fine structure transitions in the visible region provided the only means of observing actual emissions from coronal ions until after World War II, when spectrometers on rockets recorded the ultraviolet and X-ray resonance lines of these species for the first time.

It is worth noting that the selection rules for emission and absorption given here do not cover all possible types of atomic transitions. For instance, the process of Rayleigh scattering (which accounts for the blue color of the daytime sky) is a continuum scattering process, although the scattering particle is a molecule with bound states that should absorb selectively, from all that has been said so far. Rayleigh scattering turns out to be an example of a two-photon scattering process, for which different selection rules apply.

3.5   ATOMIC TRANSITIONS AND THEIR EXCITATION

Before entering into the calculation of excitation and transition rates, we summarize the basic types of atomic transition processes found to be of greatest importance in the solar atmosphere. These are

(a) *Discrete bound-bound transitions*, which produce all the emission and absorption lines in the solar spectrum. The wavelengths of these lines are given by $\lambda = hc/\Delta E$, where ΔE is the energy difference of the two levels, h is the Planck constant, and c is the light speed. For the Lyman α-line at $\lambda = 1216$ Å, $\Delta E = 10.2$ eV.

(b) *Free-bound or bound-free transitions*, which produce series limit continua such as the Balmer and Lyman continua of hydrogen. The wavelength of the emission or absorption is given again by $\lambda = hc/\Delta E$, but now $\Delta E = E_i + \frac{1}{2}m_e v_e^2$ where E_i is the ionization energy of the atom from the particular bound state, and $\frac{1}{2}m_e v_e^2$ is the kinetic energy of the free electron. For instance, the heads of the Lyman and Balmer continua lie at 912 Å and 3460 Å respectively. These wavelengths correspond to the lowest energies $\Delta E = 13.6$ eV and 3.6 eV that will ionize the hydrogen atom from the $n = 1$ and $n = 2$ levels;

(c) *Free-free transitions* between unbound (hyperbolic rather than bound elliptical) orbits. A photon is emitted or absorbed at a wavelength corresponding to the energy ΔE lost or gained respectively by the free electron moving in the Coulomb field of the atom or ion. Much of the sun's microwave emission is radiated by free-free transitions in the chromosphere and corona. This process (also called bremsstrahlung) is also responsible for a significant part of the coronal X-ray continuum emission;

(d) *Thomson scattering* of photons by free electrons. This is a continuum scattering process with no wavelength dependence. Thomson scattering of photospheric continuum off coronal electrons produces the K-corona at eclipse.

Excitation of atomic states leading to transitions can occur by collisions between the atom and other plasma particles and by absorption of photons. Electrons are usually more effective than heavier particles in collisional excitation since they travel much faster for equal energy in a Maxwellian plasma and thus collide more frequently. The inverse process of collisional de-excitation can also occur when an electron collides with an excited atom before that atom has been able to radiate away its excitation energy. In that case, referred to as a superelastic collision or a collision of the second kind, the electron carries away the energy difference between the excited and de-excited states of the atom.

If the colliding electron carries sufficient energy, it can strip electrons from the atom, thereby causing ionization. A collision between an ion and an electron of any energy can result in recombination. This results in the formation of an excited state from the electron and ion, and a photon (of free-bound continuum radiation) carries away the difference in energy between the free and bound states. This excited state then generally decays downward by a so-called cascade, emitting line radiation photons at each downward jump. The free-bound continuum and lines produced by these two-body radiative captures of electrons are referred to as recombination radiation.

Electrons can also attach to ions through a different process called dielec-
tronic recombination, whose importance in the solar atmosphere was not
recognized until 1964. In this case, the electron excites the ion (in ionization
state k) to a decaying level of the next lower ionization state $k + 1$. These
so-called autoionization states (because they can fission spontaneously into an
excited ion and electron) exist, although their excitation energy exceeds the
ionization potential of the ion. This phenomenon of autoionizing states occurs
when two electrons (instead of the usual one only) of an optically active outer
shell are excited to levels above their ground state. In helium, for instance, this
excitation requires more energy than ionization of He I to He II. The resulting
autoionizing state of He I can then decay by a radiationless process to He II in
the ground state, and a free electron.

In discussing plasma radiation, a useful distinction can often be made
between true absorption and emission, and scattering. If a photon excites an
atom that then immediately decays back to the lower state emitting a photon
of essentially identical wavelength, we describe the process as scattering. The
wavelength of the incoming and outgoing scattered photons will only be
slightly different, due to Doppler shifting through changes in the atom's
motion and to other factors (see below) contributing to the finite width of the
atomic levels. On the other hand, the incoming photon is considered to be
absorbed if, for instance, it photoionizes the atom so that some of its energy is
turned into the kinetic energy of a free electron which then goes on to collide
and thermalize with other electrons and ions in the plasma. In general,
absorption leads to conversion of some of the absorbed photon's energy into
heat, and emission converts thermal energy into radiation.

It is important to note that a (nonrelativistic) plasma can only radiate its
thermal energy by virtue of collisions. Since two-body collisions occur at rates
proportional to the product of the perturber (e.g., electron) density n_e and ion
density, emission becomes increasingly important (relative to scattering) in
higher density plasmas. We often refer loosely to emission, as in the Balmer
line emission of prominences, when we really mean a process that is mainly
scattering of photospheric light by atoms in the prominence. In this particular
case, little true emission is produced by actual collisional excitation of neutral
hydrogen from $n = 1$ to $n > 2$ by electron impacts (resonance fluorescence),
or by recombination of electrons and protons to form excited hydrogen, since
the prominence plasma is so tenuous.

In optical and ultraviolet spectra, only electrons in the atom's outermost
shell jump between levels. For instance, in the element calcium ($z = 20$), the
electronic configuration consists of two electrons in the innermost ($n = 1$)
completed shell, eight each in the next two ($n = 2$ and $n = 3$) completed shell
and sub-shell, and two electrons in the outer ($n = 4$) shell. These two outermost, optically active electrons determine the heliumlike spectrum (singlets
and triplets) of Ca I. When calcium is singly ionized to Ca II, jumps of its lone
remaining outer shell electron produce the hydrogenlike doublet spectrum. In

flares, calcium can be ionized 19 times to Ca xx. Its lone remaining electron then again produces a hydrogenlike spectrum with lines shifted by a factor of $z^2 \sim (20)^2$ into the X-ray region around 3 Å. Figure 3-5 shows some of these hard X-ray lines. In X-ray spectra, transitions of the tightly bound inner-shell electrons can also be produced by the impacts of fast-moving (hot) electrons. In solar spectroscopy, such inner-shell transitions turn out to be relatively weak compared to the usual outer-shell transitions of highly stripped ions which dominate the spectra of active regions and flares illustrated in Figs. 3-4 and 3-5.

We now proceed to apply these concepts of atomic structure and spectroscopy to the calculation of the volume emissivity and absorption coefficients of the solar plasma from a knowledge of its physical state and chemical composition.

3.6   RATES FOR RADIATIVE TRANSITIONS

Absorption of photons from the lower state m to the upper state n of an atom occurs at a rate proportional to N_m, the population of the state m (atoms cm^{-3}), to B_{mn} the Einstein coefficient for radiative absorption from m to n, and to the intensity of the radiation. The rate at which energy is removed from a beam of intensity I_λ has been discussed in terms of the linear extinction coefficient, in Section 2.2. For the case of absorption by a discrete transition from m to n at wavelength λ, this extinction coefficient can be written as

$$\kappa_\lambda = \frac{hc}{\lambda} N_m B_{mn}. \tag{3-9}$$

Transitions from the upper to lower state can occur spontaneously at the rate (cm^{-3} s^{-1}) given by the emissivity

$$\epsilon_\lambda = \frac{hc}{\lambda} N_n A_{nm}, \tag{3-10}$$

where A_{nm} is the Einstein coefficient for spontaneous emission. The decay from n to m can also be induced by radiation of intensity I_λ, and the rate of such simulated emission is

$$\epsilon_\lambda' = \frac{hc}{\lambda} N_n B_{nm} I_\lambda. \tag{3-11}$$

It can be demonstrated (from the detailed balance between upward and downward transitions between m and n that must hold in TE) that the

Einstein coefficients A_{nm}, B_{mn}, and B_{nm} are related by

$$A_{nm} = \frac{2hc}{\lambda^3} B_{nm}$$ (3-12)

and

$$g_m B_{mn} = g_n B_{nm},$$ (3-13)

where $g = 2J + 1$ is the statistical weight of the state.

Approximate values of the B_{mn} and B_{nm} can be estimated from the theory of a damped harmonic oscillator driven by the electromagnetic field of the incident intensity I_λ. Values accurate to between 10 and 50% are derived from a quantum mechanical model of the atom driven by a classical or quantized electromagnetic field. Incorporation of the quantized electromagnetic field is necessary to provide a theoretically consistent basis for estimates of the coefficients A_{nm}.

The connection between the Einstein coefficients and the theory of radiating atoms as quantized oscillators leads to an alternate form of expressing the relative strengths of atomic transitions using so-called oscillator strengths. These are related to the Einstein coefficients through

$$f_{mn} = \frac{g_n}{g_m} \frac{mc\lambda^2}{8\pi^2 e^2} A_{nm}$$ (3-14a)

and

$$g_m f_{mn} = -g_n f_{nm},$$ (3-14b)

so that f_{mn} and f_{nm} are the absorption and emission oscillator strengths respectively. These oscillator strengths are more commonly tabulated for lines of astrophysical interest than are the Einstein coefficients. This leads to the expression for the absorption coefficient from m to n, corrected for simulated emission

$$\kappa_\lambda = \frac{\pi e^2}{mc} f_{mn} N_m \left[1 - \exp\left(\frac{-hc}{\lambda kT} \right) \right].$$ (3-15)

To calculate the absorption coefficient κ_λ we need the level population N_m. In general, this requires a full simultaneous solution for all the transitions leading into and out of that state. This is clearly a forbidding prospect, since the population and de-population rates depend on the intensity I_λ as discussed above, so the problem is inherently nonlinear.

Progress can be made by noting that in thermodynamic equilibrium the atomic state populations will be determined by collisions and these will be governed by Boltzmann equilibrium. That is, we visualize our atom situated in the quasi-adiabatic cavity described in Section 2.3, colliding with other atoms whose velocities have achieved the Maxwellian distribution characterized by the cavity temperature T. These turn out to be reasonable approximations in the photosphere, where atoms collide frequently enough for the velocities to be Maxwellian and densities are high enough so that the emergent photons of similar wavelength are formed in a reasonably isothermal layer.

3.7 BOLTZMANN EQUILIBRIUM AND THE SAHA EQUATION

Assuming then that LTE holds, we can calculate the relative excitation of the two states m and n from the Boltzmann equilibrium relation

$$\frac{N_n}{N_m} = \frac{g_n}{g_m} \exp\left(\frac{-(\Delta E_n - \Delta E_m)}{kT}\right), \qquad (3\text{-}16)$$

where ΔE is the excitation energy relative to the ground state energy.

The large excitation energy $\Delta E_2 \sim 20$ eV for the $n = 2$ state of helium explains why helium lines are essentially absent in the Fraunhofer spectrum. Boltzmann equilibrium also helps to explain why the resonance lines of ions such as Ca II, Na I and Mg I can be so strong compared to the Balmer lines, although hydrogen is millions of times more abundant (but see Section 2.5.4).

Similar arguments can be extended to include ionization from bound states to the continuum k. To arrive at an expression for the ionization equilibrium populations, we consider first ionization from a ground state "0," requiring the energy $\Delta E_I = \Delta E_{0,k} + \frac{1}{2}m_e v_e^2$ where m_e and v_e are respectively the mass and velocity of the free electron produced in the ionization. Thus, we have

$$\frac{N_k}{N_0} = \frac{g_k}{g_0} \exp\left(\frac{-\left(\Delta E_I + \frac{1}{2}m_e v_e^2\right)}{kT}\right), \qquad (3\text{-}17)$$

where g_k is now the statistical weight given by

$$g_k = g_k' \times g_e, \qquad (3\text{-}18)$$

and g_k' and g_e are the statistical weights of the ground state of the ionized atom and of the free electron respectively. The statistical weight g_e is proportional to the number of phase-space elements that a free electron can occupy, which is inversely proportional to n_e the electron density per unit volume. It is given by

$$g_e = \frac{8\pi m_e^3 v_e^2}{n_e h^3}. \qquad (3\text{-}19)$$

Then, substituting for g_e in equations (3-17) and (3-18) and integrating over all the velocity states available to the electron, we obtain the equation first derived (except for the statistical weight factors) by M. Saha in 1920, namely

$$\frac{N_k}{N_0} = \frac{2}{n_e} \left(\frac{2\pi m_e kT}{h^3 z} \right)^{3/2} \left(\frac{g_k'}{g_0} \right) \exp\left(\frac{-\Delta E_I}{kT} \right). \tag{3-20}$$

Equation (3-20) can be generalized to any two successive stages of ionization k and $k + 1$, as

$$N_{k+1} = \frac{2N_k}{n_e} \left(\frac{2\pi m_e kT}{h^3 z} \right)^{3/2} \left(\frac{g_{k+1}}{g_k} \right) \exp\left(\frac{-\Delta E_I}{kT} \right), \tag{3-21}$$

where ΔE_I is now the ionization potential from the kth to $(k + 1)$st ionization state. We note that the electron density appears explicitly in the Saha equation due to the density-dependence of the free electron's statistical weight, although it does not appear in the Boltzmann equilibrium relation. That is, even in LTE, the distribution of ionization states is density and pressure sensitive, although the relative excitation of bound states within an ion depends only upon temperature.

The Saha and Boltzmann relations are derived on the assumption of thermal equilibrium. But they can be expected to hold whenever relative populations are determined by thermal gas particle collisions rather than by photon excitation and photoionization. This turns out to be a much easier condition to fulfill than strict LTE since photospheric densities are high enough that atom-electron collisions are much more frequent than atomic encounters with photons. It then hardly matters to the level populations and ionization that the radiation field may not be in perfect equilibrium with the local gas kinetic temperature. But in the chromosphere and corona, populations are no longer dominated by balance between collisional excitation and de-excitation, as discussed in Section 3.8. Then, populations and ionization balances calculated from the Saha equation and Boltzmann relation cease to resemble the observed values, even to order of magnitude.

The Saha equation enables us to understand why the lines of Ca II are so much stronger than those of Ca I in the solar photosphere (see exercises). It also is worth noting that the hydrogen ionization predicted by the Saha formula at the temperature of sunspot umbrae, or of the photospheric temperature minimum, is below 10^{-8}. The low-level ($\sim 10^{-4}$) ionization is provided by metals with relatively small ionization potentials. The sunspot umbra and the thin layer of photospheric temperature minimum thus constitute the closest approach to an electrically neutral atmosphere in the sun.

3.8 RATE EQUATIONS IN STATISTICAL EQUILIBRIUM

More generally, when LTE does not apply, as in the chromosphere and corona, we must relate the absorption coefficient κ_λ or emissivity ϵ_λ to the density and temperature through the equation of statistical equilibrium. For coronal radiations, a useful model is a simple two-level atom consisting of the ground level m and an excited state n. The population of the upper state is set by a balance between collisional excitations from m to n due mainly to electron-atom impacts, and by spontaneous radiative decay from n to m via an allowed transition. Induced emission from n to m and also photoexcitation from m to n are negligible because the intensity of coronal radiation is low.

At coronal densities, the rate of collisional de-excitation of the atom by an electron or atomic impact can also be neglected since it is much smaller than the radiative decay rate A_{nm} for an allowed transition (although not necessarily for a forbidden one). Since the radiative decay rates are so high compared to the collisional excitation rates at the low coronal densities, the population of higher excited states is negligible relative to that of the ground state. The rate equation in statistical equilibrium reduces in this simple case to

$$A_{nm}N_n = C_{mn}N_e N_m, \tag{3-22}$$

where A_{nm} (s^{-1}) is the Einstein coefficient for spontaneous emission, N_n and N_m are the number densities (cm^{-3}) of ions in the nth and mth states, and C_{mn} is the collisional excitation coefficient.

The coefficient C_{mn} $(\text{cm}^{-3}\,\text{s}^{-1})$ for collisions between electrons and ions (which are much more frequent than ion-ion collisions on account of the much higher electron thermal velocity) can be found by integrating the collision cross section $\sigma(v)$ for electron-ion collisions over a Maxwellian velocity distribution. The result of this integration is an expression for C_{mn} of the form

$$C_{mn} = 9.63 \times 10^{-6}T_e^{-1/2}(\Omega_{mn}/g_m)\exp(-\Delta E_{mn}/kT_e), \tag{3-23}$$

where T_e is the electron temperature, g_m is the statistical weight of level m, and Ω_{mn} is a dimensionless quantity, the so-called collision strength. The collision strength of an allowed transition is a slowly varying function (of order unity for coronal ions) of incident electron energy, and these have been tabulated for a few ionic species. The values (from difficult quantum mechanical calculations) are generally accurate to a factor of 2 or better.

It follows that the volume emissivity (ergs $\text{cm}^{-3}\,\text{s}^{-1}$) is given for this two-level model by

$$\epsilon_{nm} = \frac{hc}{\lambda_{nm}}N_e N_m C_{mn}. \tag{3-24}$$

In the hot corona, the electron and proton number densities N_e and N_p (produced mainly by the high ionization of hydrogen) are equal, so $N_e \sim N_p \sim N_H$. Also,

$$N_m \simeq \left(\frac{N_i}{N_x} \right) \left(\frac{N_x}{N_H} \right) N_H, \tag{3-25}$$

where N_i/N_x and N_x/N_H are respectively the fraction of atoms of element x in ionization state i and the abundance of element x relative to hydrogen. Under these conditions, equation (3-24) becomes

$$\epsilon_{nm} \simeq \frac{hc}{\lambda} C_{mn} \left(\frac{N_i}{N_x} \right) \left(\frac{N_x}{N_H} \right) N_e^2, \tag{3-26}$$

so that the emissivity (and thus the intensity $\epsilon/4\pi$) of an isothermal, optically thin plasma is proportional to the electron density squared. This is a very important relation in analysis of coronal emissions.

3.9   LINE BROADENING

So far we have treated atomic transitions between perfectly narrow energy levels of isolated, stationary atoms. Even an atom in a very low density medium at rest relative to the observer radiates a line that has a natural width set by damping of the atomic oscillator by its own radiation. But this natural width is below a milliangstrom and is thus completely negligible for purposes of solar spectroscopy. Hyperfine structure due to nuclear spin is also a negligible contributor. Isotopic contributions to hyperfine structure can produce components separated by an angstrom or more, but the strength of the lines due to less abundant isotopes is generally too low to influence the line profile noticeably. For instance, the Balmer lines have weak satellites due to the deuterium isotope ^2H, which are shifted to the blue. However, the relative intensity of the deuterium satellites is less than 1/5000th that of the ^1H lines and thus has negligible effects on the Balmer line profiles for most purposes in solar spectrophotometry.

The main effects that dominate the broadening of solar absorption and emission lines are the Doppler effect of atoms in thermal and in turbulent plasma motion, and the pressure broadening caused by atomic interactions in relatively dense plasmas ($n \gtrsim 10^{11}$ cm^{-3}). Optically thick emission lines can also be significantly broadened by self-absorption although formally this is a radiative transfer effect, unlike the atomic broadening mechanisms described above. The broadening that can be caused by Zeeman and Stark effects in magnetic and electric fields can also be significant, as has already been discussed. We have pointed out previously that particularly in the hydrogen spectrum, unresolved fine structure makes an often-overlooked but significant contribution to the widths of solar lines. We turn now to a closer look at the information on plasma conditions that can be obtained from Doppler and pressure broadening.

3.9.1 Thermal and Turbulent Doppler Broadening

Assuming a Maxwellian distribution of atomic velocities, the probability, df, that the velocity of an atom lies between v and $v + dv$ is

$$df = (\beta/\pi)^{1/2} \exp(-\beta v^2) \, dv, \tag{3-27}$$

where $\beta = m/2kT$ and m is the mass of the atom. The Doppler shift $\Delta\lambda$ caused by the velocity component u in the line of sight is given by

$$\frac{\Delta\lambda}{\lambda} = \frac{v\cos\theta}{c} = \frac{u}{c} \tag{3-28}$$

where θ is the angle between v and the line of sight. Substituting this relation (3-28) in equation (3-27), we find that the distribution of relative intensity (the line profile) as a function of wavelength shift $\Delta\lambda$ from the line center λ_0 is

$$I(\Delta\lambda) = \text{constant} \times \exp\left(-\beta \frac{c^2}{\lambda^2} \Delta\lambda^2\right). \tag{3-29}$$

This (Gaussian) line profile is plotted in Fig. 3-16. The most frequently used measure of the line width is the full-width at half-maximum intensity (FWHM), found by setting the exponential in (3-29) equal to $1/2$. In terms of $\Delta\lambda_{1/2}$ the halfwidth at half-maximum intensity, we have

$$\text{FWHM} = 2\,\Delta\lambda_{1/2} = 1.67\frac{\lambda}{c}\sqrt{\frac{2kT}{m}}. \tag{3-30}$$

Alternatively, equation (3-29) can also be solved for the thermal Doppler halfwidth at $1/e$ of its central intensity, given by

$$\Delta\lambda_D = \frac{\lambda}{c}\sqrt{\frac{2kT}{m}}. \tag{3-31}$$

When additional Doppler broadening due to turbulent macroscopic gas motions is present, a further term is added to equation (3-30) or (3-31). Assuming (for convenience) that this additional velocity field is also Maxwellian with a root mean square velocity v_{rms}, we have

$$\text{FWHM} = 1.67\frac{\lambda}{c}\sqrt{\frac{2kT}{m} + v_{\text{rms}}^2}. \tag{3-32}$$

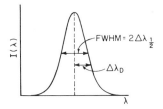

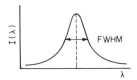

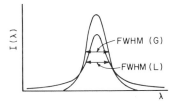

Fig. 3-16 Gaussian Doppler profile (upper), Lorentzian damping profile (middle), and their superposition (lower).

It is a common technique to evaluate the temperature and turbulent velocities separately in an atmosphere by solving equation (3-32) simultaneously for the widths of lines emitted by two atoms of greatly different atomic weight.

3.9.2 Radiation Damping and Pressure Broadening

The profile shape and width of a spectral line emitted by an isolated atom at rest can be obtained from classical mechanics by considering the atomic state to decay as a harmonic oscillator damped by its own radiation, thus obeying the equation of motion

$$\ddot{x} = -\omega_0^2 x - \gamma x, \tag{3-33}$$

where ω_0 is the angular frequency. The damping can be visualized as the loss of energy due to radiation from a charge in circular motion, at the rate γ s^{-1}, given classically as

$$\gamma = 2e^2\omega_0^2/3mc^3. \tag{3-34}$$

The solution is an exponentially damped oscillation of the form

$$x = x_0 e^{i\omega_0 t} e^{-\gamma t/2}. \tag{3-35}$$

The power spectrum (which determines the intensity profile) of an ensemble of such oscillating atoms having random phase is

$$I(\omega) = \left(\frac{\gamma}{2\pi}\right)\left[(\omega - \omega_0)^2 + \left(\frac{1}{2}\gamma\right)^2\right]^{-1}. \tag{3-36}$$

The shape of this profile is the Lorentzian illustrated in Fig. 3-16. Note how its wings are much stronger compared to the wings of a Gaussian profile of the same width.

To obtain the FWHM of this Lorentzian profile, we note that the half-intensity point falls at $\omega - \omega_0 = \gamma/2$. This yields for the full width in wavelength units

$$\text{FWHM} = (2\pi c/\omega)\gamma = 4\pi e^2/3mc^2 = 1.2 \times 10^{-4} \text{ Å}. \tag{3-37}$$

Note that this classically derived value is a constant.

Quantum theory predicts that the width of an energy level ΔE and its lifetime Δt are related through the uncertainty principle by

$$\Delta E \Delta t = \frac{h}{2\pi}. \tag{3-38}$$

If the typical lifetime of an excited level against spontaneous decay is $\Delta t \sim 10^{-8}$ s, the minimum width of the atomic transitions between two such broadened levels is given by

$$\Delta\lambda = \frac{\lambda^2}{2\pi c \, \Delta t} \sim 10^{-4} \text{ Å} \quad \text{for} \quad \lambda = 5000 \text{ Å}. \tag{3-39}$$

In general, the density of solar photospheric plasma is large enough that the radiation damping width is negligible compared to the broadening caused by electrostatic perturbation of the radiating atom's levels by other plasma particles. This effect is variously referred to as collisional or pressure broadening. Most simply, it can again be viewed as the result of the interruption of the atomic oscillator, after a characteristic time $\tau_0 = \gamma^{-1}$ taken to be the mean free time of the interacting particles. The atom is assumed to be unaffected by other particles during the time τ_0. The result is also a Lorentzian line profile with the width in wavelength units given by

$$\text{FWHM} = \frac{\lambda^2}{c}\frac{1}{\pi\tau_0}. \tag{3-40}$$

When the mean free time τ_0 is evaluated through kinetic theory, we find that the width increases linearly as the number density n and as the square root of temperature. More sophisticated treatments of pressure broadening take into account the statistical effect of the electrostatic (Stark effect) perturbations by other ions and electrons acting on the radiating atom. These treatments produce profiles that also exhibit the strong Lorentzian wings shown in Fig. 3-16, but they differ in their estimates of γ, and thus of the FWHM, for a given temperature and density.

3.9.3 Broadening by Self-Absorption

Emission lines of appreciable optical depth are also broadened by self-absorption or abundance broadening, as it is sometimes called. Essentially, at a given wavelength the plasma cannot emit at a higher intensity than that of a black body at the same temperature. For plasmas of appreciable optical depth, the line core, whose opacity is largest, reaches this maximum intensity first. Further increases of optical depth of the plasma through changes of its density or line-of-sight extent will cause points progressively further out on the wings of the profile to also achieve this maximum intensity. The effect is to produce a progressively more square profile. This effect must be kept in mind in analysis of solar emission lines whose optical thickness is uncertain.

3.9.4 Analysis of the Observed Profile of a Spectral Line

The practical problem presented in solar spectroscopy is to extract information on quantities such as the temperature, density, and macroscopic motions of the plasma from the shape of a line that is, in addition, broadened by the finite spectral resolution of the spectrometer. If the true profile of the line is $I(\lambda)$ and the spectrometer's instrumental profile is $G(\lambda)$, we find that the observed profile $O(\lambda)$ is given by the convolution integral

$$O(\lambda_0) = \int_{-\infty}^{\infty} I(\lambda - \lambda_0)G(\lambda) \, d\lambda. \qquad (3\text{-}41)$$

The instrumental profile $G(\lambda)$ represents the smoothing by the spectrometer of an infinitely narrow spectral line. It can be shown that, if $I(\lambda)$ is a Gaussian of halfwidth $\Delta\lambda_{1/2}$ (as expected if Doppler broadening dominates) and if the instrumental profile is approximated by a Gaussian of halfwidth $\Delta\lambda'_{1/2}$, then the observed profile will also be a Gaussian of halfwidth

$$\Delta\lambda_{1/2}^2(\text{obs}) = \left(\Delta\lambda_{1/2}\right)^2 + \left(\Delta\lambda'_{1/2}\right)^2. \qquad (3\text{-}42)$$

Under these circumstances, it is a simple matter to evaluate the true half-width $\Delta\lambda_{1/2}$, provided $\Delta\lambda'_{1/2}$ is known.

The instrumental profile is best determined by photoelectrically scanning or photographing a spectral line of a heavy element such as mercury to minimize thermal Doppler broadening. A line emitted by a low-pressure discharge tube can be used to avoid pressure broadening and also macroscopic motions that inescapably broaden solar lines. In most situations encountered in solar spectroscopy, the spectrometer slit is opened wide enough (to admit enough light) so that its width, rather than the fundamental resolution limit set by interference of the grating's n recombining beams, determines the instrumental profile width. For simplicity, it is customary to assume $G(\lambda)$ is Gaussian and to set $\Delta\lambda'_D = s$, where s is the slit width. Only in high-dispersion spectroscopy of photospheric lines are great efforts made to determine and eliminate the detailed shape of the function $G(\lambda)$.

If the true line profile $I(\lambda)$ is of Lorentzian form and $G(\lambda)$ is Gaussian, it can be shown that the convolution integral (3-41) yields a hybrid profile known as a Voigt function. Voigt function shapes have been conveniently tabulated as a function of the relative width of the constituent Lorentzian and Gaussian profiles, so the true profile $I(\lambda)$ can again be deconvolved from $O(\lambda)$ and $G(\lambda)$.

In the most general case, where $O(\lambda)$ and $G(\lambda)$ are of arbitrary form, $I(\lambda)$ can be obtained numerically by Fourier transform techniques using the powerful convolution theorem which states that the Fourier transform of a convolution such as $O(\lambda)$ is the product of the Fourier transform of the convolved functions $I(\lambda)$ and $G(\lambda)$. That is,

$$T(O) = T(I) \times T(G). \tag{3-43}$$

It follows that

$$I = T^{-1}[(T(O)/T(G))] \tag{3-44}$$

where T and T^{-1} denote Fourier transforms and inverse Fourier transforms respectively.

3.10 MOLECULES ON THE SUN

Seven diatomic molecules, CN, CH, CO, C_2, NH, OH, and MgH, have been identified with certainty in the photospheric spectrum. In addition, CaH, TiO, SiH, HF and H_2 have been found in umbral spectra. The electronic spectra of OH and NH contribute several bands each consisting of several hundred lines between 0.30 μm and 0.35 μm in the ultraviolet. Electronic transitions coupled with vibrations and rotations of the CN molecule give rise to several thousand molecular lines in the violet CN bands between 0.35 μm and 0.39 μm and in the red CN band between 0.78 μm and 1.16 μm. Such electronic transitions with vibrational and rotational fine structure also give rise to the numerous CH bands between 0.3 μm and 0.5 μm, the C_2 band system between 0.46 μm and 0.56 μm, and the weak MgH band observed near 0.5 μm. The fundamental and first harmonic vibrational bands of the CO molecule are prominent near 4.5 μm and 2.3 μm in the infrared. H_2 has been detected in sunspot spectra mainly through its Lyman band emission between 1200–1600 Å.

Molecular opacity probably plays a significant role in radiation transfer through upper layers of the photosphere, although it is much less important at lower levels compared to H^- and to atomic lines. The concentration of molecules is also too low to be of much importance to solar chemistry, except in a narrow layer around the photospheric temperature minimum region where the abundance of CO (which has by far the highest dissociation potential ~ 10 eV) can reach around 20–25% of the carbon and oxygen atom concentration. Even in these layers ($T \leq 5000$ K), the fraction of hydrogen in molecular form is estimated from theoretical models and ultraviolet observations to be below 10^{-3} although it is expected to reach 30% in sunspots. Nevertheless, molecular absorptions, particularly of CO, have proven useful in recent studies of the coolest regions of the photosphere (see Chapter 5), while recent calculations indicate that CH and OH make important contributions to photospheric opacity in the region between 1700–2500 Å where the sun's variable outputs have direct impact on ozone concentrations in the stratosphere of the Earth. More needs to be done in this relatively unexplored area.

ADDITIONAL READING

Atomic Structure and Spectroscopy

G. Herzberg, *Atomic Spectra and Atomic Structure*, Dover, 1945.

R. Leighton, *Principles of Modern Physics*, McGraw-Hill, 1959.

H. White, *Introduction to Atomic Spectra*, McGraw-Hill, 1934.

Observations and Interpretation of the Solar Spectrum

C. Allen, *Astrophysical Quantities*, 3rd ed.. The Athlone Press, 1976 (An authoritative compilation of astrophysical data and formulae; indispensable in solar research.)

J. Beckers, C. Bridges, and L. Gilliam, "A High Resolution Spectral Atlas of the Solar Irradiance from 380 to 700 Nanometers," Air Force Geophysics Laboratory publication AFGL-TR-76-0126 (1976).

K. Dere and H. Mason, "Spectroscopic Diagnostics of the Active Region Transition Region and Corona" in *Solar Active Regions*, F. Orrall, (Ed.), Colorado University Press, p. 129, 1981.

L. Goldberg, "Atomic Spectroscopy and Astrophysics," *Physics Today*, **41**, 38 (1988).

J. Vernazza, E. Avrett, and R. Loeser, "Structure of the Solar Chromosphere," *Astrophys. J. Suppl.*, **45**, 635 (1981).

O. White, *The Solar Output and Its Variation*, Colorado University Press, 1977.

EXERCISES

1. Calculate the relative intensities of 5500 K black-body radiation at 1500 Å and at 4500 Å. Comment on the implication of your result for the change from a solar absorption spectrum in the visible to an emission line spectrum in the far ultraviolet.

2. Calculate the relative populations of the ground level and of the $n = 2$ singlet level of He I in LTE at 6000 K, given the excitation energy of 20 eV for $n = 2$. Compare your result to the relative population of the ground and $n = 2$ levels in hydrogen. Comment on the implications for the weakness of the photospheric helium lines.

3. Calculate the relative abundances of neutral and ionized calcium (ionization potentials of 6.1 eV and 11.9 eV, respectively) at chromospheric levels where T ~ 6000 K. Use electron density values at these levels as given in Chapter 9. Comment on the implication of your result for the relative strength of Ca II and Ca I lines.

4. Calculate the typical thermal contribution to the width of Fraunhofer lines such as the Na I lines, formed at photospheric levels around $\tau_{0.5} = 0.1$. Compare this to the turbulent broadening due to granular motions on a spectrogram if granulation is not spatially resolved, and also to the pressure broadening at densities around $\tau_{0.5} = 0.1$. Use the rms granulation velocity, and photospheric density values given in Chapter 5. Compare the width of Hα to that of the H and K lines. Can you explain the difference?

5. Show that the observed profile of a spectral line is Gaussian if both the instrumental and true profiles are taken to be Gaussian.

4

Dynamics of Solar Plasmas

For the most part, the structures observed in the sun's atmosphere require dynamical explanations. On the largest spatial scales, the differential rotation of the photosphere suggests an interaction between overturning motions of solar convection and the Coriolis forces produced by the sun's rotation. Dynamical couplings between sub-photospheric plasma motions and magnetic fields have been widely studied to find an explanation of the sun's magnetic cycle.

On smaller scales, the granulation and probably also the supergranulation are best explained in terms of free convection driven by the buoyancy of plasma heated from below. In the chromosphere the structures observed in time-lapse films as spicules and fibrils contain material moving rapidly along magnetic field lines. The combination of magnetic and gas-pressure forces that is able to propel these dense plasma columns to such heights in the face of strong energy dissipation is still unknown.

Even more violent accelerations to supersonic velocities, sometimes exceeding the escape speed from solar gravity, are produced during flares and certain prominence eruptions. These sprays, surges, and coronal transients may be propelled directly by the Lorentz forces of relaxing magnetic field lines, or perhaps they are accelerated by a compressive shock driven by the flare's violent heating of chromospheric layers. Most of the material heated and lifted into the corona seems to descend as downflowing, cooling plasma at speeds that agree roughly with free-fall under solar gravity along magnetic field lines inclined from the radial direction. The downflows raise interesting questions about the thermal stability of the corona.

The importance of dynamics in understanding solar phenomena may now seem self-evident, but as recently as 20 years ago far more emphasis was placed on studies of atomic physics and radiative transfer in the static and plane-parallel layers used to represent various strata of the solar atmosphere.

Better observations at higher spatial resolution have changed this emphasis. In this chapter we discuss the equations governing moving fluids, the behavior of waves and shocks, and the dynamics of magnetized plasmas, as well as the relations governing hydrostatic equilibrium. Since the densities in the corona and solar wind can be very low, it is important to discuss both the limitations of the continuum hypothesis used in fluid mechanics and also certain transport effects that occur in magnetized plasmas when the motions of individual charged particles need to be considered.

4.1 HYDROSTATIC EQUILIBRIUM

4.1.1 Equilibrium in a Homogeneous Gravitational Field

In hydrostatic equilibrium a static balance is maintained between the force of gravity $\rho\mathbf{g}$ on a volume element dV and the difference in gas pressure dP across the external surfaces of the element. We may assume (Fig. 4-1) that the element is a cylinder whose axis is aligned with the acceleration of gravity in the radial r-direction. The cylinder's height is taken as dr, its cross-sectional area is unity, and its gas density is ρ. The balance between the gas-pressure difference across the cylinder's opposite faces and the gravitational volume force is then

$$dP = -\rho\mathbf{g}\, dr, \tag{4-1}$$

or

$$\nabla P = -\rho\mathbf{g}. \tag{4-2}$$

Here $\mathbf{g}$ is the gravitational acceleration at r. Since $\mathbf{g}$ is positive, the inward acceleration of gravity balances an outward decrease of pressure, with increasing r.

Equation (4-2) can be integrated to give the density distribution in an isothermal atmosphere at temperature T, given the equation of state. Assum-

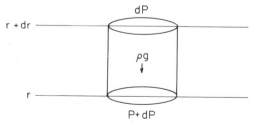

Fig. 4-1 Schematic diagram illustrating hydrostatic equilibrium.

ing an ideal gas, we can express the pressure as

$$P = \frac{\rho RT}{\mu},$$ (4-3)

where $R = 8.317 \times 10^7$ ergs deg^{-1} mole^{-1} is the gas constant and μ is the mean molecular weight, whose definition in a plasma is given in equation (4-55).

Using equation (4-3) to eliminate the pressure, we can express equation (4-2) as

$$\frac{1}{\rho}\frac{d\rho}{dr} = -\frac{\mu g}{RT},$$ (4-4)

which can be integrated for constant T, μ, and g to yield an exponential fall-off of density with increasing height r, given by

$$\rho = \rho_0 \exp\left(-\frac{\mu g r}{RT}\right).$$ (4-5)

The distance $H = RT/\mu g$, over which the density decreases by the factor $e^{-1} = 0.37$, is known as the density scale height. At the solar photosphere, with $g = 2.74 \times 10^4$ cm s^{-2}, we find that around the temperature minimum, H can be as small as 130 km. In the hot, fully ionized plasmas of the corona or deep solar interior it can assume values of order 10^5 km.

4.1.2 Self-Gravitating Atmospheres

We assumed above that g may be taken as a constant. This is often true when a homogeneous gravitational field is imposed by a body of large mass (such as the sun) upon an atmosphere such as the corona whose own gravity field is negligible. More generally, within a star held together by its self-gravity, the acceleration g is determined by the mass $M(r)$ lying within the sphere of radius r, so that its magnitude is

$$g = \frac{GM(r)}{r^2},$$ (4-6)

where $G = 6.67 \times 10^{-8}$ dyne cm^2/g^2 is the universal gravitational constant. Substituting this expression for g in equation (4-1), we obtain

$$\frac{dP}{dr} = -\frac{GM(r)}{r^2}\rho,$$ (4-7)

where ρ is the density in a spherical shell of radius r. The mass $M(r)$ can in principle be evaluated by integrating over the radial density distribution of the star, since

$$M(r) = \int_0^r 4\pi r^2 \rho(r)\, dr. \tag{4-8}$$

However, to obtain $\rho(r)$ and $M(r)$ one needs to solve the full set of coupled equations describing conservation of mass, momentum, and energy in the star together with the equation of state. As we discuss in Chapter 6, the solutions are generally obtained by numerical integration since the equations describing nuclear energy generation and the outward transport of energy by radiation and convection turn out to be too complex to admit an analytical solution unless simplifying assumptions are made.

4.1.3 The Polytropic Approximation

Early insights into the study of conditions within self-gravitating gaseous bodies were obtained already toward the end of the 19th century through use of the so-called polytropic approximation. In a polytropic variation, conditions in a gas are allowed to change within the constraint of constant specific heat of the system. To understand the implications of this constraint, we recall that the specific heat is given by

$$c = \frac{dQ}{dT}, \tag{4-9}$$

where dQ is the heat added to a unit mass of the gas. An adiabatic variation ($dQ = 0$) then corresponds to $c = 0$, while an isothermal change ($dT = 0$) corresponds to $c = \infty$. Intermediate values of c define changes in the system along paths lying between adiabats and isotherms.

To determine how thermodynamic variables are related during polytropic changes, we can express the first law of thermodynamics (the most general energy equation governing such changes) in terms of the temperature T, volume V, the specific heats at constant volume and pressure c_v and c_p, and c. The first law is

$$dQ = dU + P\, dV. \tag{4-10}$$

Since

$$dU = c_v\, dT, \tag{4-11}$$

we can differentiate equation (4-10) and substitute definition (4-9) to yield

$$c = c_v + P\frac{dV}{dT}. \tag{4-12}$$

We use the equation of state in the form

$$P = \frac{\rho RT}{\mu} = \frac{RT}{V} \tag{4-13}$$

where $V = (\rho/\mu)^{-1}$ is the specific volume. Then, recalling that $c_p - c_v = R$, we rewrite equation (4-12) as

$$\frac{dT}{T} + \frac{1}{n}\frac{dV}{V} = 0, \tag{4-14}$$

where

$$n = \frac{c_v - c}{c_p - c_v} \tag{4-15}$$

is called the polytropic index.

The equation of state (4-13) can also be used to eliminate the temperature from equation (4-14), so that

$$\frac{dP}{P} = \left(1 + \frac{1}{n}\right)\frac{d\rho}{\rho}. \tag{4-16}$$

Integration of this relation shows that during polytropic changes the pressure and density are related by

$$P = \text{const} \times \rho^{(1+1/n)}. \tag{4-17}$$

In the particular case of an adiabatic change, for which $n = 1/(\gamma - 1)$ we find the familiar relation $P/\rho^\gamma = \text{const}$. Here $\gamma = c_p/c_v$ is the usual ratio of specific heats.

The polytropic approximation is useful chiefly because the system of equations (4-3), (4-7), and (4-8) together with the polytropic relation (4-17) can be solved to give the radial profiles of P, T, M and ρ in a self-gravitating body whose energy generation and transport are imperfectly known. For $n = 0$, 1, and 5, analytical solutions can be derived. For other values of $0 \le n \le 5$, numerical solutions have been obtained and tabulated. Since for $n = 0$ the polytropic density distribution is uniform and for $n = 5$ the central condensation is infinite, such polytropes provide an approximate model in which parameters may be varied conveniently, in studies of the solar interior and also of the solar wind.

4.2 THE EQUATIONS OF MOTION

4.2.1 Euler's Equation

We first consider the motion that results when a nonuniform pressure distribution accelerates a gas of density ρ. The net force acting on a volume of gas is given by the surface integral of the pressure over a surface A bounding the volume V, namely

$$\mathbf{F} = - \int_A P \, d\mathbf{a}. \tag{4-18}$$

The negative sign indicates that the force acts in the opposite direction to the outward directed normal to the surface element $d\mathbf{a}$. This surface integral of pressure over A can be transformed to a volume integral of the pressure gradient within V by the vector relation

$$-\int_A P \, d\mathbf{a} = - \int_V \nabla P \, d\tau. \tag{4-19}$$

Thus the force per unit volume acting within the fluid is given by $-\nabla P$, and acts in the direction of decreasing pressure.

The resulting motion is described by setting the pressure gradient force equal to the product of the mass density ρ and acceleration $d\mathbf{v}/dt$

$$\rho \frac{d\mathbf{v}}{dt} = -\nabla P. \tag{4-20}$$

In the form given by (4-20), this equation describes the acceleration during a time dt of a group of particles moving together through space. One component of this total acceleration $d\mathbf{v}$ is due to the change in velocity at a fixed point $\mathbf{r}$ in space and is expressed as the partial differential $(\partial \mathbf{v}/\partial t)\, dt$. This is the only acceleration that would be measured by an observer moving with the particles. Another component is due to the difference in velocities between two points in space at a given time. This additional acceleration, which is required to evaluate the total acceleration measured by an observer fixed in space, is

$$(dr \cdot \nabla)\mathbf{v} = \left(\frac{\partial \mathbf{v}}{\partial x}\right) dx + \left(\frac{\partial \mathbf{v}}{\partial y}\right) dy + \left(\frac{\partial \mathbf{v}}{\partial z}\right) dz. \tag{4-21}$$

It follows that the total velocity change in time dt is

$$d\mathbf{v} = \left(\frac{\partial \mathbf{v}}{\partial t}\right) dt + (dr \cdot \nabla)\mathbf{v}, \tag{4-22}$$

and the total acceleration measured at a fixed point in space is

$$\frac{d\mathbf{v}}{dt} = \frac{\partial \mathbf{v}}{\partial t} + (\mathbf{v} \cdot \nabla)\mathbf{v}. \tag{4-23}$$

The resulting equation of motion under a pressure gradient alone, called Euler's equation, is

$$\rho \frac{\partial \mathbf{v}}{\partial t} + \rho(\mathbf{v} \cdot \nabla)\mathbf{v} = -\nabla P. \tag{4-24}$$

When the velocity is constant in time at each point in the flow, we have $\partial \mathbf{v}/\partial t = 0$, and the flow is steady. Otherwise the flow is described as time dependent.

Euler's equation describes the motion of an ideal fluid in which no dissipation of kinetic energy into heat takes place by viscosity nor is heat transported by thermal conduction or radiation. Although only inertia and a pressure gradient are acting, closed-form solutions are available for only a few special cases where the geometry of the flow makes the nonlinear advective term $\mathbf{v} \cdot \nabla \mathbf{v}$ identically zero. In general, when $\mathbf{v} \cdot \nabla \mathbf{v} \neq 0$, even the simple flows described by Euler's equation must be studied through numerical solutions whose behavior can be extremely sensitive to the initial conditions.

In most problems of interest in solar research, the gravitational volume force $\rho\mathbf{g}$ must also be considered, yielding the equation of motion

$$\frac{\partial \mathbf{v}}{\partial t} + (\mathbf{v} \cdot \nabla)\mathbf{v} = -\frac{1}{\rho}\nabla P + \mathbf{g}. \tag{4-25}$$

When $\mathbf{v} = 0$, we recover the hydrostatic equilibrium relation (4-2).

4.2.2 Viscous Forces and the Navier-Stokes Equation

In real fluids, the transfer of momentum occurs in part by the transport of fluid volumes having different velocity, which is expressed by the advective term in Euler's equation. But additional transfer is caused by the internal friction due to collisions between particles moving with adjacent layers of the fluid having different velocities. In the most general case of a compressible fluid the term to be added to Euler's equation to express the volume force caused by this viscosity is more complex than the force terms that govern the inviscid motion discussed so far. But in the case where the fluid motion can be considered incompressible (i.e., $\partial \rho/\partial t = 0$) the viscous force per unit volume can be shown to be proportional to the Laplacian derivative of the velocity $\nabla^2 \mathbf{v}$ and to a coefficient ν called the kinematic viscosity, whose units are cm^2

s^{-1}. The resulting (Navier-Stokes) equation of motion is given by

$$\frac{\partial \mathbf{v}}{\partial t} + (\mathbf{v} \cdot \nabla)\mathbf{v} = -\frac{1}{\rho}\nabla P + \nu \nabla^2 \mathbf{v}. \tag{4-26}$$

It follows from (4-26) that viscous forces disappear in the absence of velocity gradients within a flow. Note also that in the absence of pressure gradients, and provided the viscosity is large enough so that $\nu\nabla^2\mathbf{v} \gg \mathbf{v}(\nabla \cdot \mathbf{v})$, we find that the velocity field in a fixed volume of fluid will decay by diffusion according to the equation

$$\frac{\partial \mathbf{v}}{\partial t} = \nu\frac{\partial^2 \mathbf{v}}{\partial r^2}. \tag{4-27}$$

The time scale τ of this diffusive decay is given approximately by

$$\frac{v}{\tau} \simeq \nu\frac{v}{l^2}, \tag{4-28}$$

or

$$\tau \simeq \frac{l^2}{\nu}, \tag{4-29}$$

where l is the spatial scale over which the velocity doubles in magnitude.

The condition that $\rho(\mathbf{v} \cdot \nabla)\mathbf{v} \ll \rho\nu\nabla^2\mathbf{v}$ means that the inertial term is negligible compared to the viscous term in determining the flow. More generally, the ratio of these terms is given by the Reynolds number

$$R = \frac{v^2}{l}\left(\nu\frac{v}{l^2}\right)^{-1} = v\frac{l}{\nu}. \tag{4-30}$$

When $R \ll 1$, viscous forces dominate the flow, the nonlinear term $(\mathbf{v} \cdot \nabla)\mathbf{v}$ can be neglected, and the resulting linear equation of motion can be solved in closed form to describe laminar flows. If $R \gg 1$, the inertial term dominates and in this Eulerian regime discussed previously, the flow is usually turbulent, since viscosity is insufficient to damp out small perturbations of the laminar flow.

In the solar atmosphere the observed scales l are so large that even the smallest observable motions correspond to enormous Reynolds numbers, given the values of molecular viscosity of hydrogen, which range between 0.1 cm^2 s^{-1} at the photosphere and 10^7 cm^2 s^{-1} in the chromosphere. Under these circumstances, the observation of any laminar motions on the sun is surprising. Where such motions are observed, they imply the presence of additional, stabilizing forces due to rotation or magnetic fields or of additional sources of viscosity provided by small-scale turbulent eddy motions.

4.2.3 The Equation of Continuity

To express the condition that a flow must also conserve mass, we consider a volume V containing a parcel of fluid given by the volume integral

$$\int_V \rho \, dV,$$

where ρ is the density of the fluid. The mass of fluid flowing at velocity $\mathbf{v}$ in unit time through an element $d\mathbf{a}$ of the surface bounding V is $\rho \mathbf{v} \cdot d\mathbf{a}$. Here the magnitude of $d\mathbf{a}$ is equal to the area of the surface element, and its direction is outward along the normal to surface, so that the mass flux is positive for fluid flow out of the volume.

The total rate of flow out of the volume V is given by the surface integral $\int_A \rho \mathbf{v} \cdot d\mathbf{a}$, where A is the area of the surface bounding V. It follows that the rate of decrease of mass in V is given by

$$\frac{\partial}{\partial t} \int_V \rho \, dV = -\int_A \rho \mathbf{v} \cdot d\mathbf{a}. \tag{4-31}$$

The surface integral of the mass flux density $\rho \mathbf{v}$ can then be written as a volume integral of its divergence,

$$\int_A \rho \mathbf{v} \cdot d\mathbf{a} = \int_V (\nabla \cdot \rho \mathbf{v}) \, dV, \tag{4-32}$$

so that

$$\int_V \left[\frac{\partial \rho}{\partial t} + \nabla \cdot \rho \mathbf{v} \right] dV = 0. \tag{4-33}$$

If this equation is to hold for any volume V, the integrand must vanish, so that the equation of continuity becomes

$$\frac{\partial \rho}{\partial t} + \nabla \cdot \rho \mathbf{v} = 0, \tag{4-34}$$

or

$$\frac{\partial \rho}{\partial t} + \rho (\nabla \cdot \mathbf{v}) + \mathbf{v} \cdot \nabla \rho = 0. \tag{4-35}$$

It follows that for incompressible flow ($\rho = $ const) this reduces to

$$\nabla \cdot \mathbf{v} = 0. \tag{4-36}$$

4.2.4 The Heat-Balance Equation

Starting from the first law of thermodynamics, we describe the contributions of heat flow to the total energy balance of the gas. For a fluid element of given mass the first law relates the change in internal energy ΔU to the gain of heat ΔQ and the reversible work ΔW performed upon that mass, through the expression

$$\Delta U = \Delta Q + \Delta W. \tag{4-37}$$

If we first confine our attention to the useful situation where the work ΔW is performed only by the mechanical compressions and expansions produced as the fluid moves, we have for small reversible changes the relation

$$\Delta W = -P\,dV, \tag{4-38}$$

where V is the specific volume. Taking time derivatives, this yields

$$\frac{dU}{dt} + P\frac{dV}{dt} = \frac{dQ}{dt}. \tag{4-39}$$

If instead of the internal energy U we use the enthalpy $H = U + PV$, which includes the work done in compression and expansion of the gas, we can express the heat balance for a unit volume of isobaric fluid in terms of the directly measured quantities ρ and T as

$$\rho\frac{dH}{dt} = \rho c_p\frac{dT}{dt} = \rho\frac{dQ}{dt}. \tag{4-40}$$

In the simplest approximation, we neglect the heat flows that occur into and out of the volume by radiation and conduction, so the term dQ/dt can be set to zero, and we have adiabatic motion. Since the entropy change $dS = dQ/T$ also vanishes, such motion is often referred to as isentropic, and the energy equation can most simply be stated as

$$\rho T\frac{dS}{dt} = \rho c_p\frac{dT}{dt} = 0, \tag{4-41}$$

or

$$S = \text{constant}. \tag{4-42}$$

Although the adiabatic approximation can be useful in studying waves (where losses from a volume can be relatively small over the short period of an oscillation), it is usually a poor approximation in solar plasma dynamics since the motions are often driven (as in convection) or damped (as in shock waves) by nonadiabatic terms.

More generally, we must write the heat balance equation as

$$\rho c_p \frac{dT}{dt} = -\nabla \cdot \mathbf{F}_{cond} + \nabla \cdot \mathbf{F}_{rad} + \cdots, \tag{4-43}$$

where the right-hand side includes the divergences of all the energy fluxes of importance. The forms taken by the two leading terms on the right-hand side are considered below.

The heat flux due to thermal conduction in a gas is given by

$$\mathbf{F}_{cond} = -K\nabla T, \tag{4-44}$$

where K is the thermal conductivity. In a fully ionized plasma, the thermal conductivity in the absence of a magnetic field, or in the direction parallel to the field lines if the plasma is magnetized, is given by

$$K = 1.8 \times 10^{-12} \times \frac{T^{5/2}}{\ln \Lambda} \text{ W cm}^{-1} \text{ K}^{-1}. \tag{4-45}$$

The conductivity thus depends strongly on the temperature, and only weakly on the electron density through the factor $\ln \Lambda$, whose tabulated values range between $\ln \Lambda = 20$ at $T_e = 10^6$ K, $n_e = 10^9$ cm^{-3} in the corona, to $\ln \Lambda = 10$ at $T_e = 10^4$ K, $n_e = 10^{12}$ cm^{-3} in the chromosphere. At relatively low photospheric temperatures, the $T^{5/2}$ dependence of K makes energy transport by thermal conduction negligible for most purposes.

Using equation (4-44), the divergence of thermal conductive flux becomes

$$\nabla \cdot \mathbf{F}_{cond} = -\nabla \cdot (K\nabla T), \tag{4-46}$$

or, if K is constant,

$$\nabla \cdot \mathbf{F}_{cond} = -K\nabla^2 T. \tag{4-47}$$

The flux of thermally conducted heat is expected to be large, for instance, in the transition region between chromospheric and coronal temperatures, where $T \simeq 10^5$ K (so K is large) and the temperature gradient also attains high values. Note that the net energy deposited or extracted per unit volume, which is of interest in the local energy balance expressed by equation (4-43) requires calculation of the divergence of the conductive flux. This may be small compared to other energy balance terms in (4-43) even if the conductive flux itself is large.

The radiative flux divergence is calculated quite differently for optically thin radiation (as in the corona) and optically thick radiation (as in the solar interior). In the former case we can write $\nabla \cdot \mathbf{F}_{rad}$ in a form similar to the volume emissivity of an optically thin line given by equation (3-24), namely

$$\nabla \cdot \mathbf{F}_{rad} = n_e n_H \Phi(T) \sim n_e^2 \Phi(T) \qquad (4\text{-}48)$$

where n_e and n_H are respectively the particle densities of electrons and of protons or hydrogen atoms. The function $\Phi(T)$ expresses the temperature dependence of radiative losses from the optically thin plasma.

The calculation of this function $Q(T)$ requires consideration of all the lines and continua that contribute to radiation at a given temperature, density and chemical composition. The results have now been widely tabulated and graphed for convenient use, as illustrated in Fig. 4-2. The main points to note are the peak around 10^4 K (due mainly to Lyman lines and continuum of

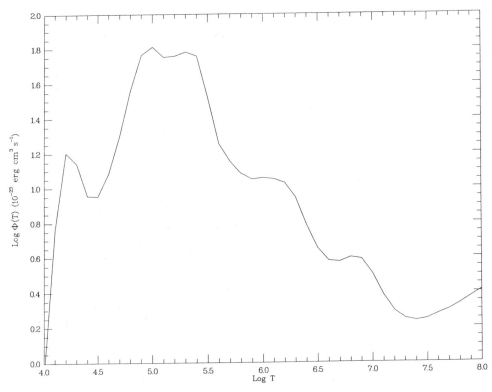

Fig. 4-2 The calculated radiative loss function for solar plasma over the temperature range between 10^4 and 10^8 K. By permission of J. Raymond.

hydrogen), the broad maximum around 10^5 K (due to strong resonance lines of highly ionized states of heavier elements such as C, N, and O), and the decrease toward coronal temperatures. The solar plasma radiative losses under optically thin conditions decrease by several orders of magnitude with increasing temperature for $T \gtrsim 3 \times 10^6$ K, in contrast to the monotonic increase as T^4 expected of an optically thick black body. We see from equation (4-48), that the radiation also increases as the square of the particle density for a fully ionized plasma.

Below the photosphere, radiation can be handled much more simply. We recall from Chapter 2 that in such an optically thick medium radiative transfer can be approximated by the diffusion equation

$$\nabla \cdot F_{\text{rad}} = \nabla \cdot (k' \nabla T) \tag{4-49}$$

where

$$k' = \frac{16\pi\sigma T^3}{3\bar{\kappa}_R}, \tag{4-50}$$

and $\bar{\kappa}_R$ is the Rosseland absorption coefficient defined in equation (2-55). Provided k' can be taken constant over the volume of interest, we then can write

$$\nabla \cdot F_{\text{rad}} = k' \nabla^2 T \tag{4-51}$$

in the equation of energy balance.

Other sources and sinks of energy in the solar plasma will be considered below and in later chapters.

4.2.5 Conservation of Total Energy

So far we have discussed only the changes in internal energy of the plasma caused by various heat-flow terms. But for a plasma moving in the sun's gravitational field, these terms must be augmented by the fluxes due to the kinetic and potential energy to obtain the equation for conservation of total energy.

The total energy of a plasma element of density ρ moving at velocity $\mathbf{v}$, and having internal energy U is $\frac{1}{2}\rho v^2 + \rho U$. It can be shown that time variations in this total energy are related to the flux of energy density in the plasma, $\rho \mathbf{v}(\frac{1}{2}v^2 + H)$, by

$$\frac{\partial}{\partial t}\left(\frac{1}{2}\rho v^2 + \rho U\right) = -\nabla \cdot \left[\rho \mathbf{v}\left(\frac{1}{2}v^2 + H\right)\right], \tag{4-52}$$

where H is the enthalpy per unit mass. To visualize the meaning of this relation, we can integrate this expression over the plasma volume and use Stokes' theorem to convert the volume integral of the divergence on the right-hand side to a surface integral. We thus obtain

$$\frac{\partial}{\partial t} \int \left(\frac{1}{2} \rho v^2 + \rho U \right) dV = -\int \rho \mathbf{v} \left(\frac{1}{2} v^2 + H \right) \cdot d\mathbf{a}. \qquad (4\text{-}53)$$

Recalling that $H = U + PV$, we can see that the time rate of change of total energy (on the left-hand side) is given by the surface integral of the flux of kinetic and internal energies, and also the flux across that surface of work done by pressure forces.

Combining these adiabatic terms with the terms of the heat equation (4-43), we arrive at the equation for conservation of total energy, which can be written as

$$\frac{\partial}{\partial t} \left(\rho U + \frac{1}{2} \rho v^2 \right) = -\nabla \cdot \left[\mathbf{F}_H + \rho \mathbf{v} \left(\frac{1}{2} v^2 + H \right) \right] + \rho \mathbf{v} \cdot \mathbf{g}, \qquad (4\text{-}54)$$

where the term $\rho \mathbf{v} \cdot \mathbf{g}$ expresses the rate of work done by or against solar gravity and $\mathbf{F}_H$ represents the heat fluxes due to conduction and radiation.

The equations describing fluid motions are completed with the equation of state (4-3) for a perfect gas. The mean molecular weight for a fully ionized hydrogen plasma is

$$\mu = \frac{n_p m_p + n_e m_e}{n_p + n_e}. \qquad (4\text{-}55)$$

Since $n_p = n_e$ and $m_p \gg m_e$, we have $\mu \simeq 0.5$. The presence of helium and small traces of heavier elements in the solar plasma leads to a value of $\mu \sim 0.6$ throughout most of the sun except very near the photosphere, where the ionization is low. The calculation of μ for a plasma of solar chemical composition is discussed more fully in Chapter 6.

4.3 THE INFLUENCE OF MAGNETIC FIELDS IN SOLAR PLASMA DYNAMICS

4.3.1 The Lorentz Force

When a plasma moves in a magnetic field, an electric current is generated. The interaction of this current of density j with the magnetic field gives rise to a magnetic volume force $\mathbf{F}_m$ on the plasma that opposes the motion. It is given in electromagnetic units (emu) by

$$\mathbf{F}_m = \mathbf{j} \times \mathbf{B}, \qquad (4\text{-}56)$$

and its geometry is illustrated in Fig. 4-3.

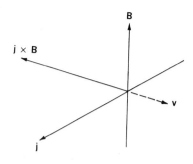

Fig. 4-3 Geometry of $\mathbf{j}$, $\mathbf{B}$, and $\mathbf{j} \times \mathbf{B}$, relative to the velocity $\mathbf{v}$.

When this Lorentz force per unit volume is included, the equation of motion becomes, in its simplest (Eulerian) form,

$$\rho\left[\frac{\partial \mathbf{v}}{\partial t} + (\mathbf{v} \cdot \nabla)\mathbf{v}\right] = -\nabla P + \rho\mathbf{g} + \mathbf{j} \times \mathbf{B}. \qquad (4\text{-}57)$$

A magnetic force is exerted on a plasma element only insofar as a current flows within that element. If a current path cannot be completed, then $\mathbf{j} = 0$ and $\mathbf{F}_m = 0$. Also, since the vector $\mathbf{j} \times \mathbf{B}$ has no component in the direction of $\mathbf{B}$, it follows that the magnetic force cannot influence local motions of a plasma along field lines, provided the plasma density is high enough for the fluid approximation to hold. For instance, the hydrostatic pressure scale height in the photosphere and chromosphere is unaltered in the presence of a magnetic field with a substantial component oriented along the acceleration of gravity.

In the study of the interaction between plasma motions and embedded magnetic field, usually referred to as magnetohydrodynamics (MHD), it is common to assume that the plasma can be treated as a continuum fluid. The motions of individual plasma particles are considered in Section (4-5), where we derive an expression for the electrical conductivity in terms of the plasma temperature and density. We also assume that $v \ll c$. Finally, we neglect any effects of rapid or small-scale plasma excitations and thus also of the displacement current $\partial \mathbf{E}/\partial t$.

4.3.2 The Importance of Self-Induction

The force $\mathbf{j} \times \mathbf{B}$ is fairly straightforward to calculate in laboratory conductors since the current density can be found from the Ohm's law relation

$$\mathbf{j} = \sigma(\mathbf{v} \times \mathbf{B}), \qquad (4\text{-}58)$$

where σ is the electrical conductivity. For conductors of larger dimensions the modification of $\mathbf{B}$ by the induced currents cannot be neglected. To take into account this self-induction in solar plasmas (where the dimensions of even the smallest resolvable structures are huge by laboratory standards), we must write the current equation as

$$\mathbf{j} = \sigma(\mathbf{E} + \mathbf{v} \times \mathbf{B}), \tag{4-59}$$

where $\mathbf{E}$ is the induced electric field given by Faraday's law

$$\nabla \times \mathbf{E} = -\frac{\partial \mathbf{B}}{\partial t}. \tag{4-60}$$

We have assumed here that electrostatic fields in the plasma are negligible ($\nabla \cdot \mathbf{E} = 0$) over the dimensions of interest. This condition on plasma neutrality is discussed further in Section 4.4.7.

To calculate the time behavior of the magnetic field and its dynamical influence on the plasma under these conditions, we can eliminate the electric field from our calculation of the changes in magnetic field by taking the curl of equation (4-59). We have then

$$\nabla \times 4\pi\mathbf{j}/\sigma = (\nabla \times \mathbf{E}) + (\nabla \times \mathbf{v} \times \mathbf{B}), \tag{4-61}$$

or

$$\frac{\partial \mathbf{B}}{\partial t} = \nabla \times (\mathbf{v} \times \mathbf{B}) - \frac{1}{4\pi\sigma}\nabla \times \nabla \times \mathbf{B}. \tag{4-62}$$

Using the vector identity

$$\nabla \times \nabla \times \mathbf{B} = \nabla \cdot (\nabla \cdot \mathbf{B}) - \nabla^2\mathbf{B}, \tag{4-63}$$

and with $\nabla \cdot \mathbf{B} = 0$, we have then

$$\frac{\partial \mathbf{B}}{\partial t} = \nabla \times (\mathbf{v} \times \mathbf{B}) + \frac{1}{4\pi\sigma}\nabla^2\mathbf{B}. \tag{4-64}$$

This induction equation describes how the magnetic intensity at a fixed point in space changes in time within a magnetized fluid of spatially uniform conductivity σ, given a velocity field $\mathbf{v}$. In principle it can be combined with the equation of motion (4-57), an energy equation of the form (4-54), and the equations of continuity and state (4-34) and (4-3) to completely determine the dynamics of a magnetized plasma.

4.3.3   The Diffusive and "Frozen-in" Approximations

Some important insights can be obtained from an order-of-magnitude analysis of the terms in the induction equation. The terms on the right-hand side can be approximated as

$$\nabla \times \mathbf{v} \times \mathbf{B} \sim \frac{vB}{l} \qquad (4\text{-}65)$$

and

$$\frac{1}{4\pi\sigma}\nabla^2\mathbf{B} \sim \frac{B}{4\pi\sigma l^2} \qquad (4\text{-}66)$$

where l is the length scale over which $\mathbf{B}$ would change by a substantial fraction of its magnitude, given the gradients of $\mathbf{B}$ in the volume under consideration. It follows that the second term (4-66) dominates if

$$\frac{B}{4\pi\sigma l^2} \gg \frac{vB}{l} \qquad (4\text{-}67)$$

or

$$R_{\mathrm{m}} \equiv 4\pi\sigma vl \ll 1, \qquad (4\text{-}68)$$

where R_{m} is called the magnetic Reynolds number. Under these circumstances, the time behavior of $\mathbf{B}$ is governed by the diffusion equation

$$\frac{\partial\mathbf{B}}{\partial t} = \frac{1}{4\pi\sigma}\nabla^2\mathbf{B}, \qquad (4\text{-}69)$$

and $1/4\pi\sigma$ is the magnetic diffusivity. The time scale for magnetic field to leak out of a conductor of dimension l is, from equation (4-69), given approximately by

$$\tau \sim 4\pi\sigma l^2. \qquad (4\text{-}70)$$

Since this relation applies to any uniform conductor, we find for a copper sphere of radius 1, $\tau \simeq 10$ s, whereas for decay of a primordial field in sun's core we have the enormous time scale $\tau > 10^{10}$ years, which is longer than the age of the sun. The main reason for the large self-induction leading to this time scale is the enormous scale of the sun; the solar plasma conductivity is at best about that of copper at room temperature (see Section 4.5.2).

For large scales of motion, we have $R_{\mathrm{m}} \gg 1$, and diffusion of magnetic field lines relative to the plasma can be neglected. Changes in $\mathbf{B}$ are then determined entirely by the transport of field lines into and out of the plasma volume through plasma motions at velocity $\mathbf{v}$. Essentially, when $R_{\mathrm{m}} \gg 1$, the self-induction is large enough for the induced electric field, given by equation 4-60, to essentially cancel the motional electric field $\mathbf{v} \times \mathbf{B}$. In this advective limit, the field lines can be visualized as "frozen-in" to the plasma. It is left to

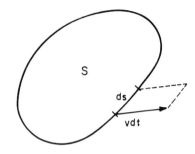

Fig. 4-4   Geometry of terms used in Exercise 4-5, on the freezing-in concept.

the exercises (refer to Fig. 4-4) to show that this advective process is described by the term $\nabla \times (\mathbf{v} \times \mathbf{B})$ in the induction equation.

Under the conditions commonly encountered on the sun, the spatial scales are so large that $R_m \gg 1$ for any perceptible motions, so that the frozen-in condition is of special relevance. When freezing-in holds, the plasma and embedded field lines move together. If, additionally, the energy density of the field given by $B^2/8\pi$ per unit volume greatly exceeds the thermal and kinetic energy density of the plasma, the field is "force-free," and the plasma frozen to it will be constrained to motions along the field lines, for which $\mathbf{j} \times \mathbf{B}$ vanishes. This tends to be the case in the chromosphere and corona, where strong magnetic fields associated with subphotospheric current systems extend out into a very tenuous medium and give rise to the conspicuously field-aligned appearance of motions in fibrils, spicules, loops, and prominences.

At the photosphere, on the other hand, the thermal energy of the much denser plasma is comparable to the magnetic energy of the fields observed in spots and faculae. At these levels, we also see motions of the magnetic structures that probably arise from buffeting by gas flows at even deeper layers. The plasma kinetic energy also dominates the magnetic energy density far out in the corona, so the ram pressure of the accelerating outflowing solar wind plasma is able to break open the magnetic field lines and stream radially out from the sun.

4.4   WAVE MOTIONS IN THE SUN

4.4.1   Types of Waves Expected and Observed

A simple analysis of the restoring forces available in the sun's atmosphere and interior suggests the presence of four main wave modes, namely sound waves, gravity waves, Alfvén waves, and plasma oscillations. There is direct observational evidence for the presence of sound waves at the photosphere and less direct evidence for their presence in the chromosphere. Certain observations in the corona and solar wind also suggest the presence of Alfvén waves, whose restoring force is the tension in magnetic field lines. Solar gravity waves, whose restoring force is the buoyancy of the fluid, are predicted to be present with greatest confidence in the sun's radiative core, but the difficulties experienced in observing them in the photosphere agree with the theoretical prediction that they should be heavily damped in the intervening convective layers. Radio observations yield information on plasma oscillations, whose restoring force is the electrostatic attraction between ions and electrons in a plasma.

4.4.2 Sound Waves

A pressure perturbation gives rise to a local pressure gradient and thus to an acceleration of fluid. When the pressure perturbation ΔP is of small amplitude relative to the ambient pressure P, some important simplifications to the equation of motion are possible, leading to an equation describing the propagation of small pressure and density signals in a compressible fluid, called sound waves.

The velocity $\mathbf{v}$ produced in such a flow is small, so that the nonlinear term $\mathbf{v} \cdot \nabla \mathbf{v}$ in Euler's equation can be neglected. Also, the flow can be considered to be adiabatic, since the perturbations in temperature and density (and thus any heat flows by thermal conduction or radiation) are negligible. Under these circumstances, it is well known that the flow obeys a wave equation of the form

$$\frac{\partial^2 \phi}{\partial t^2} = c_s^2 \nabla^2 \phi, \tag{4-71}$$

where ϕ is the velocity potential defined as

$$\mathbf{v} = \nabla \phi, \tag{4-72}$$

and

$$c_s = \left(\frac{\gamma R T}{\mu} \right)^{1/2} = \left(\frac{\gamma P}{\rho} \right)^{1/2} \tag{4-73}$$

is the sound speed. It can be shown that not only the velocity potential but also the pressure and density perturbations ΔP and $\Delta \rho$ satisfy this wave equation, which has solutions of the form

$$\phi = f_1(x - c_s t) + f_2(x + c_s t). \tag{4-74}$$

These solutions describe the propagation of a traveling wave profile given by the function f in the $-x$ and $+x$ directions at speed c_s. We note that this speed is independent of frequency (and also of amplitude, provided it is small, as we have assumed). The functional forms of f_1 and f_2 depend on how the local perturbation is generated. Most often, we are interested in the solution for a monochromatic oscillator, so we take f_1 and f_2 as sinusoids, but the perturbation could also be a single pressure pulse due to a local deposit of heat or a local compression.

The sound waves produced by a small pressure perturbation carry both energy and momentum. It can be shown that for a "plane" wave, the wave energy density per unit volume is

$$E = \rho v^2,\tag{4-75}$$

and the flux of this energy density across a plane surface transverse to the propagation direction of the wave is

$$F_w = \rho v^2 c_s.\tag{4-76}$$

Here v is the speed of the fluid caused by passage of the sound wave, so $v \ll c_s$.

A plane wave also imparts momentum to a fluid through which it propagates by virtue of this acceleration of plasma to speed v. This momentum flux density is given by

$$P_w = \rho \overline{v^2}\tag{4-77}$$

where the horizontal bar denotes a time average of v^2 over the wave period.

4.4.3   Simple Waves and Shock Formation

In many cases of interest, the pressure perturbation cannot be considered small, and the approximate solutions of the equations of motion obtained above do not hold for propagation of larger disturbances. An exact solution of the equation of motion can be obtained which describes the propagation of a traveling plane wave $f(x + ut)$ of arbitrary amplitude, still assuming that the flow is adiabatic. The properties of such a so-called simple wave differ in several important ways from the behavior of sound waves.

For one, the speed v of the fluid accelerated by the wave's passage is no longer small relative to the sound speed c_s. The total velocity $\mathbf{u}$ of any point on the wave profile at a given instant is now

$$\mathbf{u} = \mathbf{v} \pm c_s.\tag{4-78}$$

The velocity $\mathbf{v}$ is also a function of density, whose variation through the wave can be large enough to appreciably change $\mathbf{v}$. For the plasmas of interest here $du/d\rho > 0$, so the points of the wave at which ρ is higher also move faster. This dependence of the velocity on density causes the compressional parts of the wave or wave train to overtake the rarefactional parts, as illustrated in Fig.

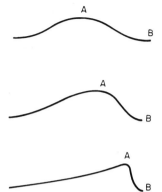

Fig. 4-5 Illustration of simple wave steepening to form a shock.

4-5. The result is the eventual formation of a thin discontinuity in density, velocity, and temperature called a shock wave.

4.4.4 Properties of Shock Waves

A shock wave is not an infinitely thin discontinuity, since changes in the fluid properties always require a finite scale comparable to the mean free path of the fluid particles. The thickness is determined by the viscosity within the shock (the viscous force determines the rate of acceleration of gas through the shock, and viscous dissipation contributes to the gas heating) and by the thermal conductivity, which determines the heat flow across the shock.

For most purposes, however, a shock can be idealized as a discontinuity, and the stationary flow conditions behind the shock (downstream) can be related to the conditions ahead (upstream) by requiring continuity of the mass, momentum, and energy fluxes through the shock. The geometry is illustrated in Fig. 4-6. If we denote with the subscript 1 the (upstream) gas that is moving into the shock and with the subscript 2 the (downstream) gas moving out of the shock, we have from the mass conservation requirement, the relation

$$\rho_1 v_1 = \rho_2 v_2. \tag{4-79}$$

For momentum conservation we require

$$P_1 + \rho_1 v_1^2 = P_2 + \rho_2 v_2^2, \tag{4-80}$$

and energy conservation is expressed as

$$H_1 + \tfrac{1}{2} v_1^2 = H_2 + \tfrac{1}{2} v_2^2. \tag{4-81}$$

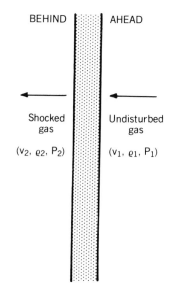

Fig. 4-6 Shock geometry.

In a perfect gas, where the enthalpy is given by

$$H = \frac{\gamma P V}{\gamma - 1},$$

(4-82)

these conservation equations lead to the following useful relations for the change in density, velocity, pressure, and temperature through the shock in terms of the shock's Mach number $M_1 = v_1/c_s$. Here, c_s is the sound speed in the gas ahead of the shock. These relations are

$$\frac{\rho_2}{\rho_1} = \frac{v_2}{v_1} = \frac{(\gamma + 1)M_1^2}{(\gamma - 1)M_1^2 + 2},$$

(4-83)

$$\frac{P_2}{P_1} = \frac{2\gamma M_1^2}{\gamma + 1} - \frac{(\gamma - 1)}{(\gamma + 1)},$$

(4-84)

and

$$\frac{T_2}{T_1} = \frac{\left[2\gamma M_1^2 - (\gamma - 1)\right]\left[(\gamma - 1)M_1^2 + 2\right]}{(\gamma + 1)^2 M_1}.$$

(4-85)

It follows that the pressure and temperature increases can be infinitely large for sufficiently large shock strengths (or Mach numbers, since the Mach number increases with the strength of the disturbance measured as $M_s - 1$)

but the density increase is limited to the value $(\gamma + 1)/(\gamma - 1)$, which is 4 for a plasma consisting of monatomic particles.

It can be shown also that a shock always moves supersonically relative to the gas ahead (i.e., $v_1 > c_s$). This means that no information about the flow disturbance can reach the gas ahead of the shock before the gas enters the shock, which explains why you cannot hear a supersonic plane coming before its shock wave breaks your windows. But the shock travels subsonically relative to gas behind it, so that disturbances arising in the downstream gas will generate waves which will eventually overtake the shock and modify its strength.

It is easy to visualize this modification of shock strength in the situation when a shock is generated by the compressive wave sent out by a piston pushing rapidly into a gas. Subsequent motions of the piston will lead to further disturbances; for instance, a rarefaction wave generated by the piston's retraction will travel behind the shock and eventually catch it and weaken it.

The strength of a shock can also increase or decrease in time under the influence of several other processes. For one, dissipation of the shock's energy into heat through viscosity and thermal conduction tends to reduce the shock's strength. The rate of this dissipation is of central importance in calculations of shock heating of the chromosphere and corona. For relatively weak shocks ($M_s - 1 \ll 1$, i.e., $\rho_1 \sim \rho_2$) a reasonable formula for the heating rate per unit volume by a train of shock waves of period τ is

$$\nabla \cdot \mathbf{F}_{sh} = \frac{2}{3} \frac{\gamma(\gamma + 1)}{\tau} P_1 (M_s - 1)^3. \qquad (4\text{-}86)$$

The dissipation and heating by weak shocks are thus seen to increase rapidly as the shock strength. The heating rate also depends inversely on the period of the wave train that generates the shocks since this period determines how many shock waves pass through a gas element per unit time.

Shocks propagating in more than one dimension are weakened simply because the original impulse is diluted over a larger surface area. A shock can also increase or decrease its strength by propagation into media of different temperature and pressure. Important effects of this kind arise in the solar atmosphere, where outward movement of shocks into increasingly rarefied plasma above the photosphere tends to increase their strength to satisfy the conservation equations (4-79)–(4-81). They weaken, however, when they begin to propagate above the temperature minimum into the rapidly increasing temperatures of the chromosphere, where the sound speed increases outward, and the disturbance tends to be "stretched out" (the inverse of the steepening illustrated in Fig. 4-5), thus decreasing its peak amplitude.

Useful insights can be obtained analytically in the asymptotic cases of weak and very strong shocks. But in general, detailed studies of shock behavior under the combined influences of geometrical effects, atmospheric gradients, and dissipation must be studied by numerical solution of the full shock equations.

4.4.5 Magnetohydrodynamic Waves

The presence of a magnetic field introduces a restoring force that leads to the possibility of a new wave mode. The magnetic volume force can be expanded as

$$\mathbf{j} \times \mathbf{B} = \frac{1}{4\pi}(\nabla \times \mathbf{B}) \times \mathbf{B} = \frac{1}{4\pi}(\mathbf{B} \cdot \nabla)\mathbf{B} - \nabla \frac{B^2}{8\pi}. \qquad (4\text{-}87)$$

The term $(\mathbf{B} \cdot \nabla)\mathbf{B}$ represents stresses due to curvature of the field lines. The term ∇B^2 can be visualized as describing stresses due to a tension $B^2/8\pi$ along the lines of force, and an equal pressure $B^2/8\pi$ transverse to them, as illustrated in Fig. 4-7.

Given this magnetic tension, we expect to find that in analogy with waves on an elastic string of mass ρ per unit length, transverse waves could be propagated in the x-direction along $\mathbf{B}$ with a velocity given by a wave equation

$$\frac{\partial^2 B}{\partial t^2} = v_a^2 \frac{\partial^2 B}{\partial x^2}, \qquad (4\text{-}88)$$

where

$$v_a = \frac{B}{(4\pi\rho)^{1/2}} \qquad (4\text{-}89)$$

is the Alfvén speed.

Figure 4-8 shows the curvature of the field lines caused by propagation of such Alfvén waves in an infinite medium. Note that the Alfvén speed varies directly with magnetic intensity and inversely with the square root of density. Thus the Alfvén speed is largest in coronal structures where $B^2 \gg 8\pi P$ and v_a considerably exceeds the sound speed under most circumstances.

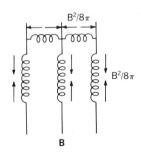

Fig. 4-7   Illustration of magnetic stresses.

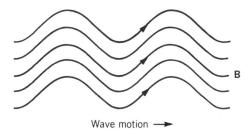

Wave motion ⟶

Fig. 4-8 Geometry of Alfvén waves in an infinite medium.

A disturbance in a magnetized plasma can also propagate across field lines, tending to compress them, as well as modifying their tension. Here the magnetic pressure $B^2/8\pi$ transverse to the field lines acts as a restoring force similar to the gas pressure. By analogy, we expect a compressional magneto-acoustic wave propagating across the field lines at a speed $B/(4\pi\rho)^{1/2}$ equal to the Alfvén speed.

In general, a perturbation of the pressure in a magnetized plasma will excite a variety of magnetoacoustic modes whose propagation speeds and directions depend on the relative values of gas and magnetic pressure in the medium. They also depend on the geometry of the medium, through the boundary conditions. A thorough understanding of these wave modes requires a full solution of the wave equation. But it is often useful to remember that the wave energy propagated away from the disturbance is distributed among the longitudinal and transverse modes in proportion to the square of their characteristic speeds. Thus, in the corona, where $v_A > v_S$, Alfvén waves are more important in determining the time scales and dynamical properties of the medium than are sound waves. Lower in the atmosphere, their relative importance changes.

4.4.6 Internal Gravity Waves

Gravity provides a restoring force that gives rise to several different types of wave motion, referred to collectively as gravity waves. Ocean waves are an example of a surface gravity wave caused by perturbation of the free surface of an incompressible fluid from its equilibrium shape, which is a plane when the gravitational force is homogeneous. Gravity waves can also propagate at the interface of two fluids whose densities are different, even if these are incompressible. In such internal gravity waves, the restoring force is derived from the buoyancy of the lighter fluid. In a compressible fluid such as the solar plasma, the most important type of gravity wave turns out to be an internal mode where the driving force is the buoyancy of a displaced parcel of fluid in hydrostatic equilibrium. We pursue this subject further in Chapter 7 since it is closely related to the onset of convection.

4.4.7   Plasma Oscillations

Oscillations arising from the electrostatic force between electrons and ions within the solar plasma have been clearly identified. The theory of such oscillations is central to the interpretation of radiowave emission from the solar corona, and these oscillations probably also contribute to the dissipation of other lower-frequency, larger-scale motions through their influence on the plasma transport coefficients.

Two main kinds of oscillations are distinguished. In electron oscillations, the electron gas component of the neutral plasma moves as a whole relative to the much heavier ions. The oscillation frequency, called the plasma frequency, can be calculated by considering the oscillations of a plasma of particle density n in the form of an infinite slab. In these oscillations a separation of a distance x normal to the plane of the slab occurs between the electrons and ions. This gives rise to a restoring electric field given in emu by $4\pi necx$. The restoring force of an electron of mass m is thus $4\pi ne^2cx$, which gives an equation of motion for the electron

$$m\frac{d^2x}{dt^2} = -4\pi ne^2cx, \tag{4-90}$$

describing harmonic oscillation at the frequency $\omega_p/2\pi$, where

$$\omega_p^2 = \frac{4\pi ne^2c}{m}, \tag{4-91}$$

is known as the plasma frequency.

The positive-ion oscillations are also longitudinal waves, called ion-sound waves. They are of two kinds, depending on whether their wavelength λ is greater or less than the distance at which the surrounding electrons effectively screen the field of a positive ion. This distance, which determines the smallest scale over which the plasma can be taken as effectively neutral, is given by the Debye length

$$h \sim \left(\frac{kT}{4\pi ne^2c^2}\right)^{1/2} \tag{4-92}$$

where T is the plasma temperature. This expression for h is derived by equating the electrostatic potential energy $-2\pi n_e e^2c^2x^2$ of an electron in the field of an unbalanced net charge $n_e e$ cm^{-3}, to the mean electronic kinetic energy $\frac{1}{2}kT$.

If $\lambda \ll h$, electrostatic waves similar to electron oscillations (but more difficult to excite) are generated. If $\lambda \gg h$, the oscillations are simply sound waves, since at these lower frequencies the restoring force is just the gas pressure, which can be transmitted on the time scale of particle collisions.

4.5   CHARGED PARTICLE DYNAMICS

4.5.1   Validity of the Continuum Approximation and of Thermal Equilibrium

Our discussion of plasma dynamics so far has treated the plasma as a continuous fluid. We have assumed that the properties of this fluid can be defined in terms of a few macroscopic state variables, such as a unique local pressure and temperature, and in terms of the molecular transport coefficients for heat, electric current, and momentum. These coefficients, called the thermal and electrical conductivity and the viscosity, constitute, along with the mean molecular weight (and thus the specific heats), the only link between the dynamics of the plasma in the continuum approximation and the microscopic properties of the charged particles. This is a valid assumption, even in the outer solar atmosphere and solar wind, since the number of particles contained in any volume of interest is always large enough to ensure that such averaged quantities can be consistently defined and expected to behave continuously from point to point.

However, at the relatively low densities found in layers above the chromosphere, the particle densities do become low enough that the mean free path between collisions can greatly exceed the dimensions of the structures observed, such as coronal loops and prominences. These dimensions are set by a magnetic field mainly associated with current systems lying below the photosphere. The collisional coupling between particles can then be quite weak, so that differences in momentum that may exist between co-existing particle populations, such as the electrons and positive ions, can persist over measurable time scales.

The assumption of thermal equilibrium, and thus the meaning of temperature in this "collisionless" regime, needs to be examined. The electrons, which travel much faster and collide far more frequently with one another than do the much heavier ions, tend to thermalize relatively rapidly to a Boltzmann velocity distribution. But a longer time is necessary before equipartition of energy is achieved by collisions between the electrons and the much heavier ions so as to equalize the electron and ion gas temperatures.

An additional complication arises when an external imposed magnetic field permeates such a low-density plasma. The transport coefficients parallel to the direction of the magnetic field lines are generally much higher than those perpendicular to the field. To understand how this anisotropy comes about and to gain some insight into the connection between individual particle motions and the MHD representation of continuum plasma motions, we briefly consider the dynamics of charged particles in a magnetic field.

4.5.2   Charged Particle Motions

The equation of motion for a particle of charge q and mass m traveling at velocity $\mathbf{v}$ in a region where an electric field $\mathbf{E}$ and magnetic field $\mathbf{B}$ are present is

$$m\frac{d\mathbf{v}}{dt} = q(\mathbf{E} + \mathbf{v} \times \mathbf{B}). \qquad (4\text{-}93)$$

If we assume for the moment that $\mathbf{E} = 0$, it is well known that the particle will travel in a circle of radius r and at a frequency ω_c, found by setting the acceleration qvB/m equal to the centrifugal acceleration v^2/r. This yields for the angular frequency

$$\omega_c = v^2/r = ZeB/m \qquad (4\text{-}94)$$

where Z is the particle charge and ω_c is called the cyclotron frequency. Note that particles of opposite charge gyrate in opposite directions. The parallel component of velocity, $v_{\parallel}$, is unaffected by the presence of $\mathbf{B}$, so the resultant motion of a particle incident upon $\mathbf{B}$ at some angle will be a helix of constant pitch around a line of force. Such free spiraling or gyration around magnetic field lines, in the absence of other forces acting in the equation of motion (4-93), is the single particle analogue of the macroscopic concept of "freezing-in" of the plasma to the magnetic field lines (but see exercises).

When other forces act on the particle, drifts across the field lines can appear. The effect of these drifts upon the validity of the "freezing-in" depends upon the nature of the force. Collisions in a thermal plasma cause the individual particles to be bumped from one field line to another, but given the isotropic nature of such collisions, a volume of gas will tend to remain "frozen in," provided the condition of high R_m is met. Electrostatic fields, gravitational fields, or inhomogeneous magnetic fields will produce systematic drifts across the field lines. The component v of motion perpendicular to $\mathbf{B}$ caused by such a force $\mathbf{F}$ is given by

$$v = \frac{\mathbf{F} \times \mathbf{B}}{qB^2}. \qquad (4\text{-}95)$$

The effect of such a force can be seen in Fig. 4-9. When the particle moves in the direction of $\mathbf{F}$, it is accelerated and its radius of gyration increases. If it moves in the opposite direction, it is slowed by $\mathbf{F}$ and r decreases. If $\mathbf{F}$ is an electrostatic field, the drift direction will be the same for positive and negative particles, since $\mathbf{F}$ acts in opposite directions on opposite charges, but their gyration directions are also in opposite senses. This will result in a bulk drift of the plasma across magnetic field lines. If $\mathbf{F}$ is gravity, ions and electrons will drift in opposite directions, which will drive an electric current across $\mathbf{B}$.

The picture given above of individual particles spiraling freely is useful, provided the density is low enough that collisions are so rare that a particle can complete many spirals before it collides and is deflected to another field line. The condition for this to hold is that the cyclotron frequency greatly

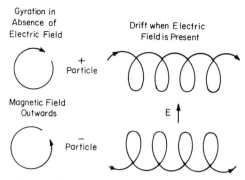

Fig. 4-9 Geometry of charged particle spiraling and drifts. Adapted from L. Spitzer, *Physics of Fully Ionized Gases*, Interscience, 1962.

exceed the particle collision frequency. The frequency w of collisions between ions and electrons in a thermal plasma is, from kinetic theory,

$$w = \pi b^2 n_i \bar{v}_t, \tag{4-96}$$

where b is the radius of the "effective" cross-section of the ion for electron-ion collisions, and $\bar{v}_t$ is the mean electron thermal velocity. Under solar conditions, for a fully ionized hydrogen plasma, we have

$$w \simeq 15 n_e T^{-3/2}. \tag{4-97}$$

Given the sharp decrease of density and rapid increase of temperature above the photosphere, it can be shown (see exercises) from comparison of the formulas for ω_c and w that effects due to the free spiraling of electrons and ions are important in the corona and solar wind and unimportant below the photosphere, where collisions are frequent.

The free spiraling of electrons creates large anisotropies in the electrical and thermal conductivities of a plasma, since these transport coefficients are largely determined by the motion of electrons relative to the much less mobile ions. When the ratio $\omega_c/w \gg 1$, the ability of electrons to move transverse to the magnetic field is severely inhibited, so both of the transport coefficients perpendicular to $\mathbf{B}$, denoted by $\sigma_\perp$ and $K_\perp$, are greatly reduced. Their values along the field, $\sigma_\parallel$ and $K_\parallel$, are unaffected by the magnetic field.

The electrical conductivity of a fully ionized hydrogen plasma is given approximately by

$$\sigma = \frac{n_e e^2}{w m_e}. \tag{4-98}$$

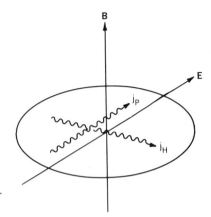

Fig. 4-10 Hall (j_H) and Pedersen (j_P) currents in a magnetized plasma.

Taking the expression (4-97) for the collision frequency, we have

$$\sigma \equiv \sigma_{\parallel} \sim 2 \times 10^{-14} T^{3/2}. \tag{4-99}$$

If an electric field **E** is applied across the magnetic field when $\omega_c \gg w$, the current drawn is given by

$$\mathbf{j} = \sigma' \mathbf{E} + \sigma'' \mathbf{B} \times \frac{\mathbf{E}}{B}, \tag{4-100}$$

where

$$\sigma' = \frac{\sigma w^2}{\omega^2 + w^2}, \tag{4-101}$$

and

$$\sigma'' = \frac{\sigma \omega w}{\omega^2 + w^2}. \tag{4-102}$$

The component of current parallel to **E** is called the Pederson current, while that perpendicular to both **B** and **E** is called the Hall current (Fig. 4-10).

Using (4-99) we see that the electrical conductivity of the solar plasma along the magnetic field is good, but no better than that of metals at room temperature. The importance of MHD in the sun is due not to the high electrical conductivity, but to the enormous spatial scales. Similar but less general results may be obtained for the thermal conductivity and viscosity parallel and perpendicular to **B**. The anisotropy of thermal conductivity is particularly important in determining heat flow in the corona as discussed in Chapter 9.

ADDITIONAL READING

Fluid Mechanics

L. Landau and E. Lifshits, *Fluid Mechanics*, Pergamon, 1959.

G. Batchelor, *An Introduction to Fluid Dynamics*, Cambridge University Press, 1967.

Plasma Physics

L. Spitzer, *Physics of Fully Ionized Gases*, Interscience, 1962.

C. Longmire, *Elementary Plasma Physics*, Interscience, 1967.

N. Krall and A. Trivelpiece, *Principles of Plasma Physics*, McGraw-Hill, 1973.

Magnetohydrodynamics

T. Cowling, "Solar Electrodynamics," in *The Sun*, G. Kuiper (Ed.), University of Chicago Press, p. 532, 1953.

T. Cowling, *Magnetohydrodynamics*, Interscience, 1957. See also 2nd edition published by A. Hilger, 1976.

E. Priest, *Solar Magnetohydrodynamics*, D. Reidel, 1982.

H. Alfvén and C. Fälthammar, *Cosmical Electrodynamics*, Oxford University Press, 1963.

EXERCISES

1. Show that the pressure and density scale heights in an isothermal atmosphere are equal.

2. Show that in a body whose density distribution is spherically symmetric the gravity force at a radius r is determined only by the mass $M(r)$ contained within the radius r.

3. Explain the difference between the formulas for P given by the equation of state (4-3) and the polytropic relation (4-17).

4. Calculate the magnetic Reynolds number for the smallest spatially resolvable motions on the sun, taking these to be on the scale $L \sim 300$ km of elemental flux tubes buffeted by horizontal granular motions of $v \sim 0.3$ km s^{-1}. What does this imply for plasma motion into and out of such small flux tubes and for larger-diameter tubes, such as those of sunspots?

5. Show that the term $\nabla \times \mathbf{v} \times \mathbf{B}$ represents the change in magnetic induction contained within the boundary s of a closed surface S. Consider that a vector element $d\mathbf{s}$ moves a distance $\mathbf{v}\, dt$ in time dt and thus sweeps out a vector element of area $\mathbf{v}\, dt \times d\mathbf{s}$, which contains an induction $\mathbf{B} \cdot (\mathbf{v}\, dt \times d\mathbf{s})$. Use this to express the time rate of change of the induction. Then

derive the required result by line integration and use of Stokes' theorem (refer to Fig. 4-4).

6. Calculate the Debye lengths of plasmas in the solar radiative core, photosphere, chromosphere, corona, and solar wind, and compare them to the typical scales of structures in these regimes. How well is the fundamental postulate of plasma neutrality satisfied in these structures? Calculate also the mean free paths for ion-ion collisions (take it to be $\lambda = (n\sigma)^{-1}$, where $\sigma = 10^{-18}$ cm$/(kT)^2$ and kT is measured in keV) in these regimes and comment upon whether the plasmas are collisional or collisionless. Finally, calculate the proton gyro radii using (4-94) but substituting the proton mass, and comment on whether the geometry of structures seen in these regimes is more likely to be determined by gas pressure gradients or by the magnetic field geometry.

7. The requirement for "freezing-in" of a collisional plasma and magnetic field requires relatively high plasma conductivity transverse to magnetic field lines. On the other hand, in a collisionless plasma, charged particles are constrained to move with field lines when collisions are entirely absent (i.e., the cross-field conductivity vanishes). Can you explain this apparent paradox?

5

The Photosphere

The image of the sun formed at the focus of a large solar telescope is an impressive sight. At the Kitt Peak Observatory's McMath tower, for instance, the photospheric disk is almost a yard across, and its features can be inspected through a pair of sunglasses without further magnification. Figure 5-1 is a white-light photograph of the solar disk which illustrates the basic structure of the photosphere. The fine detail of the dark sunspots and of the bright faculae near the limb is well seen. The remarkable sharpness of the edge of the sun's gaseous atmosphere is also evident, as is the darkening of the photosphere toward the limb. The granulation pattern can just be discerned as a low-contrast, small-scale mottling of the photospheric surface.

Figure 5-2 is a photograph in ultraviolet light, taken from a rocket-borne telescope, that shows the structure of the highest photosphere located about 400 km above the layers that contribute most of the radiation in Fig. 5-1. This is the region of the temperature minimum in the sun's atmosphere. The rather different structures seen here reveal the changes that occur over this small height difference. The pattern of granules seen in Fig. 5-1 has been replaced by a network of bright structures of relatively high contrast, and the bright faculae seen around spots are also much more evident than in the visible. These structural differences indicate the important changes in the modes of outward energy transport, and in the influence of the magnetic field, that occur within the photosphere.

In this chapter we discuss the physical structure and energy transport mechanisms of the "quiet" photosphere outside the intensely magnetized regions observed as spots and faculae. We show how relatively straightforward measurements of the limb darkening enable us to determine the dominant radiation mechanism and the physical conditions in this important layer which emits more than 99% of the sun's light and heat. Observations of the granulation and of other photospheric brightness and velocity patterns that we discuss

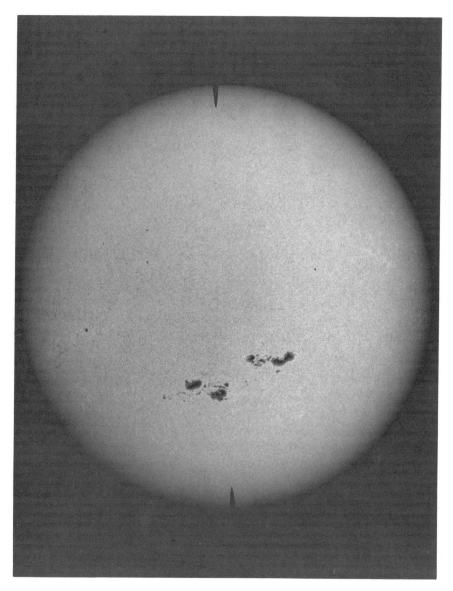

Fig. 5-1 A full-disk white-light image of the photosphere. By permission of Mt. Wilson and Las Campanas Observatories, Carnegie Institution of Washington.

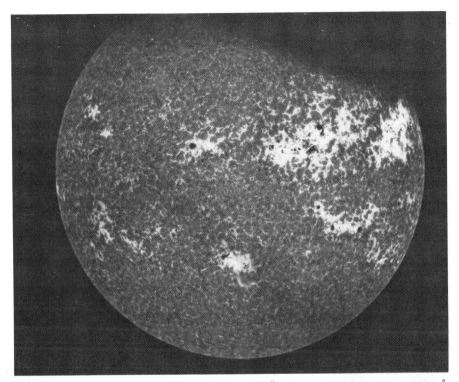

Fig. 5-2   The solar disk photographed in the 1600-Å ultraviolet continuum with a 40-Å bandpass on July 13, 1982, by a rocket-borne ultraviolet telescope. The telescope field of view does not include the entire solar disk. By permission of M. Bruner and W. Brown.

here also provide us with most of our direct information on stellar convection, whose dynamics is explored in Chapter 7. Photospheric observations also tell us most of what we know about the sun's chemical composition. The physics of sunspots and faculae is described in Chapters 8 and 11, together with other aspects of solar magnetic activity.

5.1 OBSERVATIONS OF THE QUIET PHOTOSPHERE

5.1.1 Limb Darkening

Figure 5-1 shows that the photospheric intensity falls off significantly toward the limb. As mentioned already in Chapter 2, the explanation of this limb darkening lies in the decreasing temperature of the higher photospheric layers seen as we look nearer the limb. Analysis of this effect provides a particularly direct technique for determining the photospheric temperature structure with

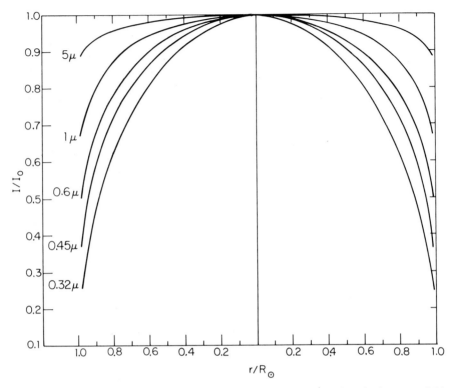

Fig. 5-3 Photospheric limb-darkening curves measured at wavelengths between 0.32 μm in the ultraviolet and 5 μm in the infrared.

depth, as explained in Section 5.2.2. The usual procedure for measuring the disk intensity profile is to stop the telescope drive and allow the centered solar image to drift across a detector, as the Earth rotates through the 30 arc minutes that correspond to the solar diameter, in two minutes of time. This method helps to ensure that all rays reaching the detector during the drift scan pass through the same parts of the optical system and also through the same atmospheric absorption.

Five intensity profiles measured at continuum wavelengths from the ultraviolet to the infrared are plotted in Fig. 5-3. The limb darkening decreases with increasing wavelength, i.e., the disk intensity profile becomes decidedly more square in the infrared. Figure 5-4 shows the behavior of limb darkening plotted against wavelength between 0.3 μm and 2.5 μm for various positions on the disk. This plot confirms the decrease of limb darkening with increasing wavelength, except around a sharp discontinuity at 3460 Å, the location of the opacity jump due to the Balmer continuum head.

Since limb darkening is caused by the photospheric temperature gradient, we might expect it to disappear in the far infrared and in the ultraviolet

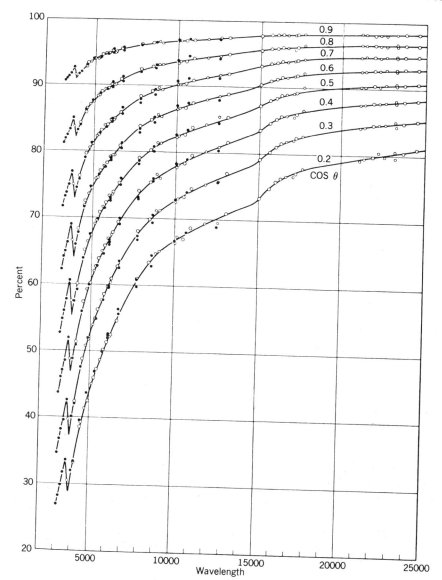

Fig. 5-4 Limb darkening as a function of wavelength at selected points of the solar disk defined by $\mu = \cos\theta$. From L. Goldberg and A. Pierce, *Handbuch der Physik*, Vol. 52, *Astrophysics III: The Solar System*, S. Flugge (Ed.), 1959. By permission of Springer-Verlag, N.Y. Inc.

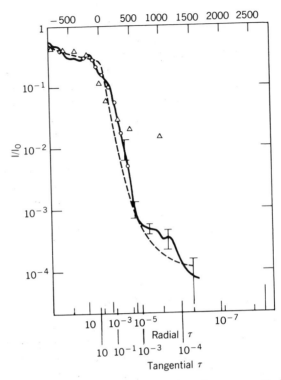

Fig. 5-5 The photospheric intensity I, observed at the extreme limb, relative to the disk–center intensity I_0. The solid curve indicates measurements in the red continuum at 6404 Å; other measurements at 6190 Å and 6563 Å are given as circles and triangles respectively. The dashed curve gives the predicted falloff of I from a photospheric model. The abscissae are radial optical depth at 5000 Å and tangential optical depth at 6404 Å. From S. Weart and J. Faller, *Astrophys. J.*, **157**, 887 (1969).

around 0.15 μm where we observe layers around T_{min}, where $dT/dh \sim 0$. A determination of the wavelength at which the disk intensity profile changes from limb darkening to limb brightening provides a technique for locating the height of the layers where the photospheric temperature minimum occurs, and considerable work has been done on the solar limb-darkening curve at far infrared and millimeter wavelengths with this objective in mind.

In practice, the profile near the limb is difficult to reconstruct exactly because of the low spatial resolution caused by telescope diffraction at these long wavelengths. The results indicate limb darkening at wavelengths of at least 24 μm, and probably out to 100 μm. Some evidence for limb brightening has been reported at $\lambda > 250$ μm, and most (but not all) measurements at $\lambda > 1$ mm show it. These results are broadly consistent with the location of the temperature minimum in photospheric models, but it becomes difficult to compare far infrared and millimeter observations with homogeneous models since the photosphere exhibits complex spatial structure at layers around the temperature minimum, as can be seen in Fig. 5-2.

The falloff of intensity at the edge of the photospheric disk is so abrupt that it is normally not resolved by even the largest solar telescopes unless special precautions are taken to remove the smearing caused by instrumental diffraction and scattered light, and atmospheric seeing. Figure 5-5 shows a plot of the intensity profile in the visible, observed during solar eclipses to reduce scattered light. The measurements indicate an intensity scale height of less than 100 km around the inflection point, so a dropoff at least as rapid as observed can be understood given the acceleration of solar gravity and the relatively low photospheric temperature (see exercises). It is worth noting that the remarkable sharpness of the limb makes it possible to measure the size and shape of the solar disk to a precision approaching 0.001%. The implications of such measurements for the sun's internal structure are discussed in Chapters 6 and 7.

5.1.2 Observed Properties of Granulation

A high-resolution picture of the granulation that is just visible in Fig. 5-1 is illustrated in Fig. 5-6. The pattern shown in this figure consists of extended bright granules of complex polygonal shape, separated by narrow dark lanes. The characteristic scale of the bright granules is about 1000 km, ranging widely from structures as small as the ~ 300-km resolution limit up to the largest features extending to over 2000 km. The average center-to-center granule spacing is 1400 km with a wide range. The average photometric contrast (as defined in Table 5-1, which summarizes granular observations) measured near disk center in white light around 5500 Å ranges between roughly 10 and 20%. This contrast increases with decreasing wavelength in the visible. These values are obtained from microdensitometry of calibrated photographs taken under exceptional seeing, in which the blurring caused by instrumental and atmospheric stray light has been corrected to a first approximation.

When a large, high-efficiency blazed grating was first installed at the spectrograph of the Mt. Wilson 150-ft tower telescope after World War II, the shortened exposures reduced seeing effects sufficiently to show the Doppler shifts of Fraunhofer lines caused by individual granules. The resulting "wiggly-line" spectrum, photographed with a modern spectrograph, is shown in Fig. 5-7. Accurate measurement of the relative line-of-sight velocities of bright granules and dark lanes is complicated by changes in the line profile across these structures, as well as by seeing and scattering that tends to reduce the velocity differences. Early studies indicated a surprisingly weak spatial correlation between the bright granular material and blueshifted parts of the wiggly-line spectrum, given that the hot material should be upflowing if

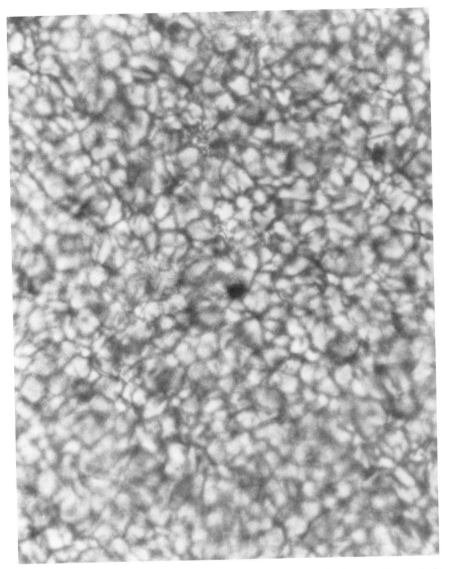

Fig. 5-6 A high-resolution photograph of granulation (and of a small pore). By permission of National Solar Observatory, Sacramento Peak, Sunspot, New Mexico.

TABLE 5-1 Observed Properties of Photospheric Granulation, Mesogranulation, and Supergranulation

	Average Diameter and (Range)	Average Center-to-Center Distance and (Range)	Disk-Center Average Contrast $C = \dfrac{I_g - I_{ig}}{\frac{1}{2}(I_g + I_{ig})}$ at (Wavelength)	Typical Velocities	Average Lifetime and (Range)	Number of Cells Covering Photosphere
Granulation	1000 km (300–2300 km) (of Bright Granules)	1400 km (400–3500 km) (of Bright Granules)	7–21% $\lambda = 5500\ \text{Å}$ $C \sim \lambda^{-1}$	Granular upflow relative to intergranule downflow: $\sim 1.8\ \text{km s}^{-1}$ Granule outflow horizontal velocity: $\sim 1.4\ \text{km s}^{-1}$	5 min (5–10 min)	$3\text{–}4 \times 10^6$
Mesogranulation	7000 km (5000–10,000 km)	Same as diameter	Not measured	60 m s^{-1} vertical velocities	2 h	10^5
Supergranulation	$3\text{–}3.5 \times 10^4$ km $(2\text{–}5.4 \times 10^4$ km)	Same as diameter	< 0.1% at visible λ	Horizontal outflow: 0.3–0.5 km s^{-1} Vertical flow at cell centers or interstices: < 0.02 km s^{-1}	20–40 h	5000

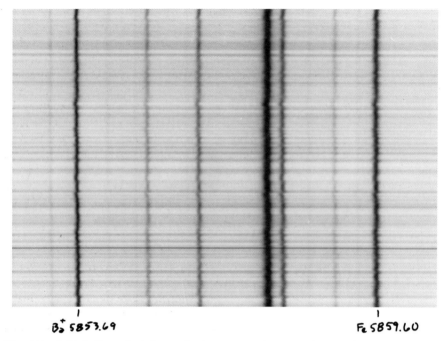

Ba⁺ 5853.69   Fe 5859.60

Fig. 5-7 A high-dispersion photospheric spectrum between the Ba I λ 5853.7 line (at left) and the Fe I λ 5859.6 line at right. Note the wiggly-line appearance due to Doppler shifts caused by granular motions. By permission of W. Livingston.

granules are convective cells. Later work showed that a substantial component of the photospheric velocity field is not associated with granulation, but is caused by a periodic 5-min oscillation. When the oscillatory velocity field is removed, the residual blueshifts coincide well with bright granules, and redshifts with the dark lanes.

The most accurate measurements of the granulation's vertical velocity field have been made by analyzing narrow-band ($\sim 1/8$ Å) filter photoheliograms taken in rapid succession. The changes in line profile between granules and intergranules can be measured from the intensity of each feature as the filter wavelength is scanned by a small amount (as little as $1/20$ Å) between successive frames. With the knowledge of the granule and lane line profile shapes, their relative Doppler shifts can be properly measured. The best measurements indicate an approximately 2 km s^{-1} relative velocity between up- and downflowing material. When radiative transfer effects are taken into account, such observations actually indicate peak-to-peak velocity differences of around 7 km s^{-1} around $\tau_{0.5} = 1$. Observations of granule displacements also indicate a substantial outflow comparable to the granular upflow speed in the hot granules.

The life pattern of an individual granule seems to consist of its formation from several smaller pre-existing components, its expansion to a maximum size of 3–5 arc sec, and its splitting up into several fragments which tend to fade away in situ (Fig. 5-8). Recent observations from the space Shuttle indicate that a typical granule lasts for about 5 minutes. Several observers have noted that the granulation evolution cycle is often repeated over several "generations" suggesting an ordering process that might last longer than the granules themselves. Such evidence for ordering needs to be investigated since it could have important implications for understanding of solar convection.

The observation that the visibility of granulation also depends on height provides some of the most important clues to the dynamics of the photosphere. Granules can be seen to at least 5″–10″ from the limb, corresponding to $\cos \theta < 0.1$. This implies that the temperature variation pattern caused by granulation persists to at least 150 km (see Section 5.4.1) above the $\tau_{0.5} = 1$ level.

The height dependence of the granular velocities (measured from Doppler shifts using lines formed at several depths) is more difficult to determine. Correction of the observations for the velocity field of 5-min oscillations is the main difficulty. It appears certain, however, that the velocity of an upward moving granule, which can persist as far as 500 km above $\tau_{0.5} = 1$ decays much less rapidly than its excess brightness. It is interesting that upflowing granular material is dark when observed in Fraunhofer lines formed relatively high in the photosphere. As discussed in Chapter 7, the explanation lies in the "supercooling" of the granular material caused by its rapid expansion during the upflow.

The granulation structure is affected by the presence of nearby strong magnetic fields of spots and magnetic network. Granules within about 1 spot radius of the outer edge of a penumbra have a roughly 20% smaller diameter. Granulation contrast in facular regions is also about a factor of 2 lower. Changes in the granulation structure, contrast, and velocity field around spots, plages, and network have also been inferred indirectly from observations of Fraunhofer line profile shapes. Models of radiative transfer in the granulation indicate that the line profile near its mid-depth portion is formed in the most rapidly upflowing bright material, and so is most blueshifted. The deepest portion of the line core is formed higher in the ballistically decelerated upflow, so its blueshift is less. The line wings, where the opacity is least, tend to be formed deepest in the cool, downflowing intergranule lanes and so are strongly redshifted. The result is a characteristically C-curved profile bisector.

This sort of analysis has shown, for instance, that the marked (~ 200–400 m s^{-1}) blueshift of Fraunhofer lines near disk center relative to their position near the limb can be understood by taking into account the decrease of granular contrast and line-of-sight velocity variation with distance from disc center. Studies of line profile shapes near network and faculae reveal significantly straighter bisectors than in areas located well away from magnetic

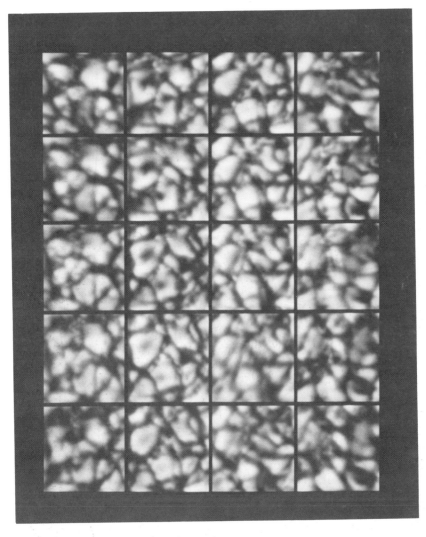

Fig. 5-8 A time sequence showing granule evolution. The field of view in each of the 20 frames is 7×7 arc sec, and the time intervals are about 1 min. Photos taken at La Palma, Canary Islands, by G. Scharmer and P. Brandt and provided by T. Tarbell.

fields. Disruption of granulation by magnetic flux tubes may be responsible. Observations with the spectrograph slit illuminated by integrated light from the whole solar disk show that the bisectors tend to be less C-curved near solar activity maximum than at minimum. This is at least qualitatively consistent with the fact that near activity maximum the higher packing density of magnetic flux tubes at the photosphere will tend to disrupt granular convection more than near activity minimum.

5.1.3 The Supergranulation and Photospheric Network

A Doppler picture of the photospheric disk, in which line-of-sight velocities appear as brightness variations, has a much less uniform appearance than Fig. 5-1. R. Leighton and his students, R. Noyes and G. Simon, first showed the power of Doppler imaging of the photosphere in the late 1950s. One of their Dopplergrams is illustrated in Fig. 5-9. It was obtained by photographically subtracting two spectroheliograms obtained consecutively in the two opposite (red and blue) wings of the Fraunhofer line (see Sections 1.3 and 1.5 for a discussion of the spectroheliograph and Leighton's techniques).

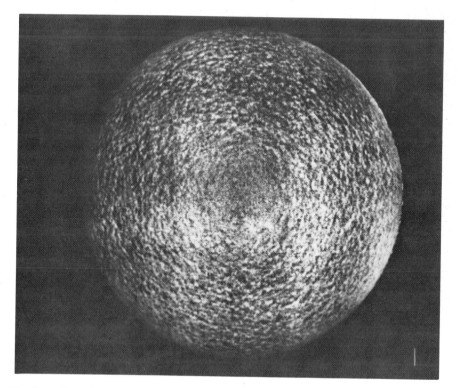

Fig. 5-9 Doppler picture of the photosphere showing supergranulation. By permission of G. Simon, National Solar Observatory, Sacramento Peak, Sunspot, New Mexico.

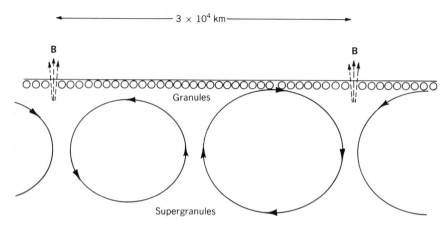

Fig. 5-10 Illustration of the approximate horizontal and vertical scales of granules and supergranules.

Examination of Fig. 5-9 shows a much larger scale pattern than granulation, with velocity amplitude diminishing toward disk center (the axisymmetric brightness gradient produced by solar rotation has been removed). This velocity pattern, called supergranulation by Leighton, consists of cells of characteristic scale around 30,000 km in which predominantly horizontal velocities of about 0.5 km s^{-1} are observed. Careful analysis showed that, in each cell, the material closer to disk center exhibits a blueshift, whereas material closer to the limb appears to be red-shifted. The geometry of the flow, illustrated in Fig. 5-10, is thus interpreted as diverging from the center of a supergranule cell and flowing to its perimeter. More recent measurements, using the observed displacement of granules carried by the supergranular flow, confirm this diverging geometry. The exact lifetime of the cells is difficult to measure directly from the velocity observations since it is of the order of a day, making it hard to obtain Doppler data sets of sufficient length.

Upflows at the cell centers and downflows at the perimeter might be expected (in analogy to granulation) if supergranules represent a larger scale of convection. Studies using line wing Doppler shifts do tend to show a significant redshift at the cell edges suggesting downflows of typically 0.1 km s^{-1} in many Fraunhofer lines, although no satisfactory evidence of cell-center upflows has been detected.

More recent work, based on analysis of full line profiles, shows that only the line wings are shifted at the cell edges and not the line cores. The change in C-curvature of the line profile shape (see Section 5.1.2) is similar to that found in models of the line shape when granulation is suppressed, so it is most likely due to the disturbance of granulation by intense magnetic flux tubes known to be located at the cell boundaries. There is no clear observational evidence for supergranule-associated up- or downflows as yet although recent studies of an intermediate scale of motion, called mesogranulation, may offer some clues. The main parameters of supergranulation (and of mesogranulation) are summarized in Table 5-1.

The absence of observable downflow does not rule out the interpretation of supergranulation as an overturning convective pattern. The uncertainty of roughly 20 m s^{-1} in the relative Doppler shift of line cores measured at cell centers and edges is large enough to mask small downflow speeds sufficient to satisfy mass conservation, provided the downflow area is some 25 times larger than the thickness of horizontal crossflow. Since the depth of the line-forming region in which the large horizontal flow is observed is only of order 100 km, this condition is quite easily met. Further comparisons of the observed velocity structures with theoretical models of supergranule cells will be necessary to clear up this question.

Well before Leighton's discovery from Doppler pictures (a velocity field of the supergranular scale and amplitude was actually reported several years earlier by A. Hart at Oxford, from a study of variation in the photospheric rotation rate), solar astronomers were aware of the so-called chromospheric network, seen particularly well in Ca K spectroheliograms (e.g., Fig. 9-3). The studies of Leighton and Simon showed that the bright Ca K points, first photographed on the disk by Hale and others, tend to be located at the outer perimeter of supergranule cells. Developments in magnetograph sensitivity in the 1960s showed that these bright Ca K points defined a photospheric magnetic network.

The photospheric network can be glimpsed in white light near the limb as a tracery of bright structures roughly outlining the supergranular cells. It becomes invisible in a photograph closer to disk center, on account of granulation noise. But careful photometry shows that the magnetic network points are between 0.1–1% brighter than the mean photosphere even at disk center and even when observed in monochromatic continuum radiated from photospheric levels near $\tau_{0.5} = 1$. The network becomes easily observable in photospheric lines and UV continua formed higher in the photosphere. It is well seen in Fig. 5-2 around the layers of the temperature minimum.

Several studies have been carried out to search for the temperature or velocity signature of larger convective scales than supergranules. No significant signal has been detected in the photospheric velocity field down to a level of a few meters per second and in the temperature field to about 0.3 K rms over spatial scales extending from supergranulation to about $0.5 R_\odot$. Detections have been reported below this level, but they must be regarded with suspicion, given the potential for low-level contamination by the relatively high supergranular velocities and by photospheric network.

The main evidence for large ("giant") scales of solar convection is based on the morphology of photospheric magnetic structures. The distribution of active regions generally shows a preferred spacing in longitude that persists over many solar rotation periods, and it may indicate the presence of large structures extending typically several hundred thousand kilometers across the solar surface. The evolution and morphology of such structures have been extensively studied and are discussed in Chapter 8. It is unclear whether they represent large scales of convection at some depth below the photosphere or, for instance, magnetic dynamo waves engendered by nonuniform solar rotation.

5.2 CONSTRUCTION OF A PHOTOSPHERIC MODEL

5.2.1 Physical Assumptions

The construction of a photospheric model builds on three basic assumptions. It is usual to adopt the geometry of a plane-parallel, homogeneous layer, since the thickness of the photosphere is much less than the solar radius and the inhomogeneities of granulation can be neglected to a first approximation. The gas pressure distribution with height h is assumed to be determined by the hydrostatic equation (4-1) and the equation of state (4-3). The photosphere can also be assumed to be in radiative equilibrium, although so-called radiative-convective models include a mixing-length representation of the convection that carries a substantial part of the heat flux through the deep photospheric layers.

The photosphere is not hot enough for thermal conduction to play an appreciable role since the plasma thermal conductivity decreases rapidly as $T^{5/2}$, yet the contribution of waves in the photospheric energy transport is of interest. Although its magnitude must be below 1% of the total flux, it might be relatively variable in time. Waves are neglected in most models because, until recently, the photospheric wave spectrum was poorly determined (it still is at the shorter periods $P < 30$ s), and the interaction of large-amplitude waves with convection and radiation is a difficult subject that requires further theoretical effort.

The iterative procedure leading to a model begins with an estimate of the vertical temperature profile $T(z)$, which is used to calculate a density distribution $\rho(z)$ using the hydrostatic relation. Given a photospheric chemical composition (or at least the abundances of the dozen elements that dominate the opacity), the opacities for this layer can be found from calculated tabulations as a function of ρ and T. In an LTE model, these opacities are based on calculations using the Saha and Boltzmann equilibrium equations to determine ionization and excitation of the important species. The LTE radiative transfer equation is solved to obtain a value of the emergent flux $F = \sigma T_{\text{eff}}^4$ corresponding to the temperature profile with optical depth. This total wavelength-integrated flux is a conserved quantity with height in a radiative equilibrium model.

The procedure used in deriving a non-LTE model is similar, but much more involved. The ionization and excitation needed to calculate the opacities must now be determined from the simultaneous solution of the coupled set of statistical equilibrium equations together with the non-LTE radiative transfer equation.

Purely theoretical models are iteratively adjusted toward the condition of radiative (or radiative-convective) equilibrium by changing the temperature profile. Each model is thus determined by a specific value of T_{eff}, the value of g, and by the assumed solar abundances. Semi-empirical models are iterated instead toward agreement with certain observable quantities such as the continuum limb darkening and the disk-center emergent fluxes over a range of continuum wavelengths. The depths and profiles of strong Fraunhofer lines also provide important constraints on $T(z)$ in the upper photosphere and are usefully compared to the predictions of models that account for the non-LTE effects in these layers. When the model is used to calculate line profiles, a distribution of turbulent velocity with depth must also be assumed along with the other model parameters.

5.2.2 Determination of the Temperature Profile from Continuum Limb Darkening

A direct determination of the temperature profile with optical depth can be obtained by inversion of the photospheric limb darkening. We begin with the expression (2-20) for the emergent intensity from a semi-infinite atmosphere

$$I_\lambda(0, \mu) = \frac{1}{\mu} \int_0^\infty S_\lambda(\tau_\lambda) \exp\left(\frac{-\tau_\lambda}{\mu}\right) d\tau_\lambda. \tag{5-1}$$

Assuming LTE and expressing the intensity relative to the disk-center intensity $I(0, 1)$, we have for the limb-darkening profile

$$\phi_\lambda(\mu) = \frac{I_\lambda(0, \mu)}{I_\lambda(0, 1)} = \frac{1}{\mu} \int_0^\infty \frac{B_\lambda(\tau_\lambda)}{I_\lambda(0, 1)} \exp\left(\frac{-\tau_\lambda}{\mu}\right) d\tau_\lambda. \tag{5-2}$$

The normalized source function can be expanded as a power series in τ_λ

$$\frac{B_\lambda(\tau_\lambda)}{I_\lambda(0, 1)} = a_\lambda + b_\lambda \tau + c_\lambda \tau^2 + \cdots. \tag{5-3}$$

The coefficients of this expansion are evaluated by substituting equation (5-3) into equation (5-2) and integrating to yield the series

$$\phi_\lambda(\mu) = a_\lambda + b_\lambda \mu + c_\lambda \mu^2 + \cdots. \tag{5-4}$$

whose coefficients can be evaluated by least-squares fitting to observations of the monochromatic limb darkening.

Once the coefficients are known, this expression gives the optical depth dependence of the normalized source function. We can then measure the intensity $I(0, 1)$ at sun center with an absolutely calibrated detector (preferably from space or at least correcting for the Earth's atmospheric transmission) to obtain $B_\lambda(\tau)$ from equation (5-3). This yields the temperature as a function of optical depth, $T(\tau_\lambda)$, at that wavelength.

5.3 DETERMINATION OF THE PHOTOSPHERIC OPACITY

5.3.1 The Empirical Technique

To determine the temperature profile with the *geometrical* depth needed to integrate the pressure in the hydrostatic equation, we require the photospheric opacity. A plot of the limb-darkening functions $\phi_\lambda(\mu)$ for widely separated wavelengths, as in Fig. 5-3, shows that the falloff of the intensity near the limb is much less pronounced in the red than in the violet. The geometrical depth z (and thus the layer temperature T_b) visible at a given μ depends upon wavelength, so the difference between limb darkening at various wavelengths is no surprise.

To extract information on the photospheric absorption coefficient, one might plot the value of μ (or equivalently of τ_λ) at which a fixed brightness temperature is observed, against the wavelength of the observation. Since a fixed T_b defines a fixed geometrical depth, the variations of μ (or τ_λ) in such a plot show the wavelength variation of the photospheric absorption coefficient.

It is most convenient to proceed by differentiating the expression for the optical depth to yield

$$\frac{d\tau_\lambda}{dT} = k_\lambda \rho(z) \frac{dz}{dT}, \qquad (5\text{-}5)$$

where we have assumed that the wavelength dependence of k_λ is the same over the range of conditions found in photospheric layers. At a layer of given geometrical depth z, the temperature gradient dT/dz and the density $\rho(z)$ will be fixed. The left-hand side of expression (5-5) can be derived directly from observations since from the Eddington-Barbier relation one can obtain $d\tau_\lambda/dT = (dT_b/d\mu)^{-1}$. It follows that a plot $d\tau_\lambda/dT$ against wavelength, obtained from limb-darkening observations, should directly yield the wavelength dependence of k_λ and also a rough idea of its absolute value. A family of such plots for a range of atmospheric temperature levels is given in Fig. 5-11.

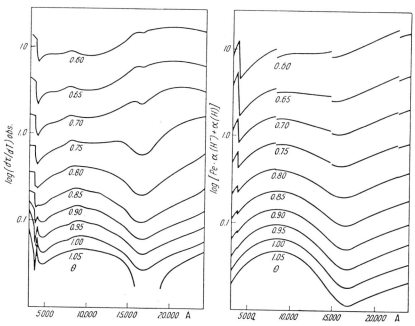

Fig. 5-11 Plots of (left panel) the wavelength dependence (in angstroms) of the observed photospheric absorption coefficient and (right panel) the behavior of the calculated coefficient. The curves are parametrized by the normalized temperature $\theta = T/5040$. From L. Goldberg and A. Pierce, *Handbuch der Physik* Vol. 52, *Astrophysics III: The Solar System*, S. Flugge (Ed.), 1959. By permission of Springer-Verlag, N.Y.

5.3.2 The Sources of Photospheric Opacity

One of the most interesting results of photospheric modeling was the identification in 1938 of the esoteric negative hydrogen ion H$^-$ as the main source of photospheric absorption and emission. Until that time, the atomic mechanism now realized to be responsible for over 95% of the sun's total radiative output was quite unknown. The bound-free continua of atomic hydrogen, the most abundant element in the sun, might be expected to dominate the photospheric opacity. But it had been recognized earlier that the discontinuities in brightness observed at the heads of the Balmer (λ 3650) or Paschen (λ 8204) continua are too small to support this. The discontinuities are expected because the sudden increase of k_λ (proceeding toward shorter wavelengths) at the continuum head reduces the geometrical depth in the atmosphere at which $\tau_\lambda = 1$ occurs, and thus the brightness temperature of T_b of the emergent radiation. The magnitude of the solar Balmer discontinuity measured at disk center is about 1.3. It becomes much larger for hotter, earlier-type stars in which atomic hydrogen opacity is dominant.

Following the initial suggestion by R. Wildt in 1938, S. Chandrasekhar and his co-workers carried out the difficult quantum mechanical calculations which showed that the wavelength dependence and magnitude of k_λ exhibited by the H^- ion, shown in Fig. 5-11, matched the observed behavior of photospheric opacity also shown in that figure. The negative hydrogen ion H^- is formed by the weak electrostatic attachment of a second electron to the imperfectly shielded nucleus of the "neutral" hydrogen atom. The electrons that form this attachment are produced by the ionization of metals such as Na, Mg, and Fe, which have relatively low first ionization potentials between 5 and 7 eV and thus are ionized even at low photospheric temperatures that leave H mainly neutral. Since the ionization potential of H^- is only 0.75 eV, it is ionized by photons of wavelengths $\lambda < 1.65$ μm.

Bound-free H^- absorption makes the dominant contribution to the photospheric absorption coefficient in the visible and near infrared although H^- ions are only about 10^{-8} as abundant as H atoms in the photosphere. At longer wavelengths, the free-free opacity of H^- dominates. The opacity of H^- plotted against wavelength in Fig. 5-11 rises steadily from the ultraviolet to a peak near 8500 Å and then decreases to a minimum around 1.6 μm before rising again into the infrared. Figure 5-13 gives a more complete picture of the principal contributions to photospheric opacity from the extreme ultraviolet to the far infrared, as discussed below.

Once the main photospheric opacity mechanism is known, the curve for $T(\tau)$ can be transformed to a curve of mass column density by starting at $\tau = 0$ and calculating the amount of material (in hydrostatic equilibrium) required to increase the monochromatic opacity by the amount τ_λ. Comparison of the results obtained from this calculation independently at various wavelengths shows good agreement in establishing the curve of $T(z)$ in the photosphere. Since the curve of $T(\tau_\lambda)$ changes shape with wavelength, the curve of $T(\tau_\lambda)$ at $\lambda = 0.5$ μm is used as the reference profile and is usually denoted $T(\tau_{0.5})$.

5.4 PHYSICAL STRUCTURE AND ENERGY BALANCE OF THE PHOTOSPHERE

5.4.1 Models of Photospheric Structure

The temperature structure of the photosphere given by a non-LTE semiempirical model is shown in Fig. 5-12. The tabulation of this model presented in Table 5-2 lists (as a function of optical depth $\tau_{0.5}$ and also geometrical depth) the plasma column density, the plasma temperature; the proton and electron densities, the total pressure including turbulence, and the mass density. We see that the photospheric temperature decreases monotonically from $T \sim 9400$ K in the deepest layers of $Z \sim -100$ km constrained by

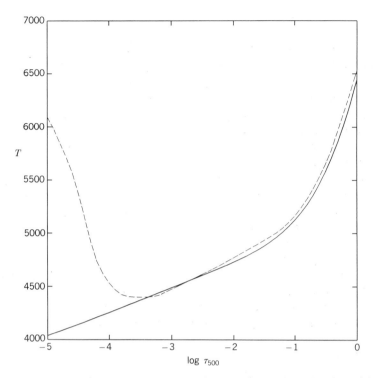

Fig. 5-12 The photospheric temperature profile from the semi-empirical model given in Table 5-2 (dashed curve) and as calculated by a theoretical radiative-convective model (solid curve). By permission of R. Kurucz.

observations, to a minimum of 4400 K roughly 500 km above $\tau_{0.5} = 1$. The position of the white-light limb on this scale falls about 350 km above $\tau_{0.5} = 1$. Above the temperature minimum region, temperatures rise rapidly to chromospheric and coronal values.

Note that the densities and pressures given in Table 5-2 span almost two orders of magnitude between the highest photospheric layers observed at the limb, and the deepest layers that contribute to white light observed near disk center. Also, the ionization of hydrogen in the photosphere is quite low. It ranges between about 2% near $\tau_{0.5} = 10$ to a few parts in 10^7 around T_{min}. Much of the contribution to the electron density comes from ionization of metals. It is only around 1% at $\tau_{0.5} = 10$, but rises to over 99% around T_{min}, and then decreases again as hydrogen ionization rises rapidly in the chromosphere.

The exponential density scale height varies from approximately 800 km around $\tau_{0.5} = 10$ to as little as 140 km near T_{min}. Still, this scale and the photon mean free path of about 30 km around $\tau_{0.5} = 1$ are vast compared to the fundamental plasma size scale on the order of a few centimeters imposed

TABLE 5-2 A Semi-empirical Photospheric Model

h (km)	m (g cm^{-2})	τ	T (K)	n_H (cm^{-3})	n_e (cm^{-3})	P_{total} (dyn cm^{-2})	ρ (g cm^{-3})
1198	3.693E − 04	4.940E − 06	6150	9.317E + 12	1.118E + 11	1.012E + 01	2.183E − 11
711	1.009E − 02	4.841E − 05	5030	3.553E + 14	1.078E + 11	2.764E + 02	8.325E − 10
503	6.513E − 02	3.763E − 04	4400	2.654E + 15	3.040E + 11	1.785E + 03	6.220E − 09
378	2.101E − 01	3.015E − 03	4610	8.186E + 15	9.128E + 11	5.755E + 03	1.918E − 08
200	9.778E − 01	4.712E − 02	4990	3.502E + 16	3.997E + 12	2.679E + 04	8.205E − 08
100	2.183E + 00	2.018E − 01	5410	7.136E + 16	1.001E + 13	5.982E + 04	1.672E − 07
50	3.160E + 00	4.130E − 01	5790	9.598E + 16	1.977E + 13	8.659E + 04	2.249E − 07
0	4.420E + 00	9.994E − 01	6520	1.187E + 17	7.676E + 13	1.211E + 05	2.780E − 07
−50	5.891E + 00	4.123E + 00	7900	1.300E + 17	6.897E + 14	1.614E + 05	3.046E − 07
−100	7.449E + 00	2.361E + 01	9400	1.357E + 17	3.856E + 15	2.041E + 05	3.180E − 07

Source: Adapted from Maltby et al., *Astrophys. J.*, **306**, 284 (1986). By permission of E. Avrett. In the notation used in this table, the number to the right of the E indicates the power of ten (positive or negative) by which the significant figures are to be multiplied.

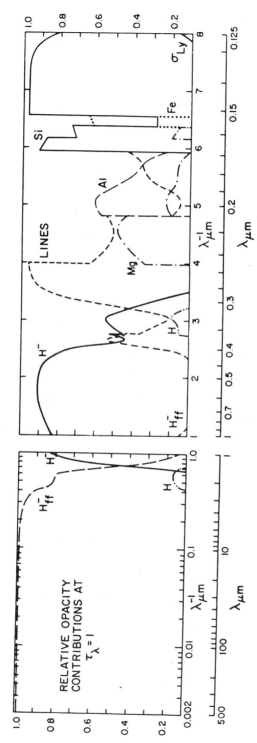

Fig. 5-13 Contributions to photospheric opacity. Abscissae are given in both wavelength (μm) and inverse wavelength. Reproduced, with permission, from J. Vernazza, E. Avrett, and R. Loeser, *Astrophys. J. Suppl.*, **45**, 635 (1981).

by the mean free path between plasma particle collisions, so that treatment of this plasma as a fluid is quite justified.

The contribution of various species to the total opacity, calculated from a similar model, is graphed in Fig. 5-13 as a function of wavelength between 0.125 μm and 500 μm. Molecular opacity is not included in this plot since it is not well known. From the far IR down to about 0.33 μm, except for a narrow region around 0.38 μm where line opacity is large, the dominant opacity source consists of free-free and free-bound H$^-$ transitions. Lines are dominant between about 0.32 μm and 0.23 μm, and important down to 0.17 μm, together with the atomic continua of Mg, Al, Si, and Fe, which dominate between 0.22 μm and 0.16 μm. Further in the UV, scattering in the Lyman α wing takes over as the main opacity source to at least 950 Å. Local dips in the total opacity occur around 1.6 μm in the IR, and around 0.36 μm in the UV.

Figure 5-14 shows the geometrical height in the atmosphere (above $\tau_{0.5} = 1$) at which $\tau_\lambda = 1$ occurs, for 0.125 μm $< \lambda <$ 500 μm. We see deepest into the photosphere around 1.6 μm, but the difference in penetration depth between that window and at 0.5 μm is no more than 35 km. The curve of depth at $\tau_\lambda = 1$ does not simply mirror the curve of opacity, because the penetration depth is determined also by the Planck function behavior as a function of wavelength. Throughout the visible and near IR we look about 450–500 km deeper into the photosphere than in the far IR and at far UV wavelengths.

5.4.2 Comparison with Observations

Over most of the visible region and near infrared, the model described here agrees with observations to within roughly 1%. Below 0.5 μm, uncertain line opacities and the increasing scatter in the observations make it more difficult to assign a unique model. In the deep photosphere ($\tau_{0.5} > 1$), where the model is constrained by measurements of limb darkening and absolute intensity in the IR between 1 and 2 μm, it is uncertain to at least ± 100 K. Higher in the photosphere around $\tau_{0.5} = 1$ it is probably accurate to better than 15 K in T_b.

In the UV, the scatter in observations is between 10 and 30%, and the model uncertainty increases rapidly to about 100 K. The temperature in the T_{min} region is determined by UV continuum intensities near 0.16 μm and IR intensities between 100 and 200 μm. It is also adjusted to agree with the temperatures found from non-LTE models of the Ca II and Mg II line core shapes and with microwave observations of free-free H emission, which yields T_b in this region. The plasma temperature errors quoted above refer to uncertainties in the temperature of an idealized homogeneous atmosphere. Uncertainties in the actual conditions in the inhomogeneous photospheric plasmas that produce the emissions used to construct the mean model presented here are much larger, perhaps 20% or more.

Interesting evidence on the spatial structure of layers around T_{min} has been obtained from measurements of the CO molecular bands near 2.3 μm and 4.5 μm. The CO molecule reaches its peak concentration in the upper photosphere

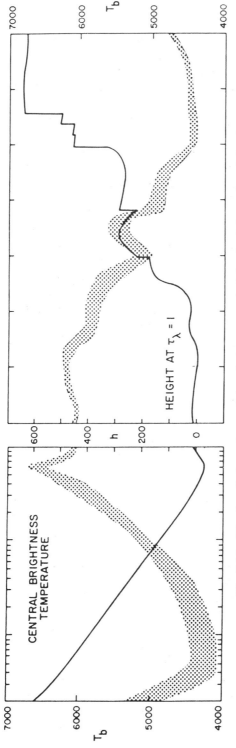

Fig. 5-14 Wavelength dependence of the relative geometrical depth of the surface defined by $\tau_\lambda = 1$ and of the brightness temperature. Abscissa as in Fig. 5-13. Reproduced, with permission, from J. Vernazza, E. Avrett, and R. Loeser, *Astrophys. J. Suppl.*, **45**, 635 (1981).

around $\tau_{0.5} \sim 10^{-2}$, and its band strengths are highly temperature sensitive to the temperatures encountered in those layers. It has been suggested that outside magnetic flux tubes, thus in the regions where nonthermal heating is likely to be relatively low, the plasma temperature can drop to values as low as 3500–4000 K, perhaps in part because of the strong cooling of CO. In general, the contribution of molecules to photospheric energy balance deserves closer attention.

5.4.3 Energy Transport Mechanisms in the Photosphere

As mentioned earlier, it is possible to compute photospheric models on the assumption of radiative or radiative-convective equilibrium, provided the behavior of the main opacity sources is sufficiently well known. Until recently, the contribution of the temperature-sensitive absorption lines to the total opacity was difficult to treat. This difficulty has now been overcome in large measure with models where the temperature and pressure sensitivity of several million lines is included.

It is interesting to compare the temperature profile of such a model, given in Fig. 5-12, with the semi-empirical model discussed above. Below roughly $\tau_{0.5} = 10^{-3}$, the theoretical and semi-empirical curves coincide to within observational errors. This suggests that radiation and convection can account for energy flow through the photosphere, but given the uncertainties it is not possible to say whether other energy transport mechanisms, such as the generation and dissipation of waves, contribute appreciably in the layers below the temperature minimum. Above $\tau_{0.5} \sim 10^{-3}$, the rapid rise of temperature to chromospheric values produces a large departure from the radiative-convective model. In these layers, it is clear that nonthermal heating, which is not included in the theoretical model, plays an important role.

The fraction of the total photospheric flux carried by convection, as computed by this theoretical model, is about 70% at $\tau_{0.5} = 4$, but drops very quickly to much less than 1% above $\tau_{0.5} = 1$. This rapid decrease suggests that the photospheric temperature profile and thus the spectrum are affected but little by the details of convective processes going on underneath.

Precise observations of time variations in the depths and shapes of Fraunhofer lines and in the continuum limb darkening should enable us to determine whether the temperature profile of the quiet photosphere varies slowly over time scales of years. As discussed in Chapter 13, the total solar irradiance measured by satellite radiometry is constant to better than $\sim 0.1\%$ over the solar cycle, so the value of T_{eff} must be constant to within roughly 2 K over that time scale. But the radial temperature structure within the photosphere could change, thus affecting the shape of the sun's spectrum and, in particular, its flux of ultraviolet continuum radiation, which is largely formed in the photospheric layers. Such changes could be caused, at least in principle, by variations in the global efficiency of heat transport caused by modification in the packing density of photospheric magnetic flux tubes or by variations in wave or other nonthermal energy transports within the photosphere. The shape of $T(\tau)$ is also influenced by the photospheric line opacities since the radiative flux obstructed by Fraunhofer lines is close to 10% of the total photospheric heat flux. This "line blanketing" effectively increases the thermal impedance of the photosphere to heat flow by about 10%, so the temperature gradient across these layers is correspondingly higher than it would be in the absence of line opacity. In principle, an instability of solar radiative output might occur if a temperature perturbation changed the line opacity so as to feed back positively on the thermal impedance of the photosphere. But precise observations of the kinds mentioned above, performed over a substantial part of the most recent 11-year cycle, show no evidence for variability of the photospheric temperature profile sufficient to cause appreciable change in the sun's ultraviolet continuum output.

5.5 THE PHOTOSPHERIC CHEMICAL COMPOSITION AND THE CURVE OF GROWTH

We consider first how one might proceed to find the concentration N (cm^{-3}) of an atomic species responsible for line absorption in a uniform isothermal layer of thickness h. Supposing that the continuum absorption outside the line is negligible and that no spontaneous or induced emission is produced from the layer itself, we find from the discussion of Section 2.2.1 that the line and continuum fluxes are related by

$$\frac{F_l}{F_c} = 1 - \exp(-\kappa_l h). \qquad (5\text{-}6)$$

Here κ_l is the line absorption coefficient given for a transition from m to n by

$$\kappa_l = \frac{\pi e^2}{mc} f_{mn} N_m. \qquad (5\text{-}7)$$

Given the oscillator strength f_{mn} of the transition and the thickness h of the layer, it is a simple matter to derive N_m. From the temperature and pressure of the layer, assuming LTE, we can also use the Boltzmann and Saha equations to relate the level population N_m to the ion concentration N_i and ultimately to the abundance N of the atomic species.

5.5.1 The Theoretical Curve of Growth

In the photosphere, this simple approach cannot be applied since even under the simplifying assumptions of the Milne-Eddington model, the residual flux

F_l/F_c in a line formed by true absorption is determined not only by the ratio $\beta = \kappa_l/\kappa_c$, which contains the concentration N that we seek, but also by the temperature structure of the photospheric layers. Here we need a technique that isolates the dependence of the line strength on the concentration of absorbing atoms. The first step is to define a convenient measure of the line strength, called the equivalent width,

$$W = \int_0^\infty A_\lambda \, d\lambda, \qquad (5\text{-}8)$$

where

$$A_\lambda = (F_c - F_\lambda)/F_c, \qquad (5\text{-}9)$$

is the line depth at wavelength λ in the line profile. By definition, W is the width (usually measured in milliangstroms) of a rectangle having the same area as contained under the line profile (Fig. 5-15). The equivalent width has the convenient property that its value is unchanged by the smearing caused by the spectrograph's instrumental profile. This makes W much easier to measure than the true line profile shape.

To find the dependence of W upon the concentration of absorbing atoms, we use the relations given in our discussion of the Milne-Eddington model (Section 2.5.3) to express the line depth as a function of β_λ. For a line formed by pure absorption in a layer where the variation of the Planck function is given as

$$B_\lambda(T) = b_0 + b_1 \tau_\lambda, \qquad (5\text{-}10)$$

we have from (2-77)

$$A_\lambda = \frac{\beta_\lambda/(1 + \beta_\lambda)}{1 + \tfrac{3}{2}(b_0/b_1)}. \qquad (5\text{-}11)$$

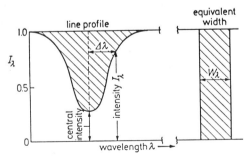

Fig. 5-15 Schematic illustration of the equivalent width. From A. Unsold, *The New Cosmos*, 1969. By permission of Springer-Verlag, N.Y. Inc.

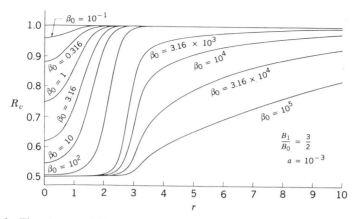

Fig. 5-16 The change of line profile shape with increasing line strength, β_0. Here $R = 1 - A_\lambda$ and r is the distance from line center in units of the Doppler width. From D. Mihalas, *Stellar Atmospheres*, 2nd ed., 1978. By permission of W. H. Freeman and Co.

Since the continuum absorption coefficient κ_c can be taken as a constant over the narrow wavelength interval of a line, we seek the profile of the line absorption coefficient κ_λ, which determines the profile of the parameter β_λ. As described in Section 3.9, the line shape is given most generally by the convolution of a Gaussian function $G(\lambda)$, whose width $\Delta\lambda_D$ is set by kinetic and turbulent plasma motions, with a Lorentzian profile $L(\lambda)$, whose width $\Delta\lambda_C$ is set by collisional broadening. The result is a Voigt function of the form

$$V(\Delta\lambda_D, \Delta\lambda_C) = \int_{-\infty}^{\infty} G(\lambda)L(\lambda - \lambda_0)\, d\lambda. \tag{5-12}$$

A convenient separation of the broadening function can be made by studying the relative importance of the Gaussian and Lorentzian functions in shaping weak, moderate, and strong lines. This change in shape with increasing β_0 (the line-center value of β) is illustrated in Fig. 5-16. We see, by comparison with the bell shape of a Gaussian and the relatively much stronger far wings of a Lorentzian shown in Fig. 3-16, that the shape of a weak line is Gaussian. As β_0 increases, the core depth in lines of moderate strength eventually saturates at a value given by equation (5-11) taken to the limit $\beta_\lambda \to \infty$

$$A_0 = \frac{1}{1 + \frac{3}{2}(b_0/b_1)}. \tag{5-13}$$

Beyond this saturation value determined by the photospheric temperature structure, lines of greater strength begin to exhibit squared cores no longer described by a Gaussian profile. Even stronger lines exhibit the increasingly prominent wide wings described by a Lorentzian profile, as well as the squared core.

In the range of β_0 corresponding to weak lines, we need only the Gaussian contribution to the convolution integral (5-12), so using equation (5-11) we can express the equivalent width as

$$W = \text{const} \times \beta_0 \int_0^\infty e^{-v^2}\left(1 + \beta_0 e^{-v^2}\right) dv, \qquad (5\text{-}14)$$

where $v = \Delta\lambda/\lambda$, with $\Delta\lambda$ the distance from the line center. The integral can be expanded and integrated to yield

$$W = \text{const} \times \beta_0\left(1 - \frac{\beta_0}{\sqrt{2}} + \frac{\beta_0^2}{\sqrt{3}} - \cdots\right) \qquad (5\text{-}15)$$

so that for weak lines the dependence of W on β_0 is linear.

For lines of moderate strength whose cores have saturated but have not yet developed Lorentzian wings, it can be shown that the equivalent width depends only weakly on the concentration of absorbers through

$$W \sim \ln \beta_0. \qquad (5\text{-}16)$$

If β_0 is increased further, the area under the wings eventually makes the dominant contribution to the equivalent width. This is the case of the Ca II H- and K- lines, for instance. The integral (5-8) is now carried out representing the Lorentzian wings as

$$L = \frac{a}{\pi^{1/2}v^2}, \qquad (5\text{-}17)$$

where a is the full width at half-maximum depth of the line. This yields

$$W = \int_0^\infty \frac{1}{1 + \left(\pi^{1/2}/a\beta_0\right)v^2}\, dv$$

$$= \text{const} \times \beta_0^{1/2}. \qquad (5\text{-}18)$$

In this so-called damping regime of β_0, the equivalent width grows as the square root of β_0.

Figure 5-17 shows the theoretical curve of growth as calculated from a Milne-Eddington model for lines formed by pure absorption, assuming a specific temperature profile determined by b_0 and b_1 and a specific damping halfwidth a. In the Doppler or linear part of the curve of growth, the width, $\Delta\lambda_D$, is unchanged so the equivalent width increases only because of an

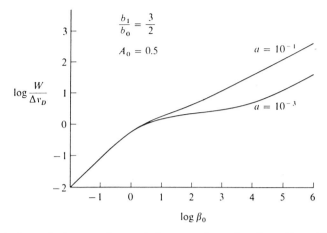

Fig. 5-17 A theoretical curve of growth for pure absorption lines. From D. Mihalas, *Stellar Atmospheres*, 2nd ed., 1978. By permission of W. H. Freeman and Co.

increase in the line's depth. It can be shown that the growth of W in the square-root regime is also independent of $\Delta\lambda_D$. In the saturation regime, however, it is only a change in $\Delta\lambda_D$ that can change W, since the line depth is fixed and the broad wings have not yet developed. Families of theoretical curves of growth similar to Fig. 5-17 have been calculated using various atmospheric models and for lines formed by pure scattering as well as pure absorption.

5.5.2 Comparison with the Empirical Curve

The abundance of an element giving rise to Fraunhofer lines of various strengths distributed along the curve of growth can be determined by comparing its theoretical curve with the empirical curve of growth constructed from the observed equivalent widths and wavelengths of these lines. The procedure is based on a separation of the abscissa, $\log \beta_0$, used in the theoretical curve into three terms, one of which contains the dependence of $\log \beta_0$ on the abundance N that we seek. The other two are dependent only upon the oscillator strength f_{mn} and on the excitation temperature T of the lower level m. From the definitions of κ_0 and β_0, we can express the value of β_0 for a transition of wavelength λ from a lower level m as

$$\beta_0 = \frac{\kappa_0}{\kappa_c} = \text{const} \times (f_{mn}\lambda) \times \left(\frac{N_m v_t}{\kappa_c}\right). \tag{5-19}$$

Here N_m is the concentration (cm^{-3}) of ions in state m and v_t is the most probable velocity of the atoms. Since N_m is related in LTE to the population

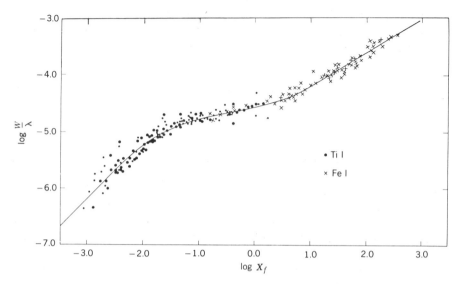

Fig. 5-18 An empirical curve of growth for solar Fe I and Ti I lines. Here $x_f =$ $\log(g_m f_{mn} \lambda)$. From K. Wright, Dominion Astrophysical Observatory, **8** (1948).

N_0 of the ground state by the Boltzmann equilibrium equation (3-16), we obtain, taking logarithms

$$\log \beta_0 = \log(g_m f_{mn} \lambda) - \log(\theta \chi_{0m}) + \log C, \qquad (5\text{-}20)$$

where $\theta = 5040/T$ and χ_{0m} is the excitation energy (in eV) of level m. The last term contains the dependence upon N_i, the concentration of the ith ionization state of the element. It can be written as

$$\log C = \log\left(\frac{N_i}{\kappa_c v_t}\right) + \log(\text{constant}). \qquad (5\text{-}21)$$

We can now plot the measured equivalent widths divided by the line wavelengths, $\log(W/\lambda)$, against the values of $\log(g_m f_{mn} \lambda)$ for each line, to construct an empirical curve of growth. Figure 5-18 gives an example of such a curve plotted for Fe I and Ti I lines. To bring the empirical and theoretical curves into coincidence, we shift them along both the abscissa and ordinate. The required shift along the abscissa directly yields the term

$$\log C = \log \beta_0 - \left[\log(g_m f_{mn} \lambda) - \theta \chi_{0m}\right]. \qquad (5\text{-}22)$$

To extract the value of N_i, we evaluate the constant (equal to $\pi^{1/2} e^2/mc$) and obtain κ_c (per hydrogen atom) from calculations of the H^- opacity or from the measurements of photospheric opacity outlined in Section 5.3. The value of v_t can be determined directly from the same comparison of theoretical and empirical curves of growth, using the shift of their ordinates. We note that the ordinate of the theoretical curve is $\log(W/\Delta\nu_D)$, where

$$\Delta\nu_D = -c\lambda^{-2}\,\Delta\lambda, \qquad \Delta\lambda_D = \frac{\lambda}{c}v_t, \qquad (5\text{-}23)$$

and c is the light speed. Thus the ordinates are displaced by

$$\log(W/\Delta\lambda_D) - \log(W/\lambda) = -\log(v_t/c), \qquad (5\text{-}24)$$

which yields the value of v_t that we seek. This usually exceeds the thermal Doppler broadening, indicating the presence of turbulent motions of some 1–2 km s^{-1}. This value is consistent with the velocity fields of granulation and of the 5 min oscillations measured through studies of spatially resolved line shifts.

It remains to relate the concentration N_i of the ion to the abundance of other states of ionization, and thus to the element abundance N, using the Saha formula (3-21). Since κ_c has been obtained in units of per hydrogen atom, we also obtain N as a value relative to the hydrogen abundance. It is customary to use as reference the hydrogen abundance set at $\log_{10} N_H = 12$ and express other element abundances as logarithms on this scale. There are fundamental limitations to the accuracy of abundances derived from curves of growth, set in part by the simple line formation model used and by inaccuracies in the oscillator strengths. For some elements, an insufficient number of lines is formed over the linear and saturation parts of the curve that are most useful in determining the horizontal and vertical shifts. Nevertheless, the technique has been widely applied to derive the photospheric chemical composition discussed below.

5.6 THE SUN'S CHEMICAL COMPOSITION

The relative abundances of solar elements were first systematically determined by H. Russell in 1929 using eye estimates of line strengths and a two-layer (i.e., Schuster-Schwarzschild) model of the photosphere. L. Goldberg and his co-workers in 1960 carried out a comprehensive study building on the curve of growth techniques described above, which were first introduced by M. Minnaert in the 1930's. Table 5-3 shows modern values of the solar abundances derived mainly from refinements of the curve of growth technique, and also from spectrum synthesis, where detailed photospheric models are iterated with different abundances to reproduce the details of the line shape and depth over a wide wavelength range of the photospheric spectrum.

The most striking result is that the sun consists mainly of hydrogen, with an admixture of about 8% helium, and much smaller concentrations of the

TABLE 5-3 The Solar Chemical Composition

Z	Element	$A_{Element}$	Comments	Z	Element	$A_{Element}$	Comments
1	H	12.0	Reference element	43	Tc		Radioactive
2	He	10.9	Prom., flare part., solar wind	44	Ru	1.9	Phot.
3	Li	1.0	Phot., spot	45	Rh	1.5	Phot.
4	Be	1.1	Phot.				
5	B	2.3	Phot.	46	Pd	1.5	Phot.
				47	Ag	0.9	Phot.
6	C	8.7	Phot.	48	Cd	2.0	Phot.
7	N	7.9	Phot., corona	49	In	1.7	Phot., spot
8	O	8.8	Phot., corona	50	Sn	2.0	Phot.
9	F	4.6	Spot				
10	Ne	7.7	Corona	51	Sb	1.0	Phot.
				52	Te		No lines identified
11	Na	6.3	Phot., corona	53	I		No lines identified
12	Mg	7.6	Phot., corona	54	Xe		No lines identified
13	Al	6.4	Phot., corona	55	Cs	< 2.1	Spot
14	Si	7.6	Phot., corona				
15	P	5.5	Phot., corona	56	Ba	2.1	Phot., chrom., spot
				57	La	1.1	Phot., spot
16	S	7.2	Phot., corona	58	Ce	1.6	Phot.
17	Cl	5.5	Phot., spot	59	Pr	0.8	Phot.
18	Ar	6.0	Corona	60	Nd	1.2	Phot.
19	K	5.2	Phot.				
20	Ca	6.3	Phot., corona	61	Pm		Radioactive
				62	Sm	0.7	Phot.

Z	Element	Value	Method
21	Sc	3.1	Phot.
22	Ti	5.0	Phot.
23	V	4.1	Phot.
24	Cr	5.7	Phot.
25	Mn	5.4	Phot.
26	Fe	7.6	Phot., corona
27	Co	5.0	Phot., corona
28	Ni	6.3	Phot., corona
29	Cu	4.1	Phot., corona
30	Zn	4.4	Phot., corona
31	Ga	2.8	Phot., spot
32	Ge	3.4	Phot., spot
33	As		No lines identified
34	Se		No lines identified
35	Br		No lines identified
36	Kr		No lines identified
37	Rb	2.6	Phot., spot
38	Sr	2.9	Phot., chrom.
39	Y	2.1	Phot.
40	Zr	2.8	Phot.
41	Nb	2.0	Phot.
42	Mo	2.2	Phot.

Z	Element	Value	Method
63	Eu	0.7	Phot.
64	Gd	1.1	Phot.
65	Tb		No f-values
66	Dy	1.1	Phot.
67	Ho		No f-values
68	Er	0.8	Phot.
69	Tm	0.3	Phot.
70	Yb	0.2	Phot.
71	Lu	0.8	Phot.
72	Hf	0.9	Phot.
73	Ta		No lines identified
74	W	0.8	Phot.
75	Re	< -0.3	Phot.
76	Os	0.7	Phot.
77	Ir	0.8	Phot.
78	Pt	1.8	Phot.
79	Au	0.8	Phot.
80	Hg	< 2.1	Phot.
81	Tl	0.9	Spot
82	Pb	1.9	Phot.
83	Bi	< 1.9	Phot.
90	Th	0.2	Phot.
92	U	< 0.6	Phot.

Source: From O. Engvold, *Physica Scripta*, **16**, 48 (1977). By permission Royal Swedish Academy of Sciences.

171

heavier elements. The abundance of helium, the second most common solar element, must be calculated from visible and ultraviolet emission lines radiated from the chromosphere and transition region since the only helium absorption line identified in the photospheric spectrum is the weak He I feature at λ 10,830. Values are also obtained from measurement of the strong D_3 emission line at λ 5876 (see Fig. 3-8) in prominences at the limb. In general, chromospheric and coronal abundance determinations are less accurate than photospheric measurements since they depend on models of temperature and density in these tenuous plasmas where non-LTE effects are very important. The helium concentration can also be estimated through calculation of photospheric models using different helium abundances. Although helium does not affect the photospheric continuum or line spectrum, its inclusion does modify the gas pressure since it increases the mean molecular weight.

Overall, solar abundances are known to a factor of 2 at best. The values for many elements, particularly those with few and often blended photospheric lines, are uncertain to an order of magnitude. Some improvements can be expected from spectra obtained with Michelson interferometers, which achieve very low levels of scattered light and thus permit the detection of weak features and accurate line profiles. For certain elements, special observations are helpful. For instance, the lines of lithium are substantially strengthened in the spectrum of spot umbrae. More accurate oscillator strengths derived from laboratory measurements are also very important. To a lesser extent, better photospheric models and more complete understanding of non-LTE effects on line formation should improve accuracy.

For about 20 of the elements, abundances have been determined also from laboratory samples of carbonaceous chondrites, a species of meteorite thought to be representative of the primordial materials from which the sun was formed. Overall, photospheric, meteoritic, and spatially averaged coronal abundances are similar, although there is some evidence that locally, such as in coronal holes, coronal abundances might differ from the photospheric values by as much as a factor of 10. Such differences might arise from differential diffusion of various species in the sharp temperature gradients of the transition region and from differential escape into the solar wind.

The plot of solar abundances given in Fig. 5-19 brings out several important features of the sun's chemical composition. As discussed in Chapter 6, the low abundances of lithium, beryllium, and boron are explained in part by nuclear reaction rates, and in part by overturning in the convection zone, which moves gases to depths where the temperature is high enough for nuclear fusion. Besides this Li, Be, and B deficiency, the most remarkable features of the solar abundance curve are (i) the steep decline in abundance with increasing atomic weight A up to $A \sim 75$, (ii) several local peaks, the most prominent in the vicinity of iron, (iii) a systematic overabundance of even-A versus odd-A nuclei, and (iv) a cyclic tendency with period $A = 4$ for $A < 40$. The presence of these regularities is ascribed to the fundamental processes of nuclear fusion that are thought to fuel the sun's output and form the basis of the theory of

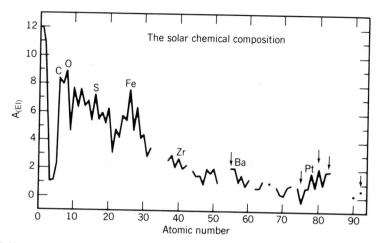

Fig. 5-19 A plot of the solar abundances given in Table 5-3. Arrows indicate upper limits. From O. Engvold, *Physica Scripta*, **16**, 48 (1977). By permission of the Royal Swedish Academy of Sciences.

stellar evolution and chemical composition of the Galaxy. Some of these nuclear processes are discussed in Chapter 6.

ADDITIONAL READING

Background

R. Bray, R. Loughhead, and D. Durrant, *The Solar Granulation*, 2nd ed., Cambridge University Press, 1984.

L. Goldberg and K. Pierce, in *Handbuch der Physik, Vol. 52, Astrophysics III: The Solar System*, S. Flugge (Ed.), Springer-Verlag, p. 1, 1959.

M. Minnaert, "The Photosphere," in *The Sun*, G. Kuiper (Ed.), University of Chicago Press, p. 88, 1953.

Observations of Structures and Motions

J. Beckers, "Dynamics of the Solar Photosphere," in "The Sun as a Star," NASA Publication SP-450, S. Jordan (Ed.), p. 11 (1981).

S. Keil (Ed.), "Small-Scale Dynamical Processes in Quiet Stellar Atmospheres," Sacramento Peak Observatory Publication (1984).

R. Muller, "The Fine Structure of the Quiet Sun," *Solar Phys.*, **100**, 237 (1985).

Theoretical Interpretation, Models, and Solar Abundances

D. Mihalas, *Stellar Atmospheres*, 2nd ed., W. H. Freeman, 1978.

J. Vernazza, G. Avrett, and R. Loeser, "Structure of the Solar Chromosphere: The Underlying Photosphere and Temperature Minimum Region," *Astrophys. J. Suppl.*, **30**, 1 (1976).

O. Engvold, "The Solar Chemical Composition," *Physica Scripta*, **16**, 48 (1977).

EXERCISES

1. Construct a simple model to explain the sharpness of the photospheric limb. Assume the photosphere is an isothermal, spherical shell. If the photospheric mass absorption coefficient is roughly proportional to $\rho^{2/3}$, show that the optical depth varies as $\rho^{5/3} \, dx$, where ρ is the density and x is a spatial coordinate. Show then that the observed intensity near the limb is given by

$$I(\tau') = S \int e^{-\tau'} \, d\tau' = S(1 - e^{-\tau'})$$

   where S is the source function in the isothermal layer and τ' is the optical depth looking tangentially to this shell. Given the density scale height expected from the temperature in the high photosphere, show that the exponential scale height of $I(\tau')$ is expected to be roughly 50 km. Compare this to the observations shown in Fig. 5-5. For a more detailed discussion, see B. Lites; *Solar Phys.* **85**, 193 (1983).

2. Assuming that roughly half of the 7 km s^{-1} velocity difference inferred near $\tau_{0.5} = 1$ between up- and downflowing granular material can be interpreted as a radial outflow relative to the sun's center of mass, calculate the ballistic trajectory of this material under solar gravity. Compare your velocity profile with the finding that only exceptional granules produce measurable Doppler shifts 500 km above $\tau_{0.5} = 1$.

3. At what wavelengths do we see deepest into the sun's atmosphere? Why? How much deeper do we see at those wavelengths than in visible light at 5500 Å? At what wavelengths do we observe the temperature minimum region? How much higher do we see at $\mu = 0.1$ than at disk center at 5500 Å?

4. Using standard curve-fitting techniques, evaluate the a, b, c, coefficients in equation (5-3) at $\lambda = 5000$ Å and 10,000 Å using the limb-darkening data given in Fig. 5-4. Use the sun-center intensities $I_\lambda(0)$ in Table 3-2 to find $B_\lambda(\tau)$. How well does the Eddington-Barbier relation apply? Compare your results at the two wavelengths and comment on differences in your two photospheric temperature models.

6

The Sun's Internal Structure and Energy Generation

Inquiry into the mysteries beneath the sun's visible surface holds a special fascination. Early ideas put forward in the late 18th century by W. Herschel held the sun to be a solid body with a cool, perhaps inhabited, surface, covered by layers of hot, luminous clouds. The source of the power required to produce the sun's light and heat output was less troubling to Herschel and his contemporaries than it might seem to us, since the laws governing energy conservation and heat flow were not recognized until over half a century later.

Subsequent developments in thermodynamics and in the kinetic theory of gases, together with new insights from the operation of convection in the Earth's atmosphere, led H. Lane in 1869 to construct the first physical model of the solar interior. Proceeding on the assumption that the sun was a sphere of ideal gas held together in equilibrium by its self-gravity, Lane found a central density about a sixth of the currently accepted value. The temperature distribution with depth in this model was based on outward heat transport by convection from a hotter interior. It was recognized that the temperature profile should then correspond to that of a rising gas element expanding and cooling adiabatically, and so would depend only upon the ratio of specific heats, γ.

In 1906, K. Schwarzschild showed that the photospheric limb darkening is more consistent with a temperature profile set by radiative equilibrium than with the adiabatic stratification assumed in the earlier models. Although the convective layers now known to exist below the photosphere are mostly in adiabatic equilibrium as Lane's model had supposed, Schwarzschild's contribution was to point out that energy might be transported by either convection or radiation in the sun. He developed a simple criterion to evaluate the absence of convection which is still in use today.

Improved dating of terrestrial rocks led to the gradual recognition that neither the infall of meteoritic material nor the contraction of the sun (both mechanisms proposed before Lane put forward his model) could fuel the sun's prodigious power output over the required time scale. Work on the transmutation of elements around the turn of the century prompted the currently accepted view that energy liberated in nuclear processes accounts for the sun's power output. It was soon recognized that the abundance of radioactive elements was too small for spontaneous fission to make an important contribution. Around 1920 A. Eddington suggested that the main process is fusion of hydrogen to helium. By the late 1930s the specific processes of nuclear burning in the sun were elucidated, notably by H. Bethe and his collaborators.

In this chapter we outline the basis for our current theoretical understanding of the solar interior and compare it with observations. Until recently, the only observations available were those of the sun's mass, diameter, luminosity, and (surface) chemical composition. In the past 20 years the new techniques of neutrino astronomy and helioseismology have enabled us to make direct measurements of conditions below the photosphere, right to the center of the sun. We are finally in a position to look into the depths of the solar interior and test our ideas on how a star generates its enormous power output.

6.1   EQUATIONS OF STELLAR STRUCTURE

6.1.1   Mechanical Equilibrium

Conditions in the solar interior can be estimated by studying the temperature and pressure profiles inside a gaseous sphere assumed to be in thermal and dynamical equilibrium. These are determined by a set of equations that describe the balance of forces acting on the gas, the distribution of mass, the internal energy generation and transport, and the equation of state.

Force balance is described by the hydrostatic equation

$$\frac{d\rho}{dr} = -\frac{GM(r)\rho}{r^2}, \tag{6-1}$$

where G is the gravitational constant, $M(r)$ is the mass inside a sphere of radius r, and ρ is the density. The photospheric rotation rate of roughly 2 km s^{-1} is too slow for centrifugal forces to be significant for most purposes in discussing the mechanical equilibrium of the solar interior.

To a good approximation the pressure in the solar interior is given by the equation of state for a perfect gas

$$P = \frac{\rho}{\mu}\frac{k}{m_\mathrm{H}}T, \tag{6-2}$$

where μ is the molecular weight of the solar plasma. The central density of around 150 g/cm^3 found in the sun is not sufficiently high for degeneracy effects to cause significant departures from this ideal gas relation. Such effects become important in more massive stars, where central densities are very high and limitations on the packing density of electrons in phase space are set by the exclusion principle. Also, the radiation pressure of the net outward flux of photons can be neglected, although it becomes important in more luminous stars.

The distribution of mass for a given density distribution ρ is given by

$$\frac{dM(r)}{dr} = 4\pi r^2 \rho. \tag{6-3}$$

Energy conservation in a star assumed to be in equilibrium requires that the total net flux of energy $L(r)$ outward across the sphere of radius r equal the total energy production through nuclear processes within it. This is expressed as

$$L(r) = 4\pi \int_0^r r^2 \rho \epsilon \, dr, \tag{6-4}$$

where ϵ is the rate of nuclear energy production per gram. Since most of the nuclear energy generation occurs in the central regions, near the photosphere we can take $L(r) = L_\odot$, the solar luminosity.

6.1.2 Energy Transport

Transport of heat in the sun's interior takes place mainly by radiation or convection. As shown in Chapter 7, the temperature gradient set up by radiative transfer through the gas is stable if it is less steep than the adiabatic gradient. Wherever this is the case in the solar interior, energy flow is governed by radiative diffusion according to the relation

$$F(r) = \frac{L(r)}{4\pi r^2} = k' \frac{dT}{dr}, \tag{6-5}$$

where k' is determined by the Rosseland mean absorption coefficient (see Section 2.4.4). This relation predicts, as we might expect, that the net outward flux $F(r)$ varies directly as the temperature gradient and inversely as the opacity, which acts as an impedance to energy flow.

When the radiative gradient is steeper than the adiabatic gradient, the gas becomes unstable to convective motions. The flux of convective energy is then proportional to the excess in temperature gradient over the adabiatic value. The excess required is only a small fraction of the gradient itself, except in the outermost layers, so the stratification can be considered effectively adiabatic over most of the convection zone depth. Differentiating the adiabatic relation between pressure and density, we thus obtain the relation

$$\frac{1}{P}\frac{dP}{dr} = \frac{\gamma}{\rho}\frac{d\rho}{dr},\tag{6-6}$$

between the pressure and density in the convection region.

To determine conditions inside the sun, we seek to integrate the two equations (6-1) and (6-3). But these contain the three dependent variables P, ρ, and $M(r)$, so we require additional relations to perform the integration with respect to r. In the convective regions of the star, this additional relation is provided by the adiabatic relation (6-6). In regions where energy transport by radiation dominates, the procedure is to use the equation of state (6-2) (which introduces a fourth dependent variable, the temperature, and relates it to P and p), together with the radiative transport equation (6-5) in the form

$$\frac{dT}{dr} = \frac{-3}{16\sigma}\frac{\bar{\kappa}_R\rho}{T^3}\frac{L(r)}{4\pi r^2},\tag{6-7}$$

to provide the required further relation between temperature and r. The luminosity $L(r)$ introduced in (6-7) can be expressed through the conservation relation (6-4), which in differential form is

$$\frac{dL(r)}{dr} = 4\pi r^2\rho\epsilon.\tag{6-8}$$

6.1.3 Boundary Conditions

The set of five basic equations (6-1)–(6-5) or (6-6) can be integrated numerically on a computer to give the radial profiles of temperature, pressure, mass, and luminosity, provided the parameters γ, μ, ϵ and $\bar{\kappa}_R$ are known over the relevant range of physical conditions. In principle, an integration might be carried outward from the center, where natural boundary conditions on mass and luminosity are $M_r = 0$ and $L_r = 0$ at $r = 0$. The starting values of central pressure and temperature, P_c and T_c, must be guessed in this case.

The integration then proceeds until the density (and pressure) decrease to zero. We cannot expect the temperature to drop to zero at the same point where $P = 0$, given the independent guessed values of P_c and T_c, so the integration is repeated with a better guess at T_c or P_c (keeping the other variable constant) until the integration produces $T = 0$ and $P = 0$ at the same value of R. In practice, a more numerically stable integration is achieved by proceeding inward from the surface and also outward from the center, meeting at an intermediate point.

The accuracy of the model can be evaluated by comparing the resultant values of R, L, and M with observations of the solar diameter, luminosity,

and mass. Improvements in the model are made by repeating the integrations ✓
using different values of P_c until the desired level of agreement is achieved.

It can be shown that for a particular chemical composition these equations
define a unique relation between the value of P_c (or T_c) and the mass, radius, ✓
and luminosity of the resulting stellar model. This relation implies that the
sun's global observable properties and internal structure are uniquely determined by the mass and chemical composition of the primordial gas that
formed it. This general result, known as the Vogt-Russell theorem (although it
has never been proved) underlies the study of stellar interiors and stellar
evolution.

The main uncertainty arises in integrating the model through the convective
region. Over most of that region, it is seen from the adiabatic relation (6-6)
that the pressure and temperature are related by

$$P = \text{const} \times T^{\gamma/(\gamma-1)},$$
(6-9)

so it follows that both T and P must go to zero at the same value of R,
regardless of how P_c and T_c are chosen. In other words, one of the physical
constraints that made it possible to produce a unique set of M, L, and R for a
given P_c is not available in the convective zone.

Removal of this constraint means that, for instance, equilibrium models of
the convective layers can be constructed which satisfy the data on $R_\odot$, $L_\odot$,
$M_\odot$, and solar chemical composition, but which leave the temperature at the
bottom of the convective layer [and thus $T(r)$ through the layer] relatively
unconstrained. To remove this ambiguity and determine which adiabat describes the convection zone requires detailed matching of the solar interior
integrations described above to semi-empirical models of the photosphere.
These models use a crude treatment of convection based on the mixing-length
approximation (see Chapter 7) to bridge the gap between layers near the
photosphere and the essentially adiabatic layers deeper in the convection zone.

This additional constraint placed by photospheric observations of the
temperature and pressure profiles enables us, in the sun, to find a smooth and
plausible fit between a convection zone model and a photospheric model.
Nevertheless, the treatment of convection is uncertain, and the depth and
structure of the solar convection zone determined in this way need to be
checked by other observations.

6.2 PHYSICAL PARAMETERS REQUIRED FOR THE SOLUTION

6.2.1 Chemical Composition

Estimates of the sun's internal chemical composition are based on measurements of the atmospheric constitution described in Sections 5.5 and 5.6.
Models of the interior show that a star with the sun's luminosity, radius, and

mass is difficult to construct from the stellar structure equations unless it contains large amounts of hydrogen and helium, in a proportion similar to that measured in the solar atmosphere. This finding supports the view that, except for the evolution of chemical structure expected from nuclear reactions, the compositions of the solar atmosphere and interior are similar. Effects such as preferential loss of certain elements from the sun or accumulation of different abundances from interstellar material onto the photosphere cannot be entirely ruled out, but they are usually assumed to be negligible.

Accordingly, evolutionary models are constructed with a first guess at the chemical composition of the infant sun similar to the present abundances in the atmosphere. The model is then allowed to evolve in time for about 4.5 billion years and compared with (mainly) the luminosity of the present-day sun. The chemical composition of successful evolutionary models gives us our best estimate of the sun's internal chemical composition.

It is customary to denote the fraction by mass of hydrogen with the symbol X, the helium fraction Y, and the fraction of heavier elements by Z, so that $X + Y + Z = 1$. Given the uncertainties in the true composition, it is usual to assume that the relative abundances of heavy elements making up the fraction Z are those observed in the solar atmosphere. Representative values used in solar interior integrations are $X = 0.71$, $Y = 0.27$, and $Z = 0.02$. The relative abundances of elements heavier than helium as found in the solar atmosphere are given in Table 5-3.

6.2.2 The Mean Molecular Weight

The radial distribution of mean molecular weight influences the structure of the solar interior through the equation of state. Each atomic species present in the solar mixture will, when ionized at the high temperatures of the solar interior, contribute N_i atoms per unit volume of atomic mass A_i, and z_i electrons. The mean molecular weight of this material per unit volume is just the total mass of these particles divided by their number. Since the electron mass can be neglected, we have

$$\mu = \frac{\Sigma N_i A_i}{\Sigma N_i (1 + z_i)}. \tag{6-10}$$

It is useful to express this relation in terms of the concentrations of the individual species by mass. In one gram of solar matter the number of free particles is given by $1/\mu m_H$, where m_H is the mass of a hydrogen atom. This quantity can also be expressed as a summation

$$\frac{1}{\mu m_H} = \sum_i \frac{N_i (1 + z_i) X_i}{N_i A_i m_H}, \tag{6-11}$$

where X_i is the number of grams of element i (given by $N_i A_i m_H$) per gram of solar material. This can be rewritten to give

$$\frac{1}{\mu} = \sum_i \frac{(1 + z_i) X_i}{A_i}.$$

(6-12)

For completely ionized pure hydrogen, we have $X_i = 1$, $1 + z_i = 2$, and $A_i = 1$, so $\mu = \frac{1}{2}$. The value of μ remains close to 0.6 in the solar interior, except in the core, where the helium abundance has been increased by hydrogen burning, and very near the photosphere, where the hydrogen ionization drops sharply.

6.2.3 The Ratio of Specific Heats

For a monatomic gas, the specific heat at constant volume at fixed ionization is $c_v = \frac{3}{2} R/\mu$, where R is the gas constant. The corresponding specific heat at constant pressure, which takes into account the $P\,dV$ mechanical work performed by the molar element of gas, is $c_p = c_v + R/\mu$, or $\frac{5}{2} R/\mu$. It follows that, in this simple situation where we are neglecting ionization changes, the ratio of specific heats is given by

$$\gamma = \frac{c_p}{c_v} = \frac{5}{3}.$$

(6-13)

In the solar interior, changes in ionization play an important role in determining the specific heats and their ratio. Ionization or recombination of hydrogen involves an energy of 13.6 eV per atom, which is roughly 16 times more than its kinetic energy at a temperature of 10^4 K. In regions of the solar interior where hydrogen ionization is changing rapidly, the value of c_v can thus be an order of magnitude higher than $\frac{3}{2} R$. Under these conditions γ approaches unity.

The value of γ determines the adiabatic temperature gradient (see Section 7.4) and thus the point of convection onset in the solar interior. Through its control of the adiabatic temperature gradient, it also determines the temperature profile of the sun's convection zone since, as we shall see later, the efficiency of convection leads to an essentially adiabatic stratification. This means that small uncertainties in c_v can appear as errors in our understanding of the sun's internal temperature profile. One example is the uncertain effect on the ionization potential of mutual electrostatic shielding by charged particles in the solar plasma.

6.2.4 The Radiative Opacity

Over most of the solar interior the temperature gradient is determined directly from equation (6-5) and thus directly by the temperature and pressure dependence of the Rosseland opacity. The plasma temperatures in this region range from about 2 million degrees at the bottom of the convection zone to about 15 million degrees at the center of the sun. Black bodies emitting in this temperature range radiate most of their flux between 1 and 10 Å in the X-ray region, so the appropriate Rosseland opacity is weighted by the flux over this wavelength interval.

In the outer regions of the radiative core, the opacity is determined mainly by bound-bound line absorptions and bound-free continua of the heavy elements. In this regime, the opacity is quite sensitive to the heavy element abundances, given their rich line spectra. In the central regions, where temperatures and densities are so high that ionization is essentially complete, Thomson scattering by free electrons and free-free transitions of electrons in the electrostatic fields of the abundant hydrogen and helium nuclei make comparable contributions to $\bar{\kappa}_R$.

The Rosseland opacity per gram of solar interior material increases by over two orders of magnitude between the sun's center and photosphere. The complex calculations leading to these opacity values are considered to be most certain (to about 5%) for electron scattering, which dominates at the high-temperature end around 10^7 K. They are least certain (no better than 10–30%) in the cooler regions, where the other processes dominate. These errors are in addition to possibly larger errors due to uncertainty in the composition of the solar interior and must be kept in mind when models of the solar interior are considered below.

6.2.5 Energy Generation Processes

It is easy to show (see exercises) that the source of the sun's energy supply cannot be derived from gravitational contraction of its primordial gas cloud, at least for any appreciable fraction of its 4.5 billion year age. Other sources of energy, such as infall of meteorites and comets, chemical reactions, or radioactivity, can also be easily dismissed. The only mechanism that appears to be capable of sustaining the solar power output over the sun's lifetime is nuclear fusion.

Considering fusion of hydrogen to helium, we find that the mass excess of four hydrogen atoms over one helium atom corresponds to an energy equivalent of 26.73 MeV or 4.283×10^{-5} ergs. From this it is easy to deduce that burning only 5% of the sun's hydrogen supply to helium would be sufficient to balance its vast energy output. The evidence for fusion reactions as the sun's principal fuel still rests in large part on their energetic plausibility, although there is now more direct evidence for nuclear reactions in the solar interior from recent, directional observations of a solar neutrino flux. Our belief is further strengthened by the identification of specific reaction chains that enable us to explain the evolution of stars similar to the sun.

6.3   NUCLEAR REACTIONS IN THE SUN'S INTERIOR

6.3.1   Factors That Determine the Dominant Reactions

The two factors which determine the most likely nuclear reactions are the abundance of the reacting species and the reaction probability at the temperatures prevailing in the solar core. The strong Coulomb repulsion between positively charged nuclei increases as the product of their nuclear charges, so only the lightest elements will have appreciable reaction probabilities. The most abundant of these are H, He, C, N, O, Ne, Mg, and Si. From the work of Bethe in 1938 we know that two nuclear reaction cycles appear to be most promising in accounting for solar energy production. Both of these have as end product the burning of four hydrogen atoms to produce one helium atom. Most (about 99%) of the solar energy is now thought to be produced by one of these mechanisms, the proton-proton chain, in which fusion occurs between the two nuclei of lowest Coulomb repulsion.

In the second mechanism, known as the carbon-nitrogen cycle, carbon and nitrogen nuclei act as catalysts in the progressive buildup of isotopes leading to formation of a stable helium nucleus as a by-product. Reactions between protons and He^4 nuclei, although more favorable from the standpoint of Coulomb repulsion, do not result in formation of any stable nuclei. Similar arguments apply to reactions between two He^4 nuclei. Reactions between He^4 and elements heavier than helium are not effective because of the Coulomb repulsion. Thus the reactions between protons and C^{12} or N^{14} are seen to be the only reactions besides the proton-proton chain which are expected to be effective in solar energy generation.

6.3.2   The Proton-Proton Chain

The first step in the p-p chain is the formation of deuterium H^2 from two protons. Since the nucleus $_2He^2$ does not exist, two protons cannot combine directly with only photon emission. Instead, deuterium is formed at a rate given by the (very low) probability of β-decay in the short interval while the two protons are in collision. The initial p-p reaction, in which a positron and a neutrino are emitted, is described as

$$_1H^1 + {}_1H^1 = {}_1H^2 + \beta^+ + \nu. \qquad (6\text{-}14)$$

Alternatively, the chain can be initiated (about $\frac{1}{4}\%$ of the time) by the so-called p-e-p reaction

$$_1H^1 + {}_1H^1 + e^- = {}_1H^2 + \nu. \qquad (6\text{-}15)$$

Both reactions proceed at an extremely slow rate, corresponding to a mean proton half-life of about 10^{10} and 10^{12} years respectively. This is by far the slowest rate of all the p-p chain reactions and thus controls the burning of the solar fires. The positron subsequently annihilates with an electron to produce two γ rays, each of 0.51 MeV. The next step proceeds much more quickly. The deuteron, $_1H^2$, produced will join with a proton within about 10 s to form He^3 and emit a γ-ray through the process

$$_1H^2 + _1H^1 = _2He^3 + \gamma. \tag{6-16}$$

Further proton capture is very unlikely since the nucleus Li^4 is unstable. By far the most likely next step is combination of the He^3 nucleus with another He^3 (previously formed by the proton-proton reaction) to produce He^4 and two protons through

$$_2He^3 + _2He^3 = _2He^4 + 2_1H^1. \tag{6-17}$$

The net effect of the p-p chain described so far is thus to produce one He^4 nucleus, two neutrinos, and γ-radiation from four protons and two electrons. The γ-radiation carries energy that is absorbed quickly and heats the sun. The neutrinos are not intercepted by solar material and escape directly into space.

The termination of the p-p chain given in (6-17) is not unique, although about 85% of the sun's energy is thought to be produced by reactions ending in that way. Alternative branches in which He^3 nuclei react with He^4 to produce beryllium and lithium isotopes are relatively unimportant energetically, but they produce neutrinos in the high energy range that is most easily detectable so far. The two alternative branches that compete with (6-17) both begin with

$$_2He^3 + _2He^4 = _4Be^7 + \gamma. \tag{6-18}$$

The $_4Be^7$ can react to produce helium either with an electron

$$_4Be^7 + e^- = _3Li^7 + \nu, \tag{6-19}$$

or with a proton

$$_3Li^7 + _1H^1 = 2_2He^4. \tag{6-20}$$

Additionally, the $_4Be^7$ can react with a proton first to form boron, and eventually helium, through the sequence

$$_4Be^7 + _1H^1 = _5B^8 + \gamma, \tag{6-21}$$

$$_5B^8 = _4Be^{8*} + \beta^+ + \nu, \tag{6-22}$$

$$_4Be^{8*} = 2_2He^4. \tag{6-23}$$

In this sequence, the asterisk represents an excited nucleus. It has also been realized quite recently that the very rare (hep) reaction

$$_2\text{He}^3 + {}_1\text{H}^1 = {}_2\text{He}^4 + e^+ + \nu \tag{6-24}$$

is of considerable interest because its neutrinos originate throughout the nuclear burning core.

The total amount of energy liberated in conversion of four hydrogen atoms to one helium atom is 26.73 MeV. The contributions made to this total at each step are given in Table 6-1, which summarizes the p-p chain. This energy appears as the direct production of γ-ray photons or as particle kinetic energy. The first reactions (6-15) and (6-16) must occur twice for each time that (6-17) occurs, so their contributions must be doubled.

Table 6-1 also gives the energy carried off by neutrinos, which is about 2% of the total. Since the neutrinos do not interact further with solar matter or radiation, they are counted as a loss to solar photon luminosity. As can be seen in Table 6-1, neutrinos are produced in the deuterium-forming step of the photon-photon chain and also in the branch that forms lithium. But the neutrinos to which the detectors constructed so far are by far the most

TABLE 6-1 Summary of p-p Chain Reactions[a]

$4^1\text{H} \rightarrow 4\text{He} + 2e^+ + 2\nu$	$Q = +26.731$	
$^1\text{H} + {}^1\text{H} + e^- \rightarrow 2\text{D} + \nu$ (p-e-p)	1.442	(1.442)
or		
$^1\text{H} + {}^1\text{H} \rightarrow {}^2\text{D} + e^+ + \nu$ (p-p)	1.442	(≤ 0.420)
$^2\text{D} + {}^1\text{H} \rightarrow {}^3\text{He} + \gamma$	5.493	
$^3\text{He} + {}^3\text{He} \rightarrow {}^4\text{He} + 2^1\text{H}$	12.859	
or		
$^3\text{He} + {}^1\text{H} \rightarrow {}^4\text{He} + e^+ + \nu$ (hep)	19.795	
or		
$^3\text{He} + {}^4\text{He} \rightarrow {}^7\text{Be} + \gamma$	1.587	(9.625)
then		
$^7\text{Be} + e^- \rightarrow {}^7\text{Li} + \nu$	0.862	$\begin{cases} 0.862 \text{ MeV}; 89.7\% \\ 0.384 \text{ MeV}; 10.3\% \end{cases}$
and		
$^7\text{Li} + {}^1\text{H} \rightarrow 2^4\text{He}$	17.347	
or		
$^7\text{Be} + {}^1\text{H} \rightarrow {}^8\text{B} + \gamma$	0.135	
and		
$^8\text{B} \rightarrow {}^8\text{Be}^* + e^+ + \nu$	15.079	(≤ 14.02)
and		
$^8\text{Be}^* \rightarrow 2^4\text{He}$	2.995	

[a] Energy released in each step is given to the right in MeV, together with the neutrino energy (in parentheses). The notation used is an alternative (one of several) to that used in the text.

sensitive are those of high energy produced in the decay of boron. As mentioned above, this reaction branch is itself of little importance to solar energy production, so we must know accurate reaction rates for the intermediate steps that form beryllium and boron if we are to relate the measurements discussed below to the rate of solar energy production, which is mainly determined by the formation of He^4 from He^3. Detectors sensitive to the lower-energy neutrinos produced directly by the primary deuterium-forming reactions, which are more directly related to solar energy production, are beginning operation in the Soviet Union and Italy.

6.3.3 The Carbon-Nitrogen Cycle

The CN cycle also plays a role in solar energy generation, although its total contribution is considered in conventional models to account for only about 1% of the total power output. This cycle begins with a reaction between a proton and a C^{12} nucleus to form a N^{13} nucleus

$$C^{12} + {}_1H^1 = N^{13} + \gamma. \tag{6-25}$$

This N^{13} β-decays to C^{13} within a few minutes

$$N^{13} = C^{13} + e^+ + \nu. \tag{6-26}$$

The stable isotope C^{13} will react with a proton under solar interior conditions to form N^{14} through

$$C^{13} + {}_1H^1 = N^{14} + \gamma. \tag{6-27}$$

This nitrogen isotope is also stable, and another proton capture turns out to be the only important process, resulting in the formation of oxygen through

$$N^{14} + {}_1H^1 = O^{15} + \gamma. \tag{6-28}$$

The O^{15} nucleus β-decays within a few minutes to a stable nitrogen isotope through

$$O^{15} = N^{15} + e^+ + \nu. \tag{6-29}$$

A proton capture on the stable N^{15} produces two nuclei, C^{12} and He^4

$$N^{15} + {}_1H^1 = C^{12} + He^4 + \gamma. \tag{6-30}$$

As in the p-p chain, the net result of the CN cycle (6-25)–(6-30) is to form a helium nucleus from four protons and two electrons, with the emission of two neutrinos. The two positrons produced in (6-26) and (6-29) annihilate electrons. The carbon and nitrogen isotopes act as catalysts; the C^{12} produced in

TABLE 6-2   Summary of the CN Cycle Reactions of Importance in the Sun

Reaction	Energy Released (MeV)[a]	
$_1H^1 + _6C^{12} \rightarrow _7N^{13} + \gamma$	1.94	
$_7N^{13} \rightarrow _6C^{13} + e^+ + \nu$	1.51	(0.71)
$_1H^1 + _6C^{13} \rightarrow _7N^{14} + \gamma$	7.55	
$_1H^1 + _7N^{14} \rightarrow _8O^{15} + \gamma$	7.29	
$_8O^{15} \rightarrow _7N^{15} + e^+ + \gamma$	1.76	(1.00)
$_1H^1 + _7N^{15} \rightarrow _6C^{12} + _2He^4$	4.96	
	25.01	(1.71)

[a] Neutrino energies in parentheses.

(6-30) is ready to re-enter the chain again in the initial reaction (6-25).

The CN cycle reactions are summarized in Table 6-2 with the energy liberated in each step. About 6% of the total is in neutrinos that are lost to the sun's photon luminosity. Some additional proton captures involving about 0.1% of the N^{15} in (6-29) go on to produce heavier isotopes such as O^{16}, F^{17}, and O^{17} and feed material back into the main CN cycle. The complete set of processes is referred to as the CNO tri-cycle. Only the processes described in the steps (6-25)–(6-30) appear to contribute appreciably to solar energy generation or to the sun's isotopic composition.

6.3.4   Nuclear Energy Generation Rates

The discussion of the p-p chain and CN cycle given above gives the amount of energy per reaction available, after neutrino losses, to produce the sun's photon luminosity. To derive the energy generation rate per gram per second, we need to evaluate the rates (cm^{-3} s^{-1}) at which the various reactions proceed. This reaction rate can be expressed as

$$P_{12} = N_1 N_2 \langle \sigma v \rangle_{12} \text{ reactions/cm}^{-3} \text{ s}^{-1} \qquad (6\text{-}31)$$

where N_1 and N_2 are the number densities of nuclei having atomic number z_1 and z_2 and atomic mass A_1 and A_2. The factor $\langle \sigma v \rangle_{12}$ is an integral of the product of the reaction cross section $\sigma(v)$, and of the particle flux averaged over the Maxwell–Boltzmann velocity distribution of the particles. The velocity is figured here in the center-of-mass system of the reacting nuclei.

This integral term is given by

$$\langle \sigma v \rangle = \left[\frac{8}{\pi m (kT)^3} \right]^{1/2} f_0 \int_0^\infty S(E) \exp(-2\pi \eta) \exp\left(\frac{-E}{kT} \right) dE, \quad (6\text{-}32)$$

where m is here the reduced mass of the nuclei, E is the center-of-mass energy, and T is the temperature. The exponential in $-E/kT$ arises from the Maxwell-Boltzmann velocity distribution noted above. The additional exponential in $-2\pi\eta$ expresses the probability of penetrating the Coulomb barrier set up by the electrostatic repulsion of the two nuclei. This Gamow barrier penetration term is given by

$$2\pi\eta = \text{const} \times \frac{z_1 z_2 A^{1/2}}{E^{1/2}}, \qquad (6\text{-}33)$$

so that reactions between light nuclei with small charges are heavily favored. The factor f_0 describes the screening effect that electrons have in diminishing the electrostatic repulsion between the colliding nuclei. This factor tends to increase the reaction rate as the square root of the particle density, with dependences also on T, z, and A. The factor $S(E)$ is a slowly varying function of energy for the reactions of interest in the solar interior.

Under solar core temperature conditions the mean energy of nuclei is about 2 keV, while the Coulomb barrier represents a potential of about 36,000 keV for a proton capture by a C^{12} nucleus. But analysis of the integral (6-32) reveals that the particle energies most effective in promoting nuclear reactions are in the range 10–15 times the mean energy. For such 20–30 keV nuclei, the penetration probability is about 10^{-35} for the proton capture by C^{12} and of order 10^{-9} for the reaction between two protons that is required to initiate the proton-proton chain. However, this high proton-reaction probability is greatly reduced by the low probability of a β-decay during the time of collision between two protons.

In equilibrium, the rates of energy generation by a sequence of nuclear reactions are equal, so that the rate of the whole sequence is set by the rate of the slowest reaction. In other words, if the factor $\langle \sigma v \rangle$ in (6-31) is low for a particular reaction, the concentrations N_1 and N_2 of the reacting species will simply increase until the product P_{12} again equals the rate of the reactions producing species 1 and 2. In the p-p chain, by far the slowest step is the proton reaction that forms deuterium. This rate, which is thought to control the sun's temperature and luminosity, is uncertain to about $\pm 5\%$ due primarily to uncertainty in the lifetime of a free neutron, which enters into the calculation of the function $S(E)$. For the CN cycle, the rate for proton capture by N^{14} (the slowest, and thus the rate determining, step in the chain) has been measured to about 1% by numerous experiments.

Theoretical values of $S(E)$ for the various p-p and CN reactions have usually been compared with multiple experiments, except for the above deuterium-forming proton reaction, which is far too slow to measure in the laboratory. But the experimental values of $S(E)$ still must be obtained at energies greatly exceeding the 20–30 keV range most effective in promoting fusion in the solar core to get practical count rates in the laboratory. Typically, these extrapolations are made from data obtained at 100–200 keV. This extrapolation can be justified provided no unknown nuclear levels exist over the intervening energy interval.

The energy generation rate ϵ $(g^{-1} s^{-1})$ is determined by the reaction rate per unit volume discussed above, by the energy generated per reaction, and by the density. When these three factors are combined, the generation rates for the p-p chain and the CN cycle can be expressed conveniently in the form

$$\epsilon = \epsilon_0 X_1 X_2 \rho T^\nu f_0. \qquad (6\text{-}34)$$

For $\epsilon_{p\text{-}p}$, $X_1 = X_2 = X$ (the hydrogen fraction). For ϵ_{CN}, $X_1 = X$, and $X_2 = X(C + N)$, (the carbon and nitrogen mass fractions). Around temperatures of 1.5×10^7 K, the temperature dependence of $\epsilon_{p\text{-}p}$ is about $T^{4.5}$, but for ϵ_{CN} it is as high as T^{20}!

6.4 THE STANDARD MODEL OF PHYSICAL CONDITIONS IN THE SOLAR INTERIOR

The equations governing the structure of the solar interior have been solved using progressively more accurate nuclear reaction rates for the p-p chain, various initial values of the sun's chemical composition, with improved opacities, and using different assumptions of the mixing length that governs the efficiency of convection in the outermost layers. The results of these computations are referred to as standard models of the solar interior. In such models, internal stresses of rotation or magnetic fields are taken to be negligible, material in the regions of the radiative energy transport is assumed to be unmixed by plasma motions, and energy generation is taken to be given mainly by the standard p-p chain reaction rates.

The physical conditions in the sun's interior predicted by a recent standard model are given in Table 6-3 and plotted in Fig. 6-1. A schematic picture of the sun's interior is shown in Fig. 6-2. The values plotted in Fig. 6-1 are computed from a model which begins with a homogeneous chemical composition given by $X = 0.71$, $Y = 0.27$, and $Z = 0.02$ and evolves to the sun's present age of 4.6×10^9 years. The density decreases smoothly from a central value of about 150 g cm^{-3} to less than 0.1% of that value at $0.75 R_\odot$, where convection takes over from radiation as the main energy transport mechanism. The depth of this convection zone, which accounts for only about 1.5% of the sun's mass, is somewhat uncertain, as discussed in more detail in Chapter 7. The density then falls by another factor 10^5 through the convection zone to the photosphere.

The temperature attains values of about 15 million degrees at the sun's center and decreases to about 1.9 million degrees at the bottom of the convection zone. The function $L_r/L_\odot$ plotted in Fig. 6-1 shows that just about all the nuclear energy generation is accomplished within a distance of $0.3 R_\odot$ from the center. Peak energy generation occurs even closer to the center,

TABLE 6-3 Conditions in the Solar Interior

$m(r)/M_\odot$	$r/R_\odot$	T	ρ	$L_r/L_\odot$	X
0.00000	0.0000	1.56E + 07	1.48E + 02	0.000	0.34111
0.00001	0.0039	1.56E + 07	1.48E + 02	0.000	0.34103
0.00005	0.0083	1.56E + 07	1.47E + 02	0.000	0.34317
0.00017	0.0120	1.56E + 07	1.46E + 02	0.001	0.34546
0.00040	0.0158	1.56E + 07	1.45E + 02	0.003	0.34885
0.00078	0.0197	1.55E + 07	1.44E + 02	0.007	0.35328
0.00135	0.0237	1.55E + 07	1.42E + 02	0.012	0.35868
0.00214	0.0277	1.54E + 07	1.40E + 02	0.018	0.36499
0.00320	0.0317	1.53E + 07	1.37E + 02	0.027	0.37217
0.00625	0.0400	1.51E + 07	1.32E + 02	0.051	0.38890
0.01080	0.0484	1.49E + 07	1.26E + 02	0.085	0.40839
0.03071	0.0708	1.42E + 07	1.08E + 02	0.217	0.46672
0.05000	0.0853	1.37E + 07	9.70E + 01	0.325	0.50536
0.10385	0.1147	1.25E + 07	7.64E + 01	0.553	0.57659
0.15000	0.1346	1.17E + 07	6.45E + 01	0.688	0.61549
0.20400	0.1551	1.09E + 07	5.40E + 01	0.798	0.64646
0.25200	0.1719	1.03E + 07	4.64E + 01	0.865	0.66550
0.30000	0.1881	9.74E + 06	3.99E + 01	0.912	0.67902
0.35000	0.2047	9.20E + 06	3.40E + 01	0.945	0.68885
0.40000	0.2212	8.70E + 06	2.88E + 01	0.966	0.69563
0.45000	0.2381	8.22E + 06	2.42E + 01	0.981	0.70024
0.50000	0.2555	7.76E + 06	2.01E + 01	0.990	0.70324
0.52000	0.2628	7.58E + 06	1.86E + 01	0.993	0.70409
0.55000	0.2739	7.32E + 06	1.65E + 01	0.996	0.70512
0.58500	0.2876	7.01E + 06	1.42E + 01	0.998	0.70621
0.65500	0.3176	6.39E + 06	1.01E + 01	1.000	0.70806
0.69000	0.3344	6.08E + 06	8.34E + 00	1.000	0.70866
0.76000	0.3737	5.44E + 06	5.32E + 00	1.000	0.70934
0.79500	0.3975	5.09E + 06	4.06E + 00	1.001	0.70952
0.86500	0.4597	4.33E + 06	2.03E + 00	1.001	0.70967
0.90000	0.5038	3.88E + 06	1.27E + 00	1.001	0.70970
0.96616	0.6559	2.64E + 06	9.94E + 13	1.000	0.70970
0.99127	0.8015	1.36E + 06	1.44E + 13	1.000	0.70970
0.99612	0.8573	9.04E + 05	5.12E + 12	1.000	0.70970
0.99869	0.9093	5.25E + 05	1.29E + 12	1.000	0.70970
1.00000	1.0000	5.77E + 03	0.000000	1.000	0.70970

Adapted from J. Bahcall and R. Ulrich, *Rev. Modern Phys.* **60**, 297 (1988).

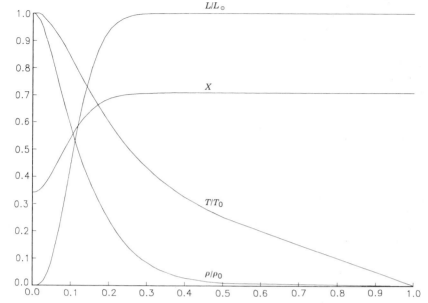

Fig. 6-1 A plot of conditions in the solar interior. The density ρ/ρ_0 and temperature T/T_0 are given relative to their sun-center values in Table 6-3, along with the net luminosity $L_r/L_\odot$ at radius r (relative to the total luminosity) and the hydrogen fraction X. Based on data given by J. Bahcall and R. Ulrich, *Rev. Modern Phys.*, **60**, 297 (1988).

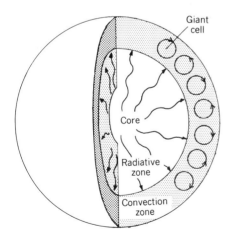

Fig. 6-2 Schematic diagram of the solar interior showing the regions of energy generation and of radiative and convective transport.

around $0.1R_\odot$. The curve of hydrogen mass fraction X indicates that the sun's core hydrogen is by now depleted by almost a factor of 2 by burning to helium.

Uncertainties in such an evolutionary model arise from questions about the initial composition and the sun's age and from errors in the opacities, nuclear reaction rates, and equation of state in the dense central regions. The values of temperature and density in the core regions are relatively insensitive to the structure of the convective envelope, as we might imagine, although the luminosity does respond weakly to the details of convective energy transport through these outer layers, as discussed further in Chapter 7.

The standard model described all that was known about the solar interior until 1968, when the first neutrino observations began to reveal a significant discrepancy between the observed neutrino flux and the much higher flux expected from the standard model. The "neutrino problem" has led to a thorough examination of the validity of the premises and parameters on which this standard model is based. The new kinds of observational techniques and their interpretation in terms of nonstandard solar models, are described in Section 6.5. These nonstandard models introduce wider options for solar energy generation and for dynamical equilibrium in the sun's interior.

6.5   OBSERVATIONAL TESTS OF THE STANDARD MODEL

6.5.1   Solar Neutrino Observations

The neutrinos produced in the solar core escape from the sun within a few seconds, traveling at the speed of light and hardly interacting with solar matter along the way. Their flux at the Earth is predicted to be quite large, about 6.6×10^{10} cm^{-2} s^{-1}. Although the probability of a reaction between a neutrino and a particle of matter is extremely small, it is not zero. Given a large detector and (mainly) extremely sensitive techniques for detecting the products of such reactions, a neutrino sensor (or even a telescope or spectrometer offering spatial and spectral resolution) is possible. These particles present an important opportunity to observe the nuclear processes in the deep solar interior.

The first solar neutrino detector was placed in operation in 1968 by R. Davis and his collaborators at the Brookhaven National Laboratory. It relies on the conversion by neutrinos of ground-state chlorine nuclei into a radioactive excited state of argon gas via the reaction

$$Cl^{37} + \nu = Ar^{37} + \beta^- \qquad (6\text{-}35)$$

The neutrino energy threshold for this reaction is 0.814 MeV, and neutrinos of sufficient energy are predicted by the standard model to be produced almost entirely by the end reaction of the p-p chain involving β-decay of B^8 to Be^8 (6-22).

In Davis's experiment (Figure 6-3) the detector is a large tank containing 2.2×10^{30} atoms of chlorine (610 tons of the cleaning fluid perchlorethylene, C_2Cl_4). About one solar neutrino-induced reaction per day is expected in this mass of chlorine. The experiment is placed 4850 ft underground in the Homestake gold mine in Lead, South Dakota, to shield it against cosmic ray particles. A typical run lasts about 80 days, during which time the Ar^{37} produced is allowed to accumulate in the detector. Inert helium gas is then passed through the tank to pick up the Ar^{37}. The purified argon gas of measured volume is placed in a proportional counter and its decays are

Fig. 6-3 The solar neutrino detector in the Homestake mine, as photographed in 1967. It has since been surrounded by a scintillation detector and cannot be photographed alone. By permission of R. Davis.

counted over a period of 8 months or more to determine the number of Ar^{37} nuclei present per unit volume. The amplitude and rise time of the counter pulses are used to discriminate between Ar^{37} decays and background.

Observations have been carried out on a regular basis since 1970. The results of about 76 runs between 1970 and 1988 are shown in Figure 6-4. The right-hand scale is in solar neutrino units (SNUs), defined as one neutrino capture per second in 10^{36} target atoms. The observed mean production rate over this time period is found to be 0.46 ± 0.03 atoms day^{-1} of which a fraction is attributed to residual cosmic ray background. If the net production rate of 0.38 ± 0.05 atoms day^{-1} is attributed to solar neutrinos, it corresponds to 2.0 ± 0.3 SNU.

The minimum rate expected from p-e-p neutrinos (which is expected independently of model details provided only that the p-p chain fuels solar luminosity) is 0.25 SNU, thus close to even the expected background rate. The neutrino flux predicted by the most recent standard solar models is 7.9 ± 0.3 SNU. The discrepancy between the flux predictions of the standard models and the low measured flux is referred to as the solar neutrino problem.

A large effort has gone into checking the nuclear reactions and cross-sections, and some improvements have been made, but none large enough to produce nearly a factor-of-3 decrease in B^8 neutrino flux. Since the B^8 neutrino flux is much more temperature dependent (T^{15}!) than the nuclear energy production rate, solar models have been constructed with a lower core temperature produced by relieving the gas pressure through introduction of a very strong internal magnetic field or of centrifugal forces caused by a high core rotation rate. But these approaches do not resolve the dilemma because the huge fields and rotation rates required violate other observational constraints on the solar gravitational quadrupole moment discussed in Section 6.5.5.

The core temperature can also be lowered for a given luminosity by reducing the radiative opacity through a low heavy-element abundance. Consistent models can be calculated with $Z \sim 0.002$ that do produce neutrino capture rates in the Cl^{37} experiment in the range 1–2 SNU. However, such models are at odds with the frequencies of acoustic modes measured in recent helioseismological observations (Section 6.5.6). Also, it is difficult to understand how the sun could have been constructed with such a "clean" hydrogen constitution. In particular, the required helium abundance is about a factor of 2 below the primordial value $Y \sim 0.2$ expected in Big Bang cosmology and raises the question of how the sun could have acquired its presently "dirty" $Z \sim 0.02$ outer layers.

The most likely astrophysical explanation of the solar neutrino problem seems to lie in possible departures from the assumptions that no mixing takes place in the standard solar model and that diffusion of nuclear species is negligible. As nuclear burning proceeds, the hydrogen depletion in the core becomes quite marked (Figure 6-1). According to relation (6-34), the energy generation rate varies roughly as $X^2 \rho T^{4.5}$, so return of hydrogen back into the

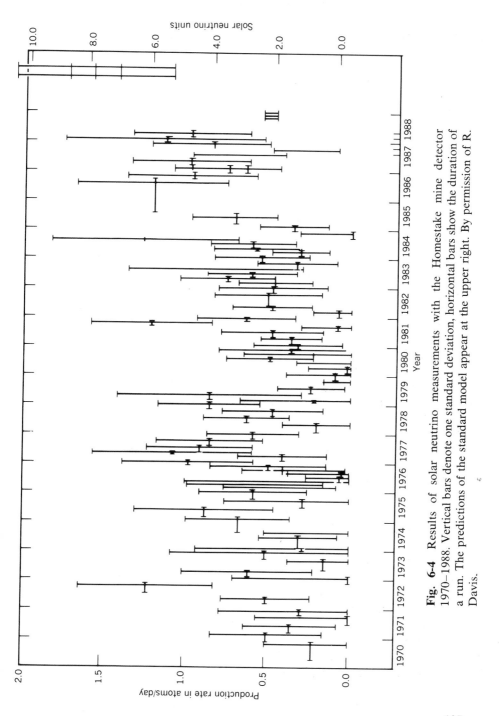

Fig. 6-4 Results of solar neutrino measurements with the Homestake mine detector 1970–1988. Vertical bars denote one standard deviation, horizontal bars show the duration of a run. The predictions of the standard model appear at the upper right. By permission of R. Davis.

195

core by episodic overturning, slow motions, or enhanced diffusion could enable the p-p chain to produce the required energy at lower core temperature and decrease the flux of B^8 neutrinos. The possibility of such dynamics in the solar core is discussed further in Section 6.5.4, although the constraints posed by helioseismology must also be considered (Section 6.5.6).

The pioneering Cl^{37} experiment now in operation has yielded important insights, and its sensitivity to temperature is desirable as a diagnostic of solar core conditions. Since the CN cycle would produce many more of the high-energy neutrinos detectable by the Cl^{37} reaction, this experiment has provided strong direct evidence against the importance of the CN cycle in the sun. However, its dependence on the high energy B^8 neutrinos makes it sensitive to details of the standard model, and it would be desirable to complement it with an experiment sensitive to low-energy neutrinos.

Rapid advances are now being made in solar neutrino observations. The low flux of B^8 neutrinos has recently been confirmed by a direct counting detector based on electron scattering of neutrinos, that is located in the Kamioka mine in Japan. This detection by a directional detector is particularly convincing. The Ga^{71} experiments now beginning operation should be particularly effective in deciding whether the low neutrino flux detected by the Brookhaven experiment implies a low solar flux or whether it might be due to uncertainties in neutrino physics. In particular, neutrinos may oscillate between the electron neutrino state discussed above, and two other states not detectable by the Cl^{37} experiment. If only $\frac{1}{3}$ of the solar neutrino flux were detectable by the Brookhaven experiment, the neutrino problem would be solved.

6.5.2   Lithium and Beryllium Abundances

In the discussion of the solar chemical composition presented in Chapter 5, it was mentioned that the abundances of Li, Be, and B were low. The generally depressed abundance of all three elements is explained in the theory of nucleosynthesis by the high reaction probability of light nuclei up to boron with protons, resulting mainly in the formation of multiple He^4 nuclei from each such reaction. However, when the solar abundances of these elements are compared to their abundances in carbonaceous chondrite meteorites (thought to represent primordial solar material), in younger stars, or in the interstellar medium, it is found that the present solar lithium abundance is about a factor of 100 lower than in the primordial material, whereas the abundances of beryllium and boron are about normal.

Observations of increased lithium depletion in cooler, low-mass stars which are expected to have deeper convection zones than the sun suggest an explanation of this depletion. It is known that lithium burns at about 2.4×10^6 K, whereas beryllium requires 3.5×10^6 K. This suggests that photospheric material is circulated to layers corresponding to lithium burning, but no deeper. Some of this mixing is expected from convection and its overshooting. It is still controversial whether the depth of the solar convection zone is sufficient or whether other, deeper motions must be invoked.

Studies of the depth of convection in the sun's early stages indicate that most of the lithium depletion should have occurred during its main sequence lifetime, so that the above reasoning can probably be applied to the present convection zone depth. This view is supported by the observation that lithium depletion tends to decrease in younger main sequence stars, implying that it is not simply a result of pre–main sequence events.

6.5.3 Stellar Structure and Evolution

Observations of the luminosities, effective temperatures, radii, and masses of stars show well-defined behavior patterns that are successfully explained by the theory of stellar structure and evolution. This agreement is the main reason for believing that the theory with which we venture into the sun's interior has some validity.

One of the principal results of stellar astronomy shows us that, when the luminosities L of stars are plotted against their colors or effective temperatures T, most stars fall into a relatively narrow band called the main sequence defined by the empirical relation

$$\frac{L}{L_\odot} = \left(\frac{T}{T_\odot} \right)^6 \tag{6-36}$$

where $L_\odot$ and $T_\odot$ denote solar values (Fig. 13-2).

It is also observed (Fig. 6-5) that the masses of these main sequence stars are related to their luminosities through

$$\frac{L}{L_\odot} = \left(\frac{M}{M_\odot} \right)^4 \tag{6-37}$$

and to their radii through

$$\frac{R}{R_\odot} = \left(\frac{M}{M_\odot} \right)^{0.7} \tag{6-38}$$

It can be shown (see exercises) that, when the appropriate dependences of opacity and nuclear energy generation on temperature and density are inserted, dimensional analysis of the equations (6-4) and (6-7) for energy production and radiative transport yields theoretical predictions that match the L-T, M-L, and M-R relations quite well, at least for stars more massive than the sun, which are expected to be in radiative equilibrium throughout. This agreement encourages some confidence in the theory of main sequence stars whose energy generation process, like that of the sun, is hydrogen burning.

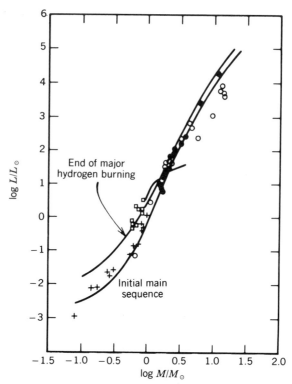

Fig. 6-5 Comparison of theoretical and empirical mass-luminosity functions. Adapted from G. Newkirk, in *The Ancient Sun*, R. Pepin and J. Eddy (Eds.), 1980, Reproduced by permission of Pergamon Press.

The further successes of this theory in explaining the evolution of young stars of different masses onto the main sequence, in predicting the ages at which they move off the main sequence into the red giant phase, their entry into the Cepheid pulsation stage, and their subsequent evolution into late life are discussed in Chapter 13. Certain aspects of this evolution have direct bearing on our understanding of the present sun. For instance, the observation that stars leave the main sequence to become red giants is explained in terms of exhaustion of hydrogen in the central regions of the star. In calculating solar models with enhanced mixing or diffusion to lower the neutrino flux, care must be taken that the mixing not be excessive or we would have difficulty in explaining the red giant phenomenon.

6.5.4 Geological and Climatological Evidence

Additional constraints on the internal structure and physics of the present sun are placed by the geological and climatological history of the Earth. The age of the sun can only be estimated on the grounds that it is at least as old as the Earth, but probably not much older. The geological evidence indicates an age for the Earth of some 4.5×10^9 years, and our ideas on how the Earth formed in the early solar nebula suggest that the sun may be 50 million years older. Nevertheless, uncertainty in the sun's age is one of the significant sources of error in solar models.

The standard solar model (and most nonstandard models also) predicts that the sun has become progressively brighter by about 20–30% while burning its core hydrogen on the main sequence. The reason is simply that, as hydrogen burns to helium, the sun's mean molecular weight will increase and the pressure scale height will decrease. The result is that the more concentrated mass will exert a higher pressure and produce a greater energy generation rate in the solar core.

Models of the present-day Earth's climate indicate that even a few percent depression in the solar constant would plunge the Earth into a deep ice age involving total global glaciation. But geological evidence from the first sedimentary rocks dating back about 3.5×10^9 years indicates that the climate has become systematically cooler. This faint sun enigma has not been fully resolved, but it is realized now that profound changes in the Earth's atmospheric composition such as its CO_2 content, in the distribution of atmospheres and oceans, and in latitudinal heat transports neglected in the simplest climate models could help to explain the discrepancy. Still, it seems difficult to reconcile such a large monotonic increase in $L_\odot$ with a steady cooling of the climate.

Finally, it is worth mentioning that episodic variations in climate marked by major ice ages may indicate dynamical processes in the solar interior. One mechanism proposed for such a connection relies on the expectation that hydrogen burning by the p-p cycle will cause He^3 to accumulate away from the sun's center. Under these circumstances an instability might arise in which material perturbed (downward, for instance) would be heated and thus would tend to generate energy at a more rapid rate than its surroundings. This would reinforce the initial higher temperature and the material would rebound, leading to growing oscillations.

Whether this particular He^3 instability operates depends on delicate computations to assess its damping by radiation, which will tend to equalize the temperature of the moving element and its surroundings. But it illustrates one way in which compositional gradients set up over hundreds of millions of years inside the supposedly stably stratified radiative core of the sun might conspire to produce mixing on much shorter time scales.

6.5.5 The Sun's Angular Momentum and Shape

Main sequence stars slightly more massive and thus hotter than the sun have higher surface rotation rates, in the range 50–200 km s^{-1}. Assuming a uniform angular rotation rate with depth, their specific angular momentum can be

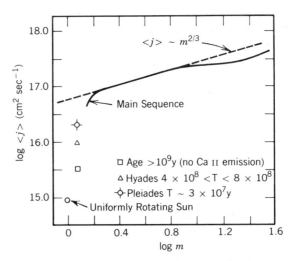

Fig. 6-6 The specific angular momentum of the sun and other stars. Adapted from R. Kraft, 1972.

calculated, and it is found to increase with stellar mass (Fig. 6-6). Cooler stars, less massive than about $1.3 M_{\odot}$, have much slower surface rotation rates (< 10 km s^{-1}), and their angular momenta per unit mass are found to be about two orders of magnitude smaller. An explanation for this sharp decrease in specific angular momentum has been given in terms of the onset of convection, which is expected from stellar interior calculations to occur at about the same point in the main sequence as the rapid decrease in rotation rate. The magnetic dynamo activity that is thought to accompany stellar convection (see Chapter 11) probably plays a role in braking at least the outer layers of the star.

A likely prospect for this braking is the loss of mass and angular momentum that accompanies a stellar wind. As discussed in Chapter 12, the wind plasma moves out along magnetic field lines to a considerable distance from the sun and, like a ballerina moving her arms out, brakes the solar rotation by exerting a torque that is transmitted back to the photosphere and deeper layers through the rigidity of the magnetic field lines. The outward flux of mass thus also carries away with it a flux of angular momentum. The efficiency of the mechanism is uncertain since a much larger mass loss rate than is observed in the present solar wind would have been required in the past to reduce the sun's angular momentum to the present low value.

Alternatively, the sun's interior may rotate much more rapidly than the photosphere, and the missing angular momentum could be stored at depth. Estimates of various hydrodynamical instabilities that might transfer angular momentum indicate that a rapidly rotating core might persist. But a high rotation rate would tend to produce an oblateness of the photospheric disk. It appears that no oblateness in excess of 1×10^{-5} (or about 10 km) is detected, which argues against a core rotating rapidly enough to explain all of the missing angular momentum.

A more direct method of detecting distortion of the sun's internal mass distribution is a measurement of the quadrupole moment J_2 of the sun's gravitational field given by

$$U = -\frac{GM}{r}\left[1 - J_2\frac{P_2(\sin\theta)}{r^2}\right], \qquad (6\text{-}39)$$

where U is the sun's gravitational potential and $P_2(\sin\theta)$ is the second Legendre polynomial expansion in the latitude. J_2 represents the difference in the equatorial and polar moments of inertia normalized to the sun's mass, and so would disappear for the perfectly spherical mass distribution expected if the sun did not rotate.

Evidence against a large J_2 value is available from measurements of the perihelion precession of Mercury's orbit—at least if the description of gravity given by general relativity is accepted. For a more sensitive measurement it has been proposed to fly a solar probe into a very close encounter with the sun and observe its detailed motion, such as has been done with geodetic satellites surveying the high-order moments of the Earth's gravity field. Estimates of the sun's quadrupole moment range over an order of magnitude $10^{-7} < J_2 < 10^{-8}$ between standard models, depending on the internal temperature distribution. It is estimated that a suitably drag-compensated solar probe could distinguish between such models. It would also be helpful in deciding whether other internal stresses such as centrifugal accelerations or very strong ($> 10^7$ G) magnetic fields might be increasing the oblateness beyond the small value expected if the sun were rotating rigidly at the photospheric rate.

6.5.6   Solar Oscillations

The solar interior acts as a resonator for sound- and gravity-mode waves. The most useful of the wave modes so far observed, from the standpoint of probing conditions in the solar interior, are the 5-min p-modes. These form standing waves in cavities that extend from just below the photosphere, right to the center of the sun, for the lowest-l modes.

The temperature structure of the solar interior can be determined using the measured frequencies of these modes. By trial and error we can achieve agreement to better than 0.5% using procedures described in Chapter 7. Such studies favor standard models with a relatively high helium abundance of $Y = 0.27$ and $Z = 0.02$, and a deep convection zone extending down to $0.7R_\odot$. In fact, the mode frequencies have now been measured to an accuracy about a factor of 10 better than this agreement. The residual discrepancies seem to arise from uncertainties in the model near the photosphere. Completely mixed models of the interior, which offer a solution to the neutrino problem, seem to be ruled out by such helioseismological data, although the degree of mixing that is allowed by uncertainties in the measurements and in their interpretation is still controversial.

The splitting of the p-modes has also been applied to the study of the solar internal rotation profile and evaluation of the sun's gravitational quadrupole moment as discussed in Chapter 7. Recent work, based on a study of low- and intermediate-degree modes, indicates that at least outside of $0.4R_\odot$ the angular rotation rate is close to the surface rate, with some evidence for a small decrease outward. A better determination of the rotation in the central core will probably require unambiguous observations of internal g-modes, which propagate mainly in the radiative core.

ADDITIONAL READING

Background on the Astrophysics of Stellar Interiors

E. Novotny, *Introduction to Stellar Atmospheres and Interiors*, Oxford University Press, 1973.

Nuclear and Atomic Processes in the Sun's Interior

W. Huebner, "Atomic and Radiative Processes in the Solar Interior," in *Physics of the Sun*, P. Sturrock et al. (Eds.), D. Reidel, p. 33, 1985.

P. Parker, "Thermonuclear Reactions in the Solar Interior," in *Physics of the Sun*, P. Sturrock et al. (Eds.), D. Reidel, p. 15, 1985.

Models of the Solar Interior, Neutrino Observations

J. Bahcall and R. Ulrich, "Solar Models, Neutrino Experiments and Helioseismology," *Rev. Modern Phys.*, **60**, 297 (1988).

M. Cherry, W. Fowler, and K. Lande, Eds., "Solar Neutrinos and Neutrino Astronomy," AIP Conference Proceedings No. 126, New York (1985).

I. Roxburgh, "Present Problems of the Solar Interior," *Solar Phys.*, **100**, 21 (1985).

EXERCISES

1. Show that the gravitational potential of a homogeneous sphere of radius R and mass M is $-\frac{3}{5}GM/R^2$. Estimate the total energy liberated in the sun's contraction to its present radius, and estimate the time over which the sun's power output might have been sustained by such a contraction assuming the present luminosity. Remember from the virial theorem that only half the released potential energy will be radiated away; the other half will heat

the star. Calculate also the rate of contraction required to balance the present luminosity, and compare it to current observational limits of about 0.05 arc sec year^{-1}.

2. Show from the equation of hydrostatic support that, to order of magnitude, the density, pressure and temperature inside the sun can be estimated to be $\rho \sim M_\odot/R_\odot^3$, $P \sim GM_\odot/R_\odot^4$, and $T \sim \mu GM_\odot/R_\odot^2$ respectively. Explain why the solar luminosity is expected to have increased with time on the main sequence.

3. Show that the approximate power law relationships between M, L, R, and T given in Section 6.5.3 for main sequence stars can be deduced from the equations of stellar energy generation and radiative transport if the nuclear energy generation rate is taken as $\epsilon_{CN} \sim \text{const} \times \rho T^{16}$ and the opacity is taken as $\bar{\kappa}_R \sim \text{const} \times \rho^{0.4} T^{-2.8}$. Does this analysis apply for the sun?

4. Using the estimates of ρ and T from exercise 2, estimate the total thermal energy of the sun. Show that the thermal relaxation time (the Kelvin-Helmholtz time) required to radiate away this energy at the present solar luminosity is of order 10^7 years. What does this imply about the response time of the sun's radius and luminosity to changes in nuclear burning rates such as during a He^3 episodic overturning? There is some evidence that the Cl^{37} detector in the Homestake mine measures an 11-year variability of the neutrino flux in anticorrelation with the solar activity level. Can such rapid variations in solar neutrino flux be ruled out?

5. Write the radiative diffusion equation using the relation (6-5) for the radiative flux, and show that the diffusion time of photons through the sun is $\tau_D \sim \rho \bar{C}_p R_\odot^2/k'$. Using representative values of the opacity, density and specific heat, show that τ_D is of the order of 10^7 years. Comment on the agreement between the radiative diffusion time scale and the Kelvin-Helmholtz time scale.

7

Rotation, Convection, and Oscillations in the Sun

The sun's rotation rate of roughly 2 km s^{-1} near the equator carries structures such as sunspots across a telescope's field of view at a rate of about 10 arc sec per hour near disk center. A displacement of this size is easily noted in even the smallest telescope, so it is not surprising that solar rotation was discovered soon after the first telescopic observations of sunspots began.

Progressively more sophisticated measurements of proper motions of solar features, and also of Doppler shifts of the solar plasma, have shown that the solar atmosphere exhibits complex motions on all observable scales. Besides the axisymmetric motion associated with solar rotation, whose angular rate decreases markedly with increasing latitude, other motions associated with convection and also with the sun's remarkable oscillation at about a 5-min period have been intensively studied. The speeds of some of these motions, such as that of granulation, considerably exceed that of solar rotation, and they are easily detected as Doppler shifts with modern spectrographs.

The dynamics of the sun's complex magnetized plasmas is far from completely understood. For instance, our ideas on the factors that determine the sun's present rotation rate are only beginning to be tested on suitably chosen samples of stars similar to the sun. Furthermore, the scales of convection mainly responsible for heat transport to the photosphere and for angular momentum distribution below the photosphere are still quite unknown. This makes it difficult to obtain a clear understanding of the photospheric differential rotation.

Nevertheless, the development of numerical models that incorporate most of the dynamical mechanisms that should operate in the sun gives us a basis for beginning a quantitative discussion. Until recently, development of these models has been hampered by insufficient observational constraints. But recently developed helioseismological techniques now make it possible to probe the sun's internal rotation rate and thermal structure. These and other advances described in this chapter have made dynamical studies of the sun's rotation, convection, and oscillations arguably the most rapidly growing part of solar research in the past ten years, with promise of much more to come.

In this chapter we survey the observational evidence on solar rotation, convection, and oscillations and put forward the results from dynamical modeling that appear to have most relevance in attempting to understand these phenomena.

7.1 OBSERVATIONS OF SOLAR ROTATION

7.1.1 Doppler Measurements

The basic features of solar rotation obtained from photospheric Doppler measurements are illustrated in Fig. 7-1, where the rotation rate is plotted against latitude. The sidereal (i.e., with respect to fixed stars rather than to the revolving Earth) rotation rate at the equator is found to be 2.84 μrad s^{-1},

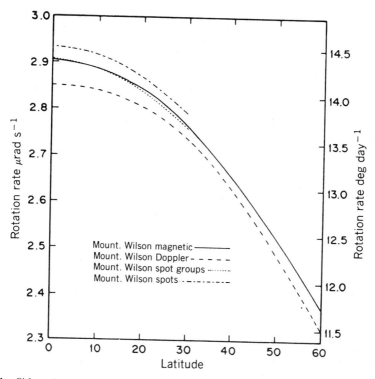

Fig. 7-1 Sidereal rotation rates from photospheric Doppler and tracer measurements. Reproduced with permission, from R. Howard, *Ann. Rev. Astr. Astrophys.*, **22**, 131 (1984), copyright by Annual Reviews, Inc.

which corresponds to a rotation period of approximately 25 days. This rate is determined to an accuracy of about 1%, judging from the agreement between the two principal sets of long-term measurements made at Mt. Wilson and at Stanford University.

Many other Doppler measurements have been made over the past 100 years, but their accuracy is generally lower because of less well understood instrumental and scattered-light errors, and shorter data sets. Scattering reduces the apparent rotation rate because less Doppler-shifted Fraunhofer line profiles from the brighter disk-center region tend to be scattered into the spectrograph slit during measurements made nearer the limb, where the line-of-sight component of rotation is largest. A large number of measurements well separated in space and time are required to achieve satisfactory precision because the strong horizontal supergranular flows contribute a large noise signal.

The most interesting feature in Fig. 7-1 is the large decrease of solar angular rotation rate with increasing latitude, amounting to about 20% between the equator and 60°. The smooth curves in this figure are obtained by fitting the data to power laws in the sine of the latitude which smooth out any abrupt latitudinal variations in the data due to either real solar effects or observational errors. Also, the north and south hemispheres are combined in these plots, although there is evidence that their rotation rates can be different. A few measurements as close as 7.5° from the poles indicate that the rotation law shown in Fig. 7-1 can be extrapolated beyond the range of the (most accurate) measurements below 60° latitude.

Short-term time variation in the rotation rate has been reported at Mt. Wilson, where the data are obtained at the 150-ft tower with a spatial resolution of about 20 arc sec. But no significant fluctuations are seen at the Stanford Solar Observatory, where observations are made with similar dispersion but at lower spatial resolution. After corrections for fluctuations due to both scattered light and instrumental effects, a day-to-day variation of up to 1% is seen at Mt. Wilson, with a variation over months of several times this value. The evidence for solar-cycle variations in solar rotation from tracer and Doppler measurements is discussed in Section 7.2.2.

Information on the radial gradient of solar rotation might in principle be obtained from Fraunhofer lines formed at different heights. Some measurements of this kind suggest an inward decrease of rotation rate in the photosphere, but their interpretation is suspect and line blends may influence the results. Any rotation gradient detectable over the accessible height range of a few hundred kilometers would imply a huge rotation rate variation with depth in the upper convection zone unless it were confined only to the photosphere, which seems unlikely. Systematic rotation rate differences between chromospheric magnetic structures and the photosphere would be easier to understand, for reasons discussed below, but the evidence here is also contradictory. A better way to study the radial gradient of solar rotation is discussed in Section 7.5.5.

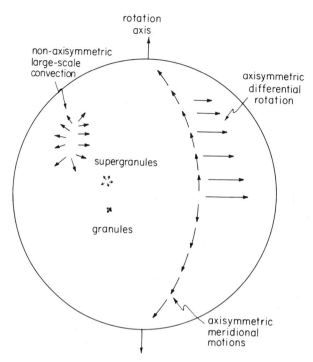

Fig. 7-2 Schematic diagram of the dimensions and geometries of some observed and theoretically predicted flow patterns at the photosphere. The size of granules is exaggerated relative to the other structures illustrated.

Doppler data have also been used to search for global meridional flows (Fig. 7-2), which in some theoretical models play an important role in transport of solar heat and angular momentum. A flow from the equator to the poles has been suggested from analyses of observations at Mt. Wilson, Kitt Peak, and elsewhere, but the interpretation of large-scale motions at such low velocities, of order 10 m s^{-1}, is still problematical, given uncertainties in separating out the contribution of the well-known large ($\sim 10^2$ m s^{-1}) blueshift of most lines near disk center. This blueshift is known to be caused by the brightness-velocity correlation of granulation, as discussed in Section 5.1.2. Small uncertainties in its relative behavior along equatorial and polar diameters can mimic meridional flows.

Analysis of residual motions (after mean rotation is removed) in the daily Mt. Wilson velocity measurements over the years 1967–1982 suggests the presence of a torsional velocity oscillation of very low amplitude. The main feature is an increase in rotation rate of about 2 m s^{-1} that moves from the poles toward the equator in 22 years. If further study is able to confirm it, this oscillation represents a tiny distortion in the differential rotation curve, Fig. 7-1, moving from right to left during a solar magnetic cycle.

7.1.2 Tracer Measurements

Solar rotation has been measured over a much longer time base than covered by the Doppler data, through observations of the daily motion of surface features. Sunspots, faculae, and magnetic field patterns have been used at photospheric levels. Chromospheric and coronal rotation rates have been obtained in the same way from the changing positions of Ca K and EUV plages, Hα filaments, and structures in the green-line and white-light corona.

All of these tracers have in common their close association with magnetic flux tubes extending downward to subphotospheric levels. The rotation rates of the different tracers thus may carry interesting evidence on the solar rotation profile below the photosphere, on the dispersion of velocities in those deeper layers, and on the transfer of momentum between the magnetic and nonmagnetic plasma components in the solar interior.

The rotation rates obtained from spot motions are also shown in Fig. 7-1 for comparison with the photospheric Doppler rate and with the rate obtained from rotation of the whole photospheric magnetic field pattern. The main point is that all the magnetic tracers seem to rotate faster by a few percent, at all latitudes, than the photospheric plasma. The fastest rotation is for small spots. The spot group rate is slower than the individual spot rate because the group rate is weighted by area, so it is biased to the relatively small number of large-area spots, which are known to rotate most slowly of all spots. The rotation rate for the magnetic patterns, including faculae and enhanced network around active regions, is about the same as that for the groups, and is about 2.92 ± 0.02 μrad s^{-1} at the equator.

The reality of the difference between the tracer and Doppler rates has been questioned on the grounds that systematic errors in the two rather different measurement techniques, such as scattered light in the Doppler observations, could explain the small difference. However, the observation of significantly different rates among the tracers themselves shows that the Doppler rate cannot be the same as all the different tracer rates.

All the rates shown in Fig. 7-1 exhibit strong differential rotation with latitude. However, study of the magnetic field rotation has revealed a tendency for long-lived field patterns at the photosphere to rotate at a much more nearly constant rate with latitude. This was shown by the same technique used to derive the magnetic field rotation rate shown in Fig. 7-1. In both cases, the Mt. Wilson daily magnetograms were divided into latitude strips, and the data in each strip were cross-correlated with data on successive days. The position of the cross-correlation peak falls at the rotation rate for that latitude. Interestingly, if the cross-correlation is carried to lags of several times the solar rotation period, the peaks obtained at different latitudes tend to fall at more nearly equal rotation periods.

It has recently been found that the sunspot rotation rate seems to exhibit a small solar-cycle modulation. Several studies of sunspot data from Mt. Wilson and Greenwich indicate that the low-latitude rotation rate speeds up at

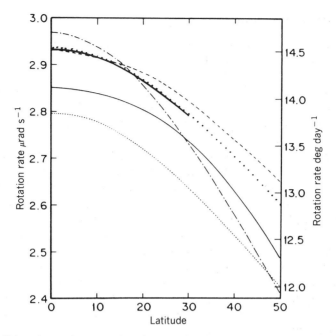

Fig. 7-3 Sidereal rotation rates for chromospheric and coronal features measured as tracers compared to photospheric Doppler rates (light solid curve). The curves refer to coronal magnetic structures (long dashes), transition region magnetic structures (dot-dashed), chromospheric magnetic network (short-dashed), photospheric faculae (dots), and sunspots (heavy solid). Reproduced with permission, from R. Howard, *Ann. Rev. Astr. Astrophys.*, **22**, 131 (1984), copyright by Annual Reviews, Inc.

activity minimum by about 0.5%. This acceleration is shown most clearly in the Mt. Wilson sunspot data for 1921–1982. A further acceleration around spot maximum is less well supported. When the Mt. Wilson Doppler measurements (made only over the shorter period 1967–1982) are averaged in the same way, they seem to exhibit the same modulations, but with about twice the amplitude.

Figure 7-3 shows the tracer rotation rates obtained for various chromospheric and coronal structures, such as Ca K plages, and green-line and white-light coronal features. The white-light structures of the outer corona, which are rooted in long-lived field patterns at the photosphere, tend to rotate less differentially than any of the other tracers or the plasma Doppler rate. Except for one uncertain measurement of the magnetic network rate, all the tracers tend to exhibit faster rotation than the Doppler rate. Care must be exercised in comparing rates obtained in quite different epochs, since the differences seen between various tracers are comparable to the solar-cycle modulation mentioned above and to its cycle-to-cycle changes.

7.2   MEASUREMENTS ON CONVECTION

7.2.1   Observations of Convection at the Photosphere

The properties of granulation, discussed in Section 5.1.2, have been identified with convection since 1930, when A. Unsold pointed out that the layers immediately below the photosphere should be convectively unstable. The observed behavior of supergranulation, described in Section 5.1.3, has also been widely associated with convection, although the evidence here is less compelling. Searches for other scales of solar convection indicate little evidence for patterns larger than supergranulation, although evidence for an intermediate mesogranular scale deserves further study.

A key question has been whether granulation represents surface turbulence generated by subphotospheric convective eddies or whether it is the overshooting of these convective eddies themselves into the photospheric layers. The first point of view originated with Unsold; the second can be traced to H. Plaskett's suggestion in 1936 that granules looked similar to the quasi-steady convective cells observed by H. Bénard in laboratory experiments on liquids heated from below.

The desire to compare solar observations with laboratory experiments and theoretical models has led to careful measurement of several basic properties of granulation, which are summarized in Table 5-1. One is the appearance of a dominant granular spatial scale. The sense of motion of granular flows is also interesting. The lifetime of granules can be compared to the predicted overturning time of convective cells of comparable scale and velocity. The brightness and velocity contrast between the hot upflowing material and the cooler, darker downflowing lanes is important in estimates of the heat flux carried by the convection. Finally, the visibility near the limb is revealing since it provides information on the height profile of the granule trajectories and the heat flux they carry.

The property of supergranular flow that suggests convection is the horizontal velocity field which is observed to diverge from the center of a cell. This is the sense of flow seen also in granulation. However, there is no evidence for a radial upflow or downflow, nor is there any measurable temperature variation across the cell. The lifetime of supergranules is roughly consistent with the overturning time expected of a convective cell having the observed spatial scale and flow speed. As discussed below, supergranules may represent the surface effect of deeper convection, or they might be caused, for instance, by a dynamical instability caused by the radial gradient of solar rotation.

7.2.2   Comparison with Laboratory Measurements

The observed properties of granulation can be compared to extensive research on laboratory convection in liquids and gases. The early results of Bénard, which inspired Plaskett to describe photospheric granules as convective cells, have since been shown to be influenced more by surface tension in the shallow (a few millimeters) layers he used than by the buoyancy forces that drive free convection. To avoid such surface effects most laboratory experiments are now performed with rigid plates as upper and lower boundaries. Also, solar convection occurs in a highly compressible, stratified gas while laboratory experiments are performed on liquids. Limitations of this kind must be kept in mind when comparing solar granulation patterns with laboratory results, but certain features of the comparison are informative.

When a liquid layer is heated from below, convection initially sets in as two-dimensional, horizontal, parallel rolls. The three-dimensional pattern of cells reminiscent of granules does not replace the elongated rolls until the temperature gradient across the layer is an order of magnitude higher than the gradient required to initiate the two-dimensional convection pattern. The existence of a three-dimensional pattern of convection cells that is steady in time is limited to fluids whose viscosity v is relatively high compared to their thermal conductivity K. The ratio of these quantities is called the Prandtl number $P = v/K$.

Although it is difficult to generalize, since the experiments use different geometries, fluids of low Prandtl number ($P \sim 1$) show little evidence of steady three-dimensional convection. For these fluids, an increase of the temperature gradient across the layer usually leads to a nonsteady (i.e., turbulent) three-dimensional convective flow pattern. Laboratory results indicate that fluids of $P \gtrsim 10^{-2}$ should pass directly into a three-dimensional turbulent regime, with very little affinity for steady three-dimensional behavior. Since $P \sim 10^{-9}$ at the photosphere, if we use molecular values of v and K, we expect to find nonsteady behavior in the granulation if the temperature gradient is sufficient to drive free convection at the observed velocities and horizontal temperature differences.

The sense of motion in convection cells observed in laboratory experiments seems to depend on the sense of temperature dependence of the fluid viscosity. In gases, where viscosity increases with temperature, the laboratory convective flows are upwards at the cell edge and down at the center. This is opposite to the sense in solar granulation, whose flow direction may thus be determined by effects of stratification that are difficult to simulate in the lab.

The factor determining the preferred scale of granulation also does not follow clearly from experiments, in which not only the density but also the Prandtl number and temperature gradient tend to be constant within the layer. The variation of these quantities with height immediately below the photosphere probably determines the granular scale. But the radial profile of these quantities in standard models does not easily explain why the upper end of the granular size distribution should be so sharply defined.

The geometry of the granular cells is uncertain. The aspect ratio of horizontal scale/vertical depth is observed to be between 5 and 10 for unsteady convection in stratified natural systems such as clouds. This would suggest a depth of about 150–300 km for granulation, which is comparable to the most highly superadiabatic layers just below the photosphere. This depth is roughly consistent with interpretation of the granule lifetime as an overturning time over a horizontal distance of $(1400/2)$ km at the mean granular speed of about 1 km s^{-1}.

7.3   DYNAMICS OF SOLAR CONVECTION AND ROTATION

7.3.1   Condition for Onset of Convection

To determine the conditions under which we expect convection, we suppose that a small element of mass within a star in radiative equilibrium has density ρ, pressure P, and temperature T, the same as its surroundings at radial distance r from the center of the star (Fig. 7-4). If the temperature of the element is increased to a value T', the gas in this element will expand rapidly to achieve a new pressure equilibrium with its surroundings. The lower-density gas of this element will then experience a buoyancy force, which will cause it to rise. We may suppose that radiative or conductive heat flows out of the element can be neglected during the short rise time that it expands. The buoyancy force will disappear when the density has dropped to the same value as that of its new surroundings after an element has traveled a distance l. If we let T_1' be its new temperature and that of its surroundings T_1, then, assuming l is small, we have

$$T_1 = T + \left(\frac{dT}{dr} \right)_R l, \tag{7-1}$$

and

$$T_1' = T' + \left(\frac{dT}{dr} \right)_{ad} l = T + \Delta T + \left(\frac{dT}{dr} \right)_{ad} l \tag{7-2}$$

Here the subscripts "ad" and "R" represent adiabatic and radiative gradients. Since we have assumed radiative equilibrium, the "R" subscript also denotes the gradient actually present in the stellar atmosphere.

Two conditions of general interest can arise, depending on the relative value of the adiabatic and radiative temperature gradients. First if $T_1' > T_1$ or equivalently, since $\Delta T > 0$, if

$$-\left(\frac{dT}{dr} \right)_R > \left(-\frac{dT}{dr} \right)_{ad}, \tag{7-3}$$

then the element continues to expand further and rise. In this situation a convection pattern is established, and the radiative gradient is unstable.

On the other hand if $T_1' < T_1$, the element begins to contract, becomes heavier, and begins to move down to its original position. In this case the

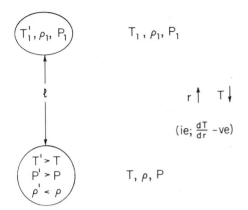

T, ρ, P

$r \uparrow \quad T \downarrow$

$(ie; \frac{dT}{dr} -ve)$

Fig. 7-4 Schematic diagram illustrating the Schwarzschild criterion for absence of convection.

radiative gradient is stable, and a displaced element will undergo oscillation about its equilibrium position.

The main conclusion is that the temperature gradient set up by radiative transport in a star can be no steeper than the adiabatic gradient in that medium before convection sets in. This criterion was first derived by K. Schwarzschild in 1906 and is named after him. Convection will tend to take over as the energy transport mechanism whenever $(-dT/dr)_R$ becomes large due to the large local opacity or strong local energy generation. The first effect is large just below the photosphere, where the opacity rises very rapidly due to increasing population of the $n = 3$ level of hydrogen since the Paschen continuum is the dominant opacity source in these layers. Convection can also set in when γ is lowered by ionization since reduction of γ lowers $(-dT/dr)_{ad}$. This effect predominates at greater depths in the convection zone.

The Schwarzschild criterion is conveniently expressed in terms of the relation between temperature and pressure for an adiabatic change

$$\frac{dP}{P} = \frac{\gamma}{\gamma - 1} \frac{dT}{T}, \tag{7-4}$$

so that

$$\left(-\frac{dT}{dr}\right)_{ad} = -\left(1 - \frac{1}{\gamma}\right)\frac{T}{P}\frac{dP}{dr}, \tag{7-5}$$

and the Schwarzschild criterion for absence of convection becomes

$$\left(1 - \frac{1}{\gamma}\right)\frac{T}{P}\left(-\frac{dP}{dr}\right) > \left(-\frac{dT}{dr}\right)_R. \tag{7-6}$$

7.3.2 Gravity Waves

When the atmospheric temperature gradient established by radiation is stable, i.e., when

$$\left(-\frac{dT}{dr}\right)_R < \left(-\frac{dT}{dr}\right)_{ad}, \tag{7-7}$$

then displacement of a gas element slowly enough to maintain gas pressure equilibrium with its surroundings leads to an oscillatory motion described as a gravity wave. The requirement for sufficiently slow displacement is met if the disturbance is much slower than the sound speed, so that acoustic waves can always maintain the pressure equilibrium. In a gravity wave the displaced elements of gas can be visualized as undeformed rods oscillating along their length at an angle θ to the local vertical. For slow motion the elements will be maintained in pressure equilibrium with their surroundings by sound waves, so that the density difference $\delta\rho/\rho$ is maintained in proportion to the temperature difference $\delta T/T$. In the adiabatic motion we consider here, this temperature difference is given by the difference in the atmospheric and adiabatic temperature gradients times the vertical displacement

$$\left[\left|\frac{dT}{dr}\right|_{ad} - \left|\frac{dT}{dr}\right|_R\right]\cos\theta. \tag{7-8}$$

The equation of motion of the element is then

$$\rho\frac{\partial^2 x}{\partial t^2} = -(\delta\rho)g\cos\theta \tag{7-9}$$

where

$$\frac{\delta\rho}{\rho} = \frac{1}{T}\left[\left|\frac{dT}{dr}\right|_R - \left|\frac{dT}{dr}\right|_{ad}\right]\cos\theta. \tag{7-10}$$

This yields an equation of motion for a simple harmonic oscillation

$$\frac{\partial^2 x}{\partial t^2} = -\omega^2 X, \tag{7-11}$$

where

$$\omega^2 = N_{BV}^2 \cos^2\theta \tag{7-12}$$

and N_{BV} is the so-called Brunt-Väisälä frequency

$$N_{BV} = \left[\frac{g}{T} \left| \frac{dT}{dr} \right|_R - \left| \frac{dT}{dr} \right|_{ad} \right]^{1/2}, \qquad (7\text{-}13)$$

which defines the natural oscillation frequency of a gas element in a stable atmosphere slowly displaced from equilibrium in a gravitational field.

As mentioned in Section 4.4.6, slow oscillations driven by buoyancy of the compressible plasma within the solar interior are of a type called internal gravity waves. A fluid element displaced slowly in a vertical direction will oscillate around its equilibrium position at the frequency N_{BV}. Typical values of N_{BV} in the interior are of order 10^{-3} rad s^{-1}, thus implying rather long internal gravity wave periods of close to an hour. Shorter surface gravity wave periods around 6 min are expected at the photosphere.

From equation (7-13) the value of N_{BV} defines the upper bound of gravity wave oscillation frequencies. More rapid oscillations in the gas bring into play pressure as a restoring force and generate sound waves.

7.3.3   Mixing Length Theory

The most elementary treatment of convection is based on the mixing length theory of heat transport by turbulence, due originally to L. Prandtl. This theory was first applied to the sun by L. Biermann, H. Siedentopf, and others in the 1930s. In this model we assume that heat is transported by the buoyancy-driven motion of fluid elements whose size is comparable to the local pressure scale height h. This element size l is called the mixing length since it is also assumed to determine the distance an element travels before merging with its surroundings and giving up (or absorbing, for downward-moving elements) its excess heat. The upward convective energy transport is due to incomplete cooling of a rising element before it merges.

The rate of convective heat transport can be calculated as follows. When an element of volume l^3 has risen a distance Δr, its enthalpy or heat content $\rho c_p l^3 T$ will differ from that of its surroundings by an amount proportional to its temperature difference and thus is given by

$$\rho c_p l^3 \left[\left(\frac{dT}{dr} \right) - \left(\frac{dT}{dr} \right)_{ad} \right] \Delta r. \qquad (7\text{-}14)$$

The convective flux F_c is the rate of transport of this enthalpy per unit area per unit time, thus

$$F_c = \rho c_p v l \left[\left(\frac{dT}{dr} \right) - \left(\frac{dT}{dr} \right)_{ad} \right], \qquad (7\text{-}15)$$

where $v = dr/dt$ is the velocity of an ascending buoyant element.

This velocity can also be calculated from the mixing length theory. If we assume that the element has been heated to a temperature in excess of its surroundings by ΔT, its density will then be less by the amount $\Delta \rho$ given by $\rho \Delta T / T$ since it is assumed to be in pressure equilibrium with the surroundings. It will then be accelerated upward by the buoyant force $g\rho l^3 \Delta T / T$. This buoyant force on an element of cross-sectional area l^2 transverse to v is balanced by a hydrodynamic drag given by $\rho v^2 l^2$. It follows from balance of the buoyant and drag forces that

$$v = \left[gl(\Delta T / T) \right]^{1/2}. \tag{7-16}$$

Since the temperature difference ΔT is difficult to predict a priori, we can express v in terms of the convective heat flux which is a more easily measured quantity. The quantity ΔT can be expressed as

$$\Delta T = \left[\left(\frac{dT}{dr} \right) - \left(\frac{dT}{dr} \right)_{ad} \right] l, \tag{7-17}$$

so, substituting for the temperature gradient difference from (7-15), we have, in terms of the convective heat flux F_c,

$$v^3 = \frac{glF_c}{\rho c_p T}. \tag{7-18}$$

This formula gives reasonable agreement (see exercises) between the sun's convective heat flux (taken to equal its photospheric radiative emittance since essentially all energy transport to the photosphere is by convection) and the observed sizes l and velocities v of photospheric granules. We also note that it predicts $v \propto l^{1/3}$, which is the distribution of velocities with turbulent element sizes that is expected from the spectrum of incompressible turbulence derived by A. Kolmogoroff in 1941.

The temperature difference ΔT between an ascending element and its surroundings then follows in terms of the heat flux carried, from equations (7-17) and (7-18), as

$$\Delta T = \frac{v^2 T}{gl}. \tag{7-19}$$

Again, adequate agreement with the brightness and temperature contrast of granulation can be achieved.

The degree of superadiabaticity required to carry heat through the solar convection zone can also be estimated from equation (7-15) using the above values of F_c, v, and l (see exercises). This superadiabaticity is a very small fraction ($< 1\%$) of the full temperature gradient dT/dr in the convection zone.

Thus the presence of convection reduces the temperature gradient from the higher value it would have assumed under radiative transport alone to the essentially adiabatic value.

Finally, it follows from the above estimates of v and l that the lifetime of a convective element should be of order $l/v \sim 10$ min, which is within the range of granulation lifetime estimates.

The mixing length theory outlined above provides a semi-empirical estimate of the efficiency of solar convection that has been calibrated against laboratory experiments and against observations of the luminosity and effective temperatures of other stars. A more precise formulation of the theory involves an empirical factor expressing the ratio $\alpha = l/H$ of the mixing length and scale height, rather than assuming $\alpha = 1$ as we have above. By changing the value of α, one then varies the efficiency of convective heat transport. The value of α is found to lie between 1 and 3 in models that are able to reproduce the correct diameter of the sun, which provides the main calibration of this parameter.

However, the theory embodies a number of uncertain basic assumptions such as the picture of convection in a stratified layer as a hierarchy of elements, each of dimension similar to the local pressure scale height, that are stacked vertically to carry heat through a star. Even if this geometry were correct, the shape of the elements, the effects of radiation on the energy transport, and of viscosity on the elements' motion, are either treated crudely or left to be specified empirically.

Mixing length theory seems to provide a good basis for estimates of the temperature stratification in the solar convection zone since all models of convection indicate that it is essentially adiabatic except very near the surface. However, the choice of adiabat depends on details of the convective efficiency near the photosphere and thus on uncertain details of the theory. These uncertainties influence the determination of how deep the convection zone actually is. More complete dynamical models, whose basis is outlined below, are required to calculate the convective motions. Such models are also needed to study the angular momentum transports that determine the sun's differential rotation and the interaction between solar rotation, convection, and magnetic fields.

7.3.4 Dynamics of Convection in a Plane Layer

The equation of motion governing convection in a viscous fluid is the Navier-Stokes equation (4-26) augmented by the gravitational body force term $g \, \Delta \rho$. This term is often written in the form $g\alpha(T - T_0)$ expressing the volume force of buoyancy. Here α is the coefficient of thermal expansion, and $T - T_0$ is the thermal perturbation from equilibrium. We then have

$$\rho \left(\frac{\partial \mathbf{v}}{\partial t} + \mathbf{v} \cdot \nabla \mathbf{v} \right) = -\nabla P + g\alpha(T - T_0) + \rho \nu \nabla^2 \mathbf{v}. \qquad (7\text{-}20)$$

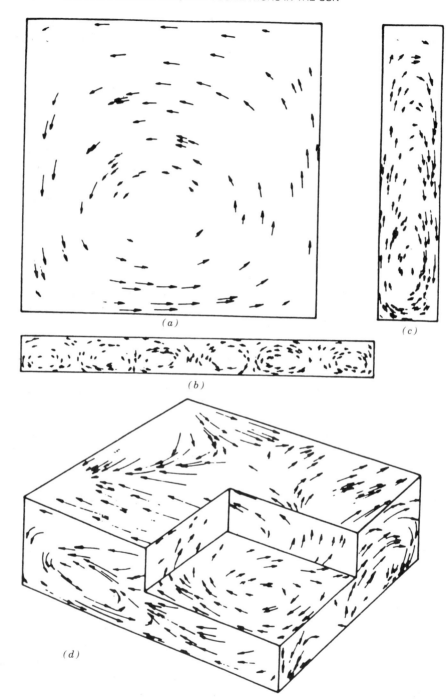

Fig. 7-5   Streamlines of numerically simulated two- and three-dimensional convection in rectangular containers of different height-to-width ratio. Reproduced with permission, from E. Graham, *J. Fluid. Mech.*, **70**, 689 (1975).

This equation of motion is supplemented by the continuity condition (4-34), the equation of state (4-3), and an energy balance relation which, in its most general form, is given by equation (4-54).

This set of equations contains all the physics involved in driving and damping convective motions in a nonrotating fluid. However, the complex nonlinear couplings implied in the above equations have made their general solution intractable until quite recently, when solutions were obtained by numerical techniques whose results are presented below. Since laboratory experimental results also do not extend to fully compressible convection in stratified layers, these numerical integrations are unique in affording a look at how solar convection might really behave.

Most of the results from this modeling are for two-dimensional convection in a polytropic atmosphere of adjustable (horizontal/vertical dimension) aspect ratio. The other variable quantities of most immediate interest are the density variation across the layer, the Prandtl number, and the vigor of convection measured by the Rayleigh number, which measures the ratio of the buoyancy force to the viscous drag in equation (7-20).

Typical two-dimensional circulation patterns found in these simulations are shown in Fig. 7-5. The key points to note are (i) a single circulation cell can fill the full depth of the layer whether the aspect ratio of the box is large or small, (ii) the aspect ratio of the circulation is about 2 : 1 (i.e., a cell is about half as deep as its full width), and (iii) the velocities at the bottom of the layer are about the same as at the top. These basic results were found to be more or less independent of the viscosity and temperature gradient over an order-of-magnitude range of Prandtl and Rayleigh numbers and were found also in three-dimensional simulations.

The finding that a single convection cell entirely filled the full depth of the layer is clearly at odds with the main postulate of the vertically stacked hierarchy of cells used in mixing length models. However, it must be kept in mind that for computational reasons, the simulations cannot yet handle convection in the regime of low Prandtl number and high Rayleigh number that is most likely to hold in stellar convection. Under the conditions of low molecular viscosity (and thus high Reynolds number) that are likely in stellar convection, motions are expected to be turbulent and may well be satisfactorily approximated by the mixing length approach.

7.3.5 Models of Granulation

The observations that granules form a cellular pattern with at least a well-defined upper limit on spatial scale, and that we see a good positive correlation between brightness and upward velocity are the main reasons for believing that granulation is free convection in a layer heated from below.

Interesting insights into the dynamics of granules can be obtained from semi-empirical models in which a simplified form of the hydrodynamic equations is solved using empirical information on the spatial and temporal scales of the flows, and on the properties of the radiation field. For instance, if we adopt a spatial scale for the granular flow a priori and adopt the observed brightness fluctuation at $\tau_{0.5} = 1$, we can predict the depth dependence of rms velocity, temperature, and pressure for a given temperature variation imposed at the base of the flow.

The results obtained from such a semi-empirical model can be adjusted to fit observed velocity and temperature profiles quite well. Interestingly, the rms horizontal velocities due to the outward transport of heat are expected to be about twice as large as the rms vertical velocities. Accurate observations of the granular velocity field's height dependence are difficult since the 5-min oscillatory velocity field must be removed. Probably most notable is the finding that the temperature fluctuations fall off much more rapidly than the velocities. That is, the upflowing convection loses its heat excess very rapidly by radiation within a few kilometers above $\tau_{0.5} = 1$, but the momentum of the plasma causes the granule to overshoot to several hundred kilometers above that level. Granular velocities are generally not detected at 500 km above $\tau_{0.5} = 1$.

Fully dynamical, time-dependent models with adequate treatment of radiative transport are required to obtain proper insight into the depth dependence of the driving forces behind granulation and to obtain any insight at all into the evolution of granules. Some results from a three-dimensional model of compressible convection are presented in Fig. 7-6, showing the temperature, vertical velocity, and pressure fields of simulated granules at several atmospheric levels.

The zero level in that model corresponds roughly to $\tau_{0.5} = 1$. One clear result of such simulations is that the true temperature differences between the hot, upflowing granules and the cool downflowing lanes are very large—about 5000 K. The corresponding intensity fluctuation one sees (even with infinite spatial resolution) is much smaller because of the strong temperature dependence of photospheric opacity. That is, in the hot granules, the opacity is higher, so we look at higher, cooler layers, and in the lanes we look deeper. The result is that around $z = 0$, the model predicts an intensity variation of only around 25–30% in green light.

A second surprising finding is that above the $z = 0$ layer, the temperature actually tends to anticorrelate with the upward velocity because of adiabatic cooling caused by rapid expansion of the rising gas. At deeper layers below $z = 0$ the most visible features are downflows confined to relatively small areas in a generally unstructured background of upflow. Better observations of granulation and its evolution will be required to determine how well such a simulation corresponds to the solar phenomenon it is intended to represent.

7.3.6 Dynamics of Supergranulation

Supergranulation does not exhibit the good correlation between upflows and excess brightness that weigh heavily in our identification of granules as an

TEMPERATURE   VERTICAL VELOCITY   LOG PRESSURE

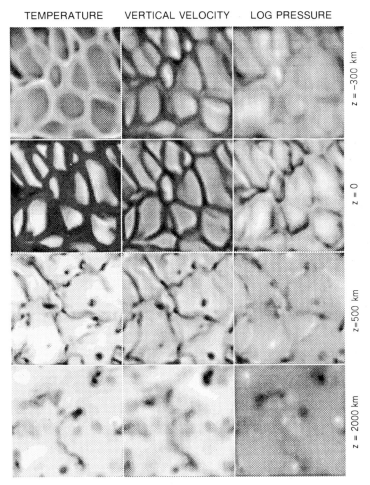

Fig. 7-6 Plots of the temperature, vertical velocity, and pressure fields in a numerical simulation of granulation, evaluated at four height levels in the photosphere. Distance downward into the sun is positive. Higher temperatures are indicated by lighter tones, as are also approaching velocities and high pressures. By permission of A. Nordlund.

efficient mode of free convection. As pointed out in Section 5.1.3, the absence of observable up- or downflows in supergranules does not necessarily violate the requirement that mass flux be conserved in an overturning circulation pattern. But the absence of an observable temperature gradient across a supergranule requires explanation if this motion is to act as an efficient mode of heat transport.

The most remarkable aspect of supergranulation is the discrete scale of the horizontal flow pattern. One possible explanation is suggested by the depths of first and second ionization of helium, which occur very roughly at 3500 km and 15,000 km. Given the recent evidence for a discrete mesogranular scale of about 7000 km in addition to the supergranular scale of 30,000 km, it is tempting to suppose that the local decrease in the adiabatic gradient at these depths increases the local superadiabaticity sufficiently to initiate flows with depths of about half of these two horizontal scales.

Other explanations have focused on the efficiency of convective modes and on the role played by magnetic fields. But more recent laboratory work and also the numerical simulations of compressible convection in stratified layers weaken the evidence that the most efficient modes necessarily dominate. A magnetic influence has been proposed on the argument that the volume-filling factor of magnetic flux tubes is likely to decrease rapidly outward near the photosphere, and the depth of this field-free zone defines a cavity in which supergranular convection can proceed relatively unhindered. Arguments of this kind are suggestive but not compelling.

The absence of a supergranular temperature variation at the photosphere could be due simply to a low heat transport efficiency of the basic convective pattern, if it is driven from deep layers where the degree of superadiabaticity is small. On the other hand, it might mean that temperature gradients present at the top of the pattern are smoothed horizontally by turbulent eddies such as granulation. It is worth considering that supergranulation may not represent free convection at all. It could be caused by a Kelvin-Helmholtz instability due to radial shear in the sun's rotation profile near the photosphere. Such ideas need to be investigated in more detail.

7.3.7 Angular Momentum Transport in a Rotating Spherical Shell

The pronounced differential rotation with latitude observed at the photosphere seems to be the result of convective flows driven radially by the buoyancy force and deflected horizontally by the Coriolis force due to solar rotation. The Coriolis force is given by an additional term $2\rho\omega \times \mathbf{v}$ in the right-hand side of equation (7-20). It deflects outflowing gas motions in the westward direction of rotation in the sun's northern hemisphere and eastward in the southern hemisphere. The specifics of this mechanism are still uncertain since the scales of convection responsible for heat flow toward the photosphere are unknown. It is thus difficult to determine whether the Coriolis effects act on slow global scale axisymmetric circulations in the meridional plane, on intermediate-scale eddies, or on a hierarchy of small turbulent eddies.

Models indicate that two basic processes contribute to the angular momentum transports in a rotating, convecting shell. The balance between them, integrated over all the scales of motion, should determine the radial and latitudinal profiles of angular velocity. These two processes are illustrated in Fig. 7-7. The first occurs if axisymmetric meridional circulations are present, as illustrated in the upper half of the figure. In the absence of any other angular momentum transport, a circulation in either sense will tend to spin up

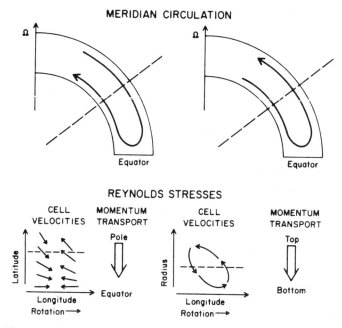

Fig. 7-7 Some mechanisms of angular momentum transport that may operate in the sun. By permission of P. Gilman.

the poles and the interior because the fluid carries angular momentum. Starting from solid body rotation, these regions will be spun up until the flux of ωr^2 is equal in both radius and latitude.

If a second process (see below) tends to enforce solid body rotation, then the circulation shown in Fig. 7-7 in which fluid flows downward at the equator could drive an equatorial acceleration. The reason is that, for equal velocity and cross-section of the flow in the meridional plane, the flux of angular momentum per unit mass, ωr^2, across the dashed line will be larger toward the equator than away from it because of the larger moment arm r in the outer branch of the circulation.

The second mechanism depends on the existence of nonaxisymmetric convective motions. When correlations are present between north-south or up-down motions and east-west motions, net fluxes of angular momentum in the latitudinal or radial directions are produced, without requiring a net mass flux. These correlations, which are illustrated in Fig. 7-7, are called Reynolds stresses. They play an important role in the dynamics of (particularly) turbulent fluids. The figure shows how velocity vectors typically found in numerical solutions can produce such angular momentum transports. The tilts of these vectors leading to Reynolds stresses are produced by Coriolis forces acting upon buoyancy-driven flows.

Note that neither buoyancy forces, which are strictly radial, nor pressure gradients, which must average to zero around the solar circumference, can themselves influence the sun's axisymmetric rotation profile. The actual solar profile of $\omega(r, \theta)$ depends on the total effect of the meridional circulations and Reynolds stress transports applied over the spectrum of convection scales in the sun.

Models have been constructed to calculate the growth and dynamics of the most unstable convective modes and explicitly compute the resultant Reynolds stresses. A large viscosity ν is invoked in many of these models on the assumption that random small-scale motions not resolvable in the simulation can be expected to carry momentum in the same way, but much more efficiently, than molecular motions alone.

The velocity pattern at low latitudes in such a high viscosity, nonaxisymmetric model resembles a "cartridge belt" configuration of convective rolls elongated in latitude, and distributed around the equator. An interesting feature is that the whole cell pattern tends to rotate prograde in the sun's frame at a rate a few percent faster than photospheric plasma rotation. The equatorial rotation produced can be more rapid than at higher latitudes, with a tendency in that case to decrease with depth.

Models of this kind have been constructed which obey the observational constraint of a very small pole-equator temperature difference and produce the observed sign and magnitude of latitudinal differential rotation. The plasma near the bottom of the convection zone at the equator is then found to rotate at about the same rate as the surface does at latitude 40°. This decreased rotation rate of about 10% is not seen in the studies of 5-min mode splitting described in Section 7.5.5. It is not yet clear whether the internal rotation indicated by these recent observations can be well described by adjustment of parameters and boundary conditions in such models, or whether a new look at the problem is required.

A second class of models is based on the idea that either the eddy viscosity or the energy transport efficiency (or both) of solar convection is likely to be anisotropic and exhibits a preferred direction relative to the radial direction or the solar rotation axis. Models based on this idea show that differential rotation can be produced without necessarily invoking high eddy viscosity. However, they have so far been less complete in their treatment of the heat flux and thus of the resulting pole-equator temperature difference.

A useful distinction between these two rather different approaches to the problem of convection in a rotating spherical shell will require some new observational approaches to measurement of the effective viscosity and thermal conductivity (and thus the Prandtl, Rayleigh, and Taylor numbers) of solar convection. Until this is achieved, models can only be suggestive of mechanisms operating in the solar interior, although they are very useful in classifying the kinds of effects that are likely to be important in such a complex system.

7.4 OBSERVATIONS OF SOLAR OSCILLATIONS

7.4.1 The 5-min Oscillations

Oscillatory motions on the sun were first reported in Doppler imaging observations of the photosphere by R. Leighton in 1960. The spatial distribution of this oscillatory velocity field can be seen in Fig. 7-8. At a given location, the velocity field exhibits a quasi-sinusoidal variation with an amplitude of hundreds of m s^{-1} and a period of around 5 min. The increased modulation seen near disk center in observations such as those in Fig. 7-8 indicates the motions are roughly radial.

Early attempts to study the distribution of the oscillations in the plane of horizontal wavenumber k_h against angular frequency ω showed some evidence of a discrete mode structure, rather than a continuous distribution. Here $k_h = 2\pi/\lambda_h$, where λ_h is the horizontal component of the wavelength, and $\omega = 2\pi\nu$ is the angular frequency in rad s^{-1}, where ν is measured in cycles sec^{-1} (hertz).

It was not until 1975 that the rich mode structure of the 5-min oscillations was resolved by the Doppler observations of F. Deubner which covered extended areas of the sun ($\sim 0.5R_\odot$) over several hours of observing. The distribution of power obtained from the most recent observations is shown in Fig. 7-9(a). Most of the oscillatory power is concentrated in the range of frequencies between 2.5 and 4.5 mHz and below wavenumbers of about 0.8 Mm^{-1} (i.e., at wavelengths greater than 8×10^3 km). The data used to construct this figure were measurements of the oscillatory brightness variation in a photospheric line, rather than the velocity oscillation measured by Leighton and by Deubner.

Observations of the 5-min modes of lowest wavenumber, having spatial scales comparable to the solar radius, were not obtained until 1979, by a group from the University of Birmingham, U.K. observing at Teneriffe in the Canary Islands and at Pic du Midi in the Pyrenees. Their technique was actually intended to detect global-scale solar velocity oscillations of much longer periods than 5 min. Disk-integrated solar light was used, which ensured that the high-k_h oscillations, consisting of many elements oscillating across the disk, would average out. The technique used to detect the tiny (m s^{-1}) velocities was optical resonance scattering, rather than the standard grating spectrographs used in previous studies. This novel type of spectrometer is particularly stable against drifts in the reference wavelength.

In a resonance spectrometer, a beam of sunlight is passed through a cell filled with a vapor having an absorption line corresponding to a strong Fraunhofer line, in this case the 7699 Å potassium line. The amount of light scattered onto a detector pointed transverse to the beam direction is determined by the exact position of the solar line relative to the line absorbed by the cell vapor. The cell is placed in a strong magnetic field oriented along the

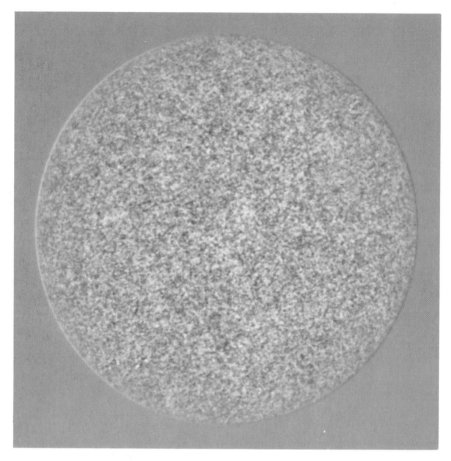

Fig. 7-8 Full-disk photospheric Dopplergram, obtained in the Ca K-line, of 5-min oscillations. Solar rotation and supergranular velocity fields have been removed. Based on data taken at the South Pole. By permission of T. Duvall.

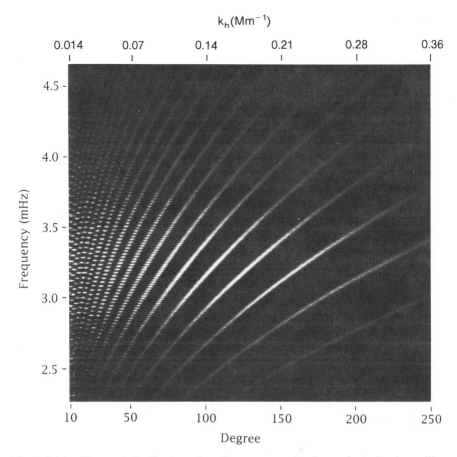

$$k_h(\text{Mm}^{-1})$$

Fig. 7-9 (a) Observed distribution of oscillatory power in photospheric 5-min oscillations plotted in the plane of frequency vs. wavenumber (top scale) or degree (bottom scale) plane. Derived from South Pole observations such as those illustrated in Fig. 7-8. By permission of T. Duvall.

incoming sunlight, so the cell line is Zeeman-broadened. By comparing the intensity of the scattered beam while looking sequentially at opposite senses of circularly polarized light, small time changes in the position of the solar line can be detected.

An example of the power spectrum of velocity oscillations thus observed in disk-integrated sunlight is shown in Fig. 7-10(a). It consists of a number of separate peaks about equally spaced in frequency. These observations were made over a continuous period of 120 h at the South Pole, where the sun stays well above the horizon in the summer. The discrete lines in the power spectrum resolved in this long continuous measurement set are just about

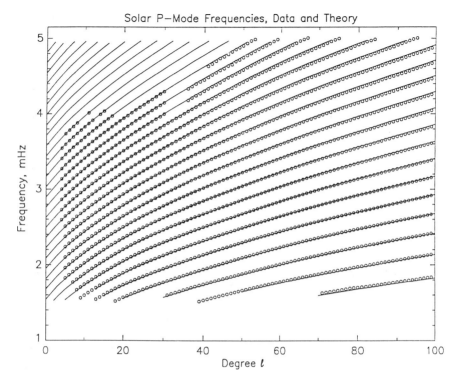

Fig. 7-9 (b)   The predicted frequencies (solid lines) calculated from a standard model of the solar interior by J. Christensen-Dalsgaard for comparison with the observed frequencies (circles) plotted in panel (a). By permission of K. Libbrecht.

uniformly spaced in frequency by $\Delta\omega = 68 \ \mu\text{Hz}$. The oscillatory power in the individual peaks corresponds to very small velocities—a few tens of cm s^{-1}.

A similar spectrum of disk-integrated 5-min modes of low k_h has also been observed as brightness oscillations in time variations of the total solar irradiance from space (Fig. 7-10b) and from balloons. The combined power of all the 5-min modes is about 10^{-5} of the solar constant. Figure 7-10(b) is based on 10 *months* of continuous observations and thus is about twice as long as the longest quasi-continuous velocity oscillation data base so far obtained.

7.4.2   Oscillations of Longer and Shorter Periods

Many detections of other oscillatory power peaks at periods longer than the 5-min modes have been reported. The power at frequencies below about 0.3 mHz (or periods exceeding 50 min) seems to be accounted for by errors expected from atmospheric transmission variations across the solar disk. These modulate the relative contribution of different photospheric areas to the Doppler signal. Given the large spatial gradients of the line-of-sight compo-

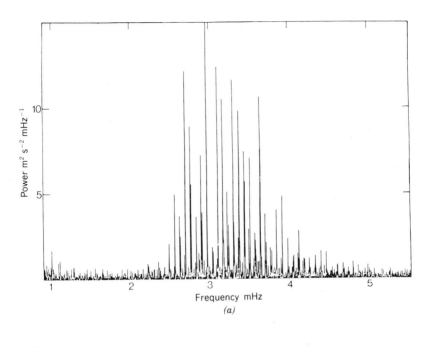

(a)

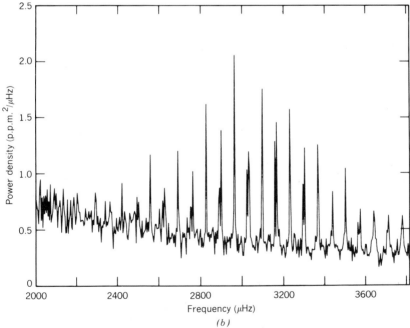

(b)

Fig. 7-10 Spectra of the photospheric 5-min oscillation observed in disk-integrated sunlight as (a) a velocity oscillation and (b) as a brightness oscillation in the total solar irradiance. Panels (a) and (b) are reproduced by permission of E. Fossat and M. Woodard respectively.

nent of the solar rotation across the disk, it is then easy to produce noise at the level below a few m s^{-1}.

Other techniques have been used to study the long-period oscillations. The first pioneering studies focused on oscillations in the sun's diameter. Measurements were carried out with a highly sensitive and stable interferometer which sensed the position of the limb using a rapid radial scan across the few tens of arc seconds inside and past the inflection point of limb intensity. The procedure and the data reduction technique used minimize the effects of atmospheric seeing on the limb position. The sensitivity to limb position changes is given as a few milli-arc seconds and thus a few kilometers on the sun!

The interpretation of the results given by the authors is that a flat white-noise spectrum is seen at frequencies exceeding 6 mHz, but that peaks at periods exceeding about 2–3 min are real. This claim has been questioned on the grounds that this power can be explained as a combination of the 5-min modes (which should cause a detectable signal in the diameter measurement) and a rise of atmospheric noise due to slowly changing atmospheric refraction properties across the disk. This ambiguity due to instrumental and especially atmospheric drifts has so far made the long-period modes a difficult subject, despite continued work to demonstrate their reality by analyses of phase coherence over many oscillatory periods.

Peaks of oscillatory power around 180 and 240 s seem to be observed in Doppler observations of chromospheric lines such as Hα or Ca K. Also, the 5-min modes are shifted to somewhat higher frequencies when a k_h-ω diagram such as Fig. 7-9(a) is constructed from observations in chromospheric lines. Observations of oscillations at frequencies much below 100 s are difficult because the wavelength becomes comparable to the photon mean free path in the upper photospheric line formation region. Detection of short-period waves would be of great interest for reasons of chromospheric heating discussed in Chapter 9. However, so far, results are contradictory.

7.5 INTERPRETATION OF SOLAR OSCILLATIONS

7.5.1 Resonances in the Sun

The discovery of the 5-min velocity oscillation in the photosphere was followed by a surge of theoretical work attempting to explain why oscillatory power around 5-min should be so heavily preferred over other periods. Interesting studies were carried out on wave generation, damping, and trapping, mainly in shallow near-photospheric layers. The fruitful idea that the 5-min modes of high k_h (the only ones then observed) might represent standing acoustic waves trapped in cavities extending well into the interior was put forward by R. Ulrich in 1970 and independently by J. Leibacher and R. Stein in 1971.

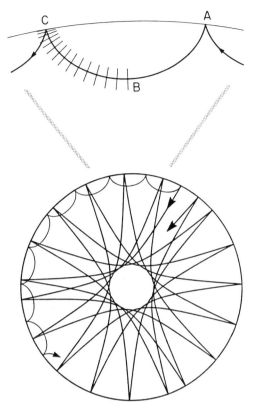

Fig. 7-11   Ray paths followed by standing waves of horizontal wavelengths roughly 1.5×10^6 km, 4.5×10^5 km, and 1.5×10^5 km. Note that the wave vector is radial at the photosphere and that the longer waves penetrate deeper. By permission of J. Toomre.

This interpretation was strikingly verified when F. Deubner first showed in 1975 that the distribution of power in the k_h-ω plane fell along well-defined narrow ridges whose shape and position agreed to a few percent with the rich mode structure calculated from the trapped acoustic wave hypothesis by H. Ando and Y. Osaki. This discovery ranks as one of the cleanest and most exciting results of astrophysics. A recent comparison of the observed and calculated frequencies is shown in Fig. 7-9(b). The uncertainties in the measured frequencies are considered to be typically about one part in 10^4 near the center of this figure.

In general, a perturbation of the hydrostatically supported plasma in the solar interior will generate a spectrum of acoustic (pressure, or p-mode) or gravity (g-mode) waves, with the dominant restoring force depending upon location in the sun and on the perturbation frequency. Acoustic waves can only propagate vertically at frequencies above their low-frequency cutoff given by

$$\omega_{ac} = \frac{v_s}{2H} = \frac{g}{2}\left(\frac{\gamma\mu}{RT}\right)^{1/2} \tag{7-21}$$

where H is the pressure scale height defined in Section 4.1.1.

At lower frequencies, below ω_{ac}, only gravity waves will propagate. The meaning of this acoustic cutoff frequency is readily visualized. If a force is applied slowly to a gravitational atmosphere, the atmosphere will simply lift and then relax back under gravity after the force has disappeared—this can be described as a gravity wave. If the disturbance is more abrupt, the gas is compressed before it can experience bodily acceleration, and a sound wave propagates.

From formula (7-21) the acoustic cutoff frequency can be seen to increase outward with r (i.e., at lower temperature and higher g). The increase of ω_{ac} causes acoustic waves of frequencies less than the photospheric ω_{ac} value (corresponding to a period of about 3 min) to be reflected back into the solar interior. The outward increase of ω_{ac} thus provides the upper boundary of the cavity required to trap acoustic waves in the sun.

The bottom boundary for sound wave reflection is provided by the condition for their total internal reflection. When an acoustic wave propagates into the solar interior, it moves into layers of increasing sound speed since $v_s \propto T^{1/2}$, so it will be refracted away from the local vertical unless it propagates exactly in the radial direction. Eventually the wave vector becomes deflected away from the vertical to the point where it can no longer propagate any deeper, leading to total internal reflection.

Thus nonradial acoustic waves are trapped in cavities whose upper and lower boundaries are determined by the frequency of the wave and by the oblique angle to the radial at which it propagates. The ray paths followed by waves of three different horizontal wavelengths are shown in Fig. 7-11. The depth of the cavity in which a given p-mode resonates is determined by the level at which total internal reflection occurs. This occurs at a depth given roughly by

$$\delta = \omega^2/(\gamma - 1)gk_h^2. \tag{7-22}$$

It follows that p-modes of high k_h are reflected at relatively shallow inner boundaries, while p-modes of longer horizontal wavelength penetrate more deeply. The depths of the resonant cavities for p- and g-modes are illustrated in Fig. 7-12.

Similar cavities exist for gravity waves. In the sun these are determined by the behavior of the Brunt-Väisälä frequency defined in Section 7.3.2 since a gravity wave's frequency must lie below the value of N_{BV} when N_{BV} is real. A gravity wave is prevented by reflection from propagating into a region where $\omega > N_{BV}$ or N_{BV} is imaginary. A cavity for such waves exists in the sun's

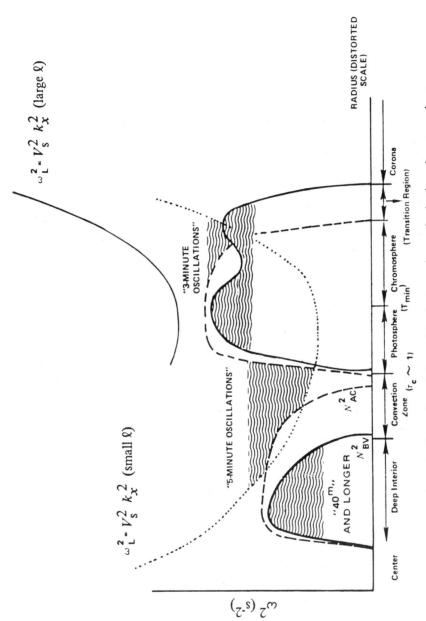

Fig. 7-12 Schematic illustration of the cavities for p- and g-modes in the solar atmosphere and interior (by permission of J. Leibacher).

233

radiative core (see Fig. 7-12), where the outer boundary is set by the convection zone, in which N_{BV} is imaginary, while near the sun's center $N_{BV} \to 0$ since $g \to 0$ as the mass $M(r) \to 0$. The internal gravity waves trapped in this deep cavity have periods greater than about 50 min. They are expected to have relatively low amplitude at the photosphere since they are evanescent in the convection zone, as explained in Section 7.3.2. However, the conditions on gravity wave reflection given above suggest that cavities appropriate for trapping surface gravity waves also exist around the temperature-minimum region of the photosphere and between the chromosphere and corona, as illustrated in Fig. 7-12.

7.5.2 Oscillation Modes of the Solar Interior

The first type of oscillation observed in stars was the radially symmetric large-amplitude contraction and expansion ("breathing") of Cepheid-type variables. The period of the sun's fundamental radial mode is easy to estimate. This mode is trapped between the center (a node must occur at $r = 0$ to satisfy radial symmetry) and the surface, where the mode is reflected from the pressure discontinuity since the pressure scale height becomes much smaller than the wavelength ($\lambda \sim R_\odot$) of the mode. Thus the solar interior contains a quarter wave length of the fundamental. The travel time $4R_\odot/v_s$ which determines the oscillation period can be found from an estimate of the mean sound speed v_s in the solar interior. It can be shown from dimensional arguments (see Chapter 6 exercises) that

$$\bar{P} \sim \left(\frac{GM_\odot}{R_\odot^2}\right)\rho R_\odot, \qquad (7\text{-}23)$$

so that

$$v_s = \left(\gamma \bar{P}/\bar{\rho}\right)^{1/2} \sim \left(\gamma GM_\odot/R_\odot\right), \qquad (7\text{-}24)$$

and the period of the fundamental radial mode is

$$\tau \sim 4\left(R_\odot^3/\gamma GM_\odot\right)^{1/2} \sim \left(G\bar{\rho}\right)^{-1/2} \qquad (7\text{-}25)$$

or about 60 min.

More generally, the normal modes of mechanical vibration of the sun include nonradial as well as radial oscillations. These normal modes are described by a solution of the equations of stellar structure, perturbed by a small change from the equilibrium solution that describes the steady state of the star. From our discussion above of resonant cavities in the solar interior, we are led to seek solutions of an eigenvalue problem in the form of standing waves trapped in the solar interior.

T. Cowling showed already in 1941 that perturbations of the linearized equations for a nonrotating, spherically symmetric star produce oscillations that can be described in terms of spherical harmonic solutions of a wave equation. The classifications of pressure (p-), gravity (g-), and fundamental (f-) modes were first made in his paper. In analogy with the quantum mechanical solutions for the motion of an electron in a central force field, the solar oscillations represent the allowed eigenfunctions in the potential well determined by the solar gravity field.

The oscillation in space and time of a quantity such as the density is then described as

$$\rho' = \rho'_{nl}(r)Y_{lm}(\theta, \phi)\exp(-i\omega t), \tag{7-26}$$

where $\rho'_{nl}(r)$ gives the radial structure of the mode of order n and degree l (the radial structure has the form of a standing wave in r) and $Y_{lm}(\theta, \phi)$ is a spherical harmonic function defined as

$$Y_{lm}(\theta, \phi) = P_l^m(\cos \theta)\exp(im\phi). \tag{7-27}$$

Here $P_l^m(x)$ is the associated Legendre polynomial given as

$$P_l^m(x) = (-1)^m (1 - x^2)^{m/2} \frac{d^m}{dx^m} P_l(x). \tag{7-28}$$

A given mode is thus defined in terms of three indices. The order n defines the number of nodes in the radial wave function excluding those at the center or outer surface, so the f-mode has $n = 0$. The degree $l = 0, 1, 2, 3, \ldots, (n - 1)$ is the number of nodal lines on the solar surface, and we have

$$\lambda_h l \sim 2\pi R_\odot, \tag{7-29}$$

so that modes of high degree correspond to short horizontal wavelengths at the photosphere. Since $k_h = 2\pi/\lambda_h = \sqrt{l(l + 1)}/R_\odot$, the 5-min oscillations of high horizontal wavenumber k_h plotted in Fig. 7-9 correspond to modes of l up to about 250 and are referred to as high-degree p-modes. The 5-min oscillations whose spectrum is given in Fig. 7-10 represent p-modes of low degree ($l \leq 3$). The order of both the high- and low-l p-modes has been determined to correspond to n between 1 and 20 from observations described below.

The index $|m| \leq l$ describes the distribution of the nodal lines on the surface (i.e., parallel to the equator or to meridians relative to a coordinate system defined by the axis of solar rotation. Figure 7-13 illustrates cases of

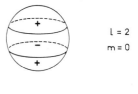

L = 2
m = 0

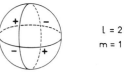

L = 2
m = 1

Fig. 7-13 Schematic illustration of oscillation modes corresponding to $l = 2$ spherical harmonics with $m = 0, 1, 2$ (by permission of E. Fossat).

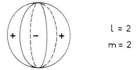

L = 2
m = 2

$m = 0, 1, 2$ for $l = 2$. In the absence of rotation, modes of different m are degenerate, i.e., of the same frequency. In the presence of rotation, modes of the same n, but of different m, are split into $2l + 1$ components in frequency due to their differing spatial oscillation pattern relative to the axis of rotation.

The reason for the observed rotational splitting is easy to understand. A standing wave arises from the reflection of a propagating wave trapped between boundaries placed an integral number of half-wavelengths apart. It follows that the pattern of standing waves established at the photosphere by the solar oscillations can be regarded as composed of two such identical waves moving in opposite directions. When the medium is in motion, as in solar rotation, the waves are carried with it at velocity v, so that the wave propagating in the sense of motion has its observed frequency Doppler shifted upward by an amount

$$\frac{\Delta\omega_+}{\omega} = \frac{v}{v_p}, \tag{7-30}$$

where v_p is the wave's phase speed. The oppositely directed wave undergoes a decrease of frequency $\Delta\omega_-$ of the same magnitude, so the two waves' frequencies are split by an amount $2\,\Delta\omega$.

The magnitude of this splitting depends on the mode. It is relatively large for the high-m p-modes and less for the low-m modes. The result is a slow but measurable drift of the whole standing wave pattern across the sun (in the direction opposite to solar rotation) due to the slight mismatch of frequencies between the oppositely propagating waves.

7.5.3   Excitation and Damping Mechanisms

The reader will wonder why most of the solar p-mode power is concentrated in a narrow band around 5-min, given all the p-mode periods that might be excited below that of the f-mode. So far there is no clear answer to this question. To make progress we calculate how and where in the sun the oscillations are excited and damped. External excitation of the p-modes could occur through generation of acoustic noise by the convection zone motions. This would be an intermittent process, and we might expect to see time variations in the amplitude of at least the high-l modes in response to changes in this buffeting.

Self-excitation might occur through a steady process such as the so-called kappa-mechanism. This operates when the opacity of a gas increases with its temperature. When such a gas is compressed adiabatically (and thus heated), its opacity rises so that it extracts more energy from radiation flowing through it. This extra heat is available in subsequent expansion of the gas and acts to amplify any initial oscillatory motion. The excitation of internal gravity waves might occur by nuclear burning instabilities in the solar core (see Section 6.5).

Either of the above p-mode excitation mechanisms, or both, might operate in the sun, and a damping mechanism is required to limit their amplitude. This is supplied in part by radiation, since the wave's compression and heating lead to radiative losses (which are neglected for acoustic waves in Section 4.4.2, where the amplitudes are assumed small), and in part by viscosity. Non-linear interactions between modes may also prove important. The amplitude of a mode is determined by a balance between excitation in some region of the solar interior and damping over a volume of the solar interior that in general will be different from the region of excitation.

The excitation and damping of the modes can be studied by analyzing the power spectra for evidence of time decay and re-excitation in the amplitude of individual peaks, and also from the width of individual modes in the k_h-diagram. Results reported from the study of low-l p-modes indicate exponential decay of individual spectrum peaks with an e-folding time of several days and suggest intermittent excitation of such global modes every few days. This evidence is in agreement with the studies of the width of modes of given n and l and suggests that the modes maintain coherence over many cycles and can justifiably be referred to as global oscillations. Similar behavior is found for the high-l modes from measurement of their width in the k_h–ω plane, although the global nature of these modes is less clear, since many more cycles of coherence are required, at their shorter horizontal wavelength.

7.5.4   Comparison of the Observed and Calculated Properties of the p-Modes

The basic parabolic shape of p-mode ridges seen in Fig. 7-9 is determined by the condition that the cavity contain an odd number of quarter-wavelengths, so that

$$\left(n + \frac{1}{2}\right)\pi = \int k_z \, dz = \omega \int \frac{dz}{v_s}. \tag{7-31}$$

Using relation (7-22), we obtain

$$\left(n + \frac{1}{2}\right)\pi = \left(\frac{2\omega^2}{\gamma - 1}\right)gk_h \tag{7-32}$$

or

$$\omega_n^2 \simeq \left(n + \frac{l}{2}\right)gk_h \tag{7-33}$$

which yields the predicted $\omega \propto k_h^{1/2}$ dependence first observed by Deubner.

The p-mode ridges which are separated temporally in ω consist of individual modes in the spatial k_h coordinate as well, as seen in Fig. 7-9. The first observations of individual p-modes resolved in both ω and k_h were the low-l p-modes (Fig. 7-10) detected in full-disk integrated light using resonance cells. These modes of low l have larger spacing in k_h and can be seen as individual power spectrum peaks, even in one-dimensional power spectra in ω such as Fig. 7-10. Their frequencies (in mHz) in the case of high order n are given approximately by

$$\nu_{n,l} \simeq (n + l/2 + k)\,\Delta\nu \tag{7-34}$$

Here k and $\Delta\nu$ are constants computed from the solar model, and the separation between adjacent n-values should be $\Delta\nu \sim 135 \ \mu$Hz. It follows that

$$\nu_{n+1,l} - \nu_{n,l} \sim \Delta\nu \tag{7-35}$$

and

$$\nu_{n,l+2} \sim \nu_{n+1,l}. \tag{7-36}$$

It can be shown that except at very low n-values, successive modes of equal l are equidistant in frequency. This is what one observes in the power spectrum of these modes, illustrated in Fig. 7-10. Further, relation (7-36) leads us to expect two sequences of odd or even l with almost coincident frequencies. The odd-l frequencies lie about midway between even-l modes.

A convincing identification of the l-values of these low-degree p-modes can be made by dividing the power spectrum of Fig. 7-10 into equal segments of $\Delta\nu = 136 \ \mu$Hz and superposing them to increase the signal-to-noise ratio. This is done in Fig. 7-14, which shows that the positions of peaks agree very well with their calculated positions. That is, if we take the $l = 0$ peak as reference, the even-series peaks of $l = 2$ and the weaker $l = 4$ follow quite closely at the

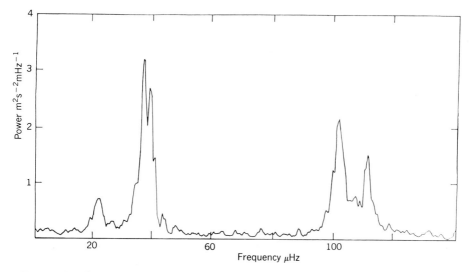

Fig. 7-14 The frequency range between 2.4 and 4.8 mHz for low-l modes, obtained by superposed averaging (by permission of E. Fossat).

slightly higher predicted frequencies, while the first odd-l peak, $l = 1$, lies about midway between $l = 0$ and the frequency of the peak of next higher order n and $l = 0$. The odd-series $l = 3$ peak also appears weakly at slightly higher frequency than the $l = 1$.

In general, the identification of the mode responsible for a given line in the power spectrum of the low-l p-modes, or for a given ridge in the k_h-ω plot of the higher-l modes, has been carried out by trial-and-error comparison between the calculated and observed spectra, as illustrated above for the low-l modes. When the k_h-ω ridges of $l \leq 100$ were resolved into individual modes, it became possible to evaluate the order n of each ridge simply from the number $n - 1$ of individual l-values that it contains. The radial order n can be deduced for the low-l modes from the decrease in l for the lowest-n modes.

7.5.5 Oscillations as a Probe of the Solar Interior

The close agreement between the calculated and observed ridges in the k_h-ω diagram encourages the view that analysis of the residual differences might be used to improve our understanding of solar interior models. The technique has been used so far to test the thermal structure of the convection zone, to determine its depth, and to place limits on the radial profile of rotation in the solar interior.

The direct approach is to calculate the mode structure in the k_h-ω diagram expected for a standard solar interior and compare it with observations. Optimal fit to the data is then sought by varying the parameters of the model by trial and error. This is the method used most effectively so far and Fig. 7-9(b) shows some of the most recent results. A more elegant approach is to solve the inverse problem where the relations that determine the modes of a standard model are inverted to give a solution for the temperature structure directly from the mode spectrum. This is the technique used by seismologists, who have at their disposal time-of-travel information derived from individual pulses (earthquakes). In the solar situation the inverse approach is less well defined because only spectral (not phase) information is available so far. In principle, pulse response information might be obtained if the oscillations that are expected to be excited by a flare could be measured, but this has not yet been achieved.

Application of the direct approach to study of the convection zone temperature structure is based on the relation of the mode period τ, cavity depth δ, and sound speed v_s given by

$$\tau = 2 \int^{\delta} \frac{dz}{v_s}. \tag{7-37}$$

In the first comparisons it was found that the observed frequencies were a few percent lower than the values calculated from a standard convection zone model with mixing ratio $l/H = 1.0$. The best fit is now found at $l/H = 2$, which implies more efficient convection and thus a slower temperature decrease (smaller superadiabatic gradient) in the outermost layers. The result is a model with a lower temperature throughout the convection zone and also a deeper convection zone extending down to $0.7R_\odot$. Both effects will, from relation (7-37), tend to lower the frequency of the calculated modes.

None of the models is able to reproduce the observed frequencies to the precision of better than 0.1% now achieved in the observations. Analysis of the errors suggests that the models are incomplete in the outer few percent of the solar interior, the only region common to all the modes studied. This is also the region where crudely parameterized, spherically symmetric models might be expected to produce largest errors since temperature gradients are highest. Also, magnetic obstacles in the form of active regions tend to produce temperature inhomogeneities that reflect, refract, and absorb the acoustic waves differently than does the nonmagnetic medium.

The measurement of the solar internal rotation from p-mode splitting is based on the principle outlined in Section 7.5.2, using also the fact that p-modes of progressively lower l penetrate deeper so their splitting is indicative of conditions nearer the center of the sun. Given the contribution functions in depth of the individual modes, the radial profile of solar rotation can then be reconstructed from a sufficient number of p-mode splitting measurements. The first attempt to apply this technique used high-l modes of $|m| = l$. Since the p-mode splitting in frequency is proportional to m, these shifts were relatively easy to measure, but since the high-l modes only penetrate to shallow layers, they give information only on rotation quite close to the photosphere. The results indicated an increase in angular rotation rate of about 5% in the first 15,000 km. This result is in good agreement with the analysis of tracer rotation (see Section 7.1.2), and it is consistent, to within errors, with more recent analyses using a wider range of l-values.

Considerable attention has focused on analysis of the low-degree p-modes, which provide information on rotation in much deeper layers. However, their splitting is much smaller since m is greatly reduced. No consistent picture has emerged yet, and there is doubt that the small splitting of these modes can be measured without spatial resolution given the finite lifetime (and Q-value) of the 5-min modes, which determines their width.

Several analyses of low- to intermediate-degree p-modes ($l < 100$) with spatial resolution across the disk have recently been carried out. These analyses indicate that within a shell extending between $0.4R_\odot$ and almost $1R_\odot$ from the sun's center, the rotation rate exhibits no significant radial gradient. The dependence of angular rotation rate on latitude within this shell appears to follow about the same differential rotation law as the photosphere. There is some evidence that the plasma within $0.4R_\odot$ of the sun center rotates rigidly at a rate intermediate between the photospheric rotation rates of the pole and equator. The most interesting result of these measurements is that they show no sign of the tendency to rotation at constant angular velocity on cylinders that is predicted by the most completely studied models (Section 7.3). Such a rotation law would imply a substantial inward decrease of angular rotation speed that lies outside the error bars of these observations.

A clear detection of g-modes would greatly improve diagnostic techniques for the deep interior, since p-mode frequencies are influenced mainly by conditions relatively near the photosphere. So far, the difficulties of detecting the g-modes against the noise of instrumental drifts, and of atmospheric and solar noise sources in the low-frequency regions where their signal is expected, have precluded their clear measurement. Longer data records are required to make use of the long coherence times predicted for these modes.

ADDITIONAL READING

Observations and Theory of Solar Rotation

R. Howard, "Solar Rotation," *Ann. Rev. Astron. Astrophys.*, **22**, 131 (1984).

E. Schröter, "Solar Differential Rotation," *Solar Phys.*, **100**, 141 (1985).

P. Gilman, "Observations and Theories of Solar Convection, Global Circulation and Magnetic Fields" in *Physics of the Sun*, P. Sturrock et al. (Eds.), D. Reidel, p.   , 1985.

B. Durney, "On Theories of Rotating Convection Zones" in "Solar Instrumentation: What's Next?" R. Dunn (Ed.), Sacramento Peak Observatory Publication (1981).

J. Hart et al., "Laboratory Experiments on Planetary and Stellar Convection Performed on Spacelab 3." *Science* **234**, 61 (1986).

Solar Convection

R. Bray, R. Loughhead and C. Durrant, *The Solar Granulation*, Cambridge University Press, 1984.

A. Nordlund, "Solar Convection," *Solar Phys.*, **100**, 209 (1985).

Helioseismology

J. Leibacher, R. Noyes, J. Toomre and R. Ulrich, "Helioseismology," *Scientific American*, **253**, 48 (1985).

J. Leibacher and R. Stein, "Oscillations and Pulsations," in *The Sun as a Star*, NASA publication SP-450, S. Jordan, (Ed.), p. 263 (1981).

F. Deubner and D. Gough, "Helioseismology: Oscillations as a Diagnostic of the Solar Interior," *Ann. Rev. Astron. Astrophys.*, **22**, 593 (1984).

T. Brown, B. Mihalas and E. Rhodes, "Solar Waves and Oscillations," in *Physics of the Sun*, P. Sturrock et al. (Eds.), D. Reidel, p. 177, 1986.

K. Libbrecht, "Solar and Stellar Seismology," *Space Sci. Rev.*, **47**, 275 (1988).

EXERCISES

1. Compare the velocities and temperature differences observed in granulation with those calculated assuming a value of $\alpha = 1$ for the mixing length parameter. Also calculate the superadiabatic temperature gradient expected at $\tau_{0.5} = 1$ and also at the convection zone base with the above assumption. Compare your values with the total temperature gradients at these locations, determined from photospheric and convection-zone models in Chapters 5 and 6.

2. Estimate the relative magnitude of the Coriolis force term and inertial term for typical solar flows such as granules, supergranules, and possible larger-scale convective flows of $v \sim 10$ m s^{-1}. Show that rotation is unimportant in granulation dynamics but significant at larger scales. Show also that for a steady, inviscid flow in a rotating fluid the equation of motion reduces to $2\omega \times v = -\nabla P/\rho$, which implies constant angular velocity along cylindrical surfaces parallel to the rotation axis. This is known as Proudman's theorem and explains the tendency to constant angular velocity along cylinders seen in the models of solar internal rotation. (See, for instance, G. Batchelor, *An Introduction to Fluid Dynamics*, Cambridge University Press, 1967, p. 558.)

3. The effect of stellar rotation and centrifugal acceleration upon a star's internal energy function was discussed by G. Sweet [*Mon. Not. Royal Astr. Soc.*, **110**, 548 (1950)]. The slightly lower pressure in the equatorial plane relative to the poles gives rise to a slow circulation, even in a purely radiative atmosphere. Consult this reference and comment on the speed of the flow for the sun and its likely importance for mixing the solar interior.

4. It is tempting to reason that the radial profile of angular velocity in a rotating gaseous atmosphere (such as that of the Earth or sun) should be given by conservation of angular momentum in molecular motions. Show that if that were the case, the sun's profile would be given by $w(r) \propto r^{-2}$. Observations show that the Earth's atmosphere rotates roughly rigidly $[\omega(r) \sim \text{const}]$ for reasons that were not made clear until quite recently [see F. Bretherton and J. Turner, *J. Fluid Mech.* **32**, 449 (1968)]. Discuss the effects operating, and their relevance to the sun's internal rotation profile, given the primary mechanisms that seem to determine the profile of internal rotation in the sun.

5. Refer to Fig. 7-9 and use equation (7-22) to calculate the reflection depths of a few points on each p-mode ridge. Thus construct lines of equal penetration depth (isobaths) on this diagram. What l values would be appropriate to observe the influence of active regions on the convection zone structure in a shallow ($\leq 10^4$ km) layer near the photosphere? Is this a practical observation, given the relative spatial scale of, e.g., spots and of these high-l modes?

8

Observations of Photospheric Activity and Magnetism

The study of sunspot structure and of the statistical behavior of spot groups constituted the first research on solar activity. Early observers also noticed that bright, irregular patches called faculae are seen in the vicinity of spots near the limb. In later filter or spectroheliograph observations made in chromospheric lines the white-light faculae were seen to be overlaid by bright areas called plages which are visible across the whole disk. These narrow-band observations also revealed elongated dark filaments in the vicinity of the spots and plages.

The extended areas where spots, faculae, plages, and filaments occur together are called active regions. These are also the locations of the powerful explosions observed as flares. Even well away from active regions, narrow-band observations in chromospheric lines show huge prominences, intricately structured atmospheres of relatively cool gas often standing hundreds of thousands of kilometers high in the corona. The transient phenomena in active regions and the prominences are referred to collectively as solar activity.

It is now well established that the solar magnetic field is the underlying cause of solar activity. But many fundamental problems, such as the reasons for spot coolness and for facular brightness, still elude a satisfactory solution. The mechanism responsible for the sun's remarkable magnetic cycle is also not well understood, although a number of fundamental effects that must play a role in amplifying the field and switching its polarity have been identified. The discovery of starspots, faculae, and magnetic cycles on other stars has brought new observations to bear on the topic of solar activity and opened new possibilities for applying the advances in understanding gained from studying these phenomena on the sun.

In this chapter, we discuss the structure of spots and faculae (the two manifestations of solar activity observed at the photosphere) and the observed behavior of the photospheric magnetic field underlying the activity phenomenon. The behavior of chromospheric and coronal activity such as filaments, prominences, and flares, and their magnetic fields are described in Chapters 9 and 10. The solar activity cycle is discussed in Chapter 11, along with the dynamics of solar magnetism.

8.1 SUNSPOT OBSERVATIONS

8.1.1 Structure of the Umbra and Penumbra

A typical sunspot observed in white light consists of the darkest central area called the umbra surrounded by a less dark, roughly annular region, called the penumbra. Sunspots vary widely in their structure, as can be seen from the examples in Fig. 8-1, and even large spots can be found without penumbrae altogether. Also, the shapes of umbrae and penumbrae are often highly irregular, so that the prototype of a round umbra surrounded by a regular annular penumbra is the exception rather than the rule.

The umbral diameter of the largest spots can exceed 20,000 km, with a more typical value about half that. Its area is on average about 20% of the total area of the spot, although the dispersion in this number is considerable. The smallest dark structures, below about 2500 km in diameter and lacking penumbrae, are called pores. These are noticeably less dark than spots, and their photometric contrast is more similar to that of granulation lanes. Some pores can be seen in Fig. 8-1. They have typical lifetimes of a few tens of minutes, much shorter than spots.

The structure of some spots includes lanes of bright material intruding into the umbra, called light bridges. A prominent example can be seen in Fig. 8-2. The intrusion occurs over many hours or days, and it is remarkable that the spot is stable enough to sustain itself during their formation. With sufficient exposure to bring out the umbra and care to avoid photospheric scattered light, photographs of spots show granulation within the umbral area, as illustrated in Fig. 8-2(a). Such umbral granules are about 30% smaller than the photospheric variety, but they have been observed to last for hours, and their mean lifetime is three to four times that of photospheric granules. Their presence indicates that the kilogauss-intensity vertical magnetic field of the umbra does not entirely suppress convection. Even smaller bright structures called umbral dots can also be seen on overexposed umbral photos such as Fig. 8-2(a). Their diameters are only a few hundred kilometers, and their lifetimes are a few tens of minutes. Several dozen can be present in an umbra at a given time.

The penumbra typically appears in low-resolution pictures as a gray annulus surrounding the black umbra, extending the spot diameter to 50,000 km or more in the very largest spots. Its general shape is often irregular (see Fig. 8-1) and depends in part on the evolutionary stage of the spot, as will be discussed later, but its outer edge is always remarkably sharp. At the spatial resolution

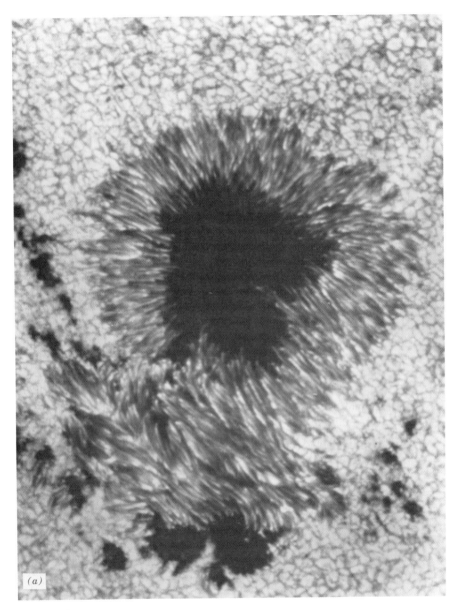

Fig. 8-1 High-resolution white-light images of (a) a large sunspot and neighboring pores and (b) a group of smaller spots. Both pictures obtained at the Pic du Midi Observatory. By permission of R. Muller.

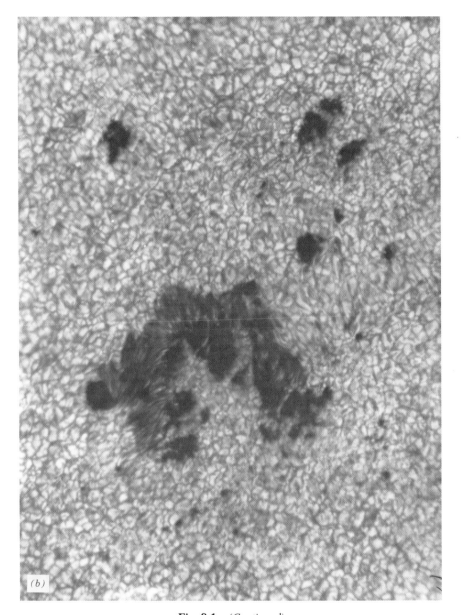

(b)

Fig. 8-1   *(Continued)*

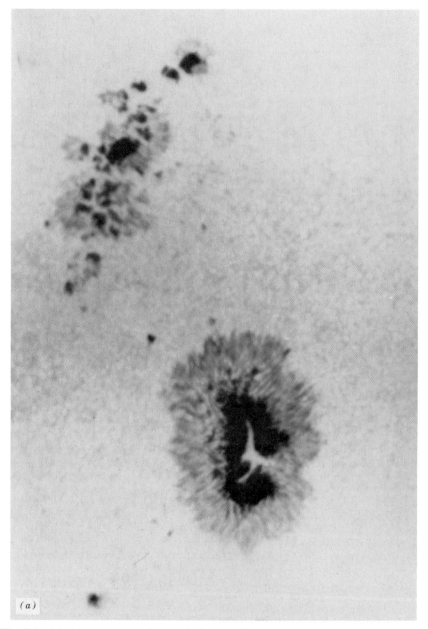

(a)

Fig. 8-2   (a) A large umbra with a light bridge. The same umbra is shown overexposed in (b) to illustrate umbral granules. By permission of P. McIntosh.

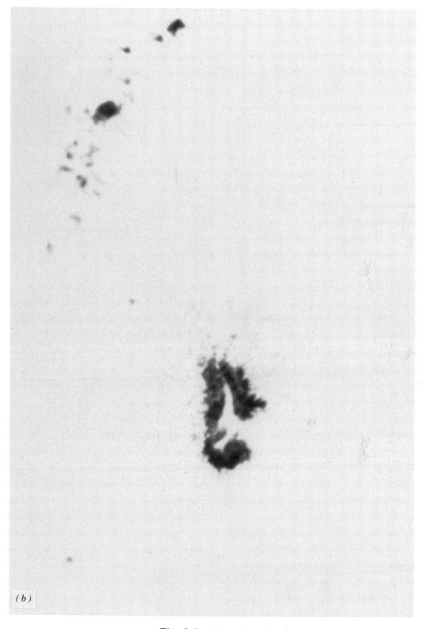

(b)

Fig. 8-2 *(Continued)*

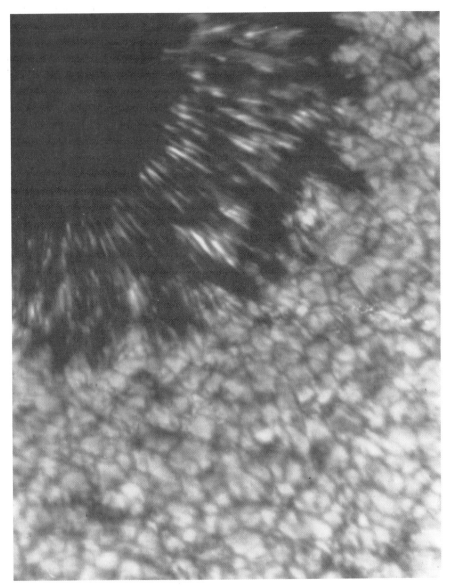

Fig. 8-3 Detail of a sunspot penumbra and of nearby granulation. Big Bear Solar Observatory.

of Fig. 8-1 the penumbra is resolved into mainly radial fine structures that appear to be elongated dark and bright filaments. Under enlargement of the very best pictures of spots, such as Fig. 8-3, the bright filaments are shown to consist of alignments of tiny, somewhat elongated bright grains. These penumbral grains occupy roughly half the penumbral area and drift inward toward the umbra during their lifetime of roughly 1 h. As they approach the umbra, they increasingly resemble the umbral dots, suggesting a common nature as small regions where convection is able to penetrate between magnetic field lines.

This interpretation is supported by observations that the dots and grains have brightness similar to or exceeding the photosphere and exhibit upward motion (as do bright granules) and lower magnetic field intensity than their surroundings. The geometry of the dark photospheric structures in penumbrae is less clear. The best pictures indicate that at least some of the dark structure consists of radial filaments that in some cases overlie the bright grains. But there also seems to be a general dark background of penumbral material.

8.1.2 Birth and Evolution of Spot Groups

A spot is born by the darkening and growth in diameter of a pore, turning it into an umbra. Only a small fraction of the many pores observed near active regions turn into spots, and over half of these spots, in turn, last less than 2 days. Usually spots of opposite polarity form close together in time, as one might expect from the condition that $\nabla \cdot \mathbf{B} = 0$. Their birth seems to take place in the magnetic network, and the two spots of a bipolar pair are spatially separated by a relatively fixed distance of about one supergranule diameter.

Subsequent growth of spots that are longer lived than about a day takes place mainly by coalescence of smaller spots and development of the penumbra, although a rudimentary penumbra can sometimes be found on even the smallest spots. For those spots that do grow, the increase in size can be extremely rapid, and their maximum area can be achieved within a day, although 4–5 days is more typical. A large group covers about 0.05% of the disk area. One of the largest ever recorded, in 1947, covered over 0.6% of the disk. The largest spots tend to form in the side of the bipolar group that is preceding (p) in the direction of solar rotation. Follower (f) spots tend to be smaller and greater in number.

Evolution of the longer-lived spot groups proceeds through a series of morphological stages which are still conveniently described by the venerable Zurich classification system. Class A describes either a small single spot or group of spots a few degrees across. Class A spots have no penumbrae, and the group shows no structure of the kind described below. Class B describes a bipolar group of spots lacking penumbrae. Structurally, the spots are concentrated in opposite ends of the bipolar group oriented roughly east-west. Classes C and D contain groups similar to B, but Class C has at least one spot with penumbra, and in D the largest spots show penumbrae.

Classes E and F contain large bipolar groups extending over 10° and 15° in longitude respectively, with the large spots having penumbrae and with many small spots between them. This represents the maximum size and complexity of a group. The increasing extent in longitude of a growing group is achieved mainly through rapid forward motion (by about one spot diameter per day) of the p-spot. The f-spots tend to stand still in longitude or move somewhat backward. In Classes G, H, and J the group decreases in size and structure, first losing the small spots intervening between the large group members, then leaving only a unipolar spot with possibly complex penumbral and umbral structure, and finally even that structure decreases, leaving a single roundish spot with penumbra.

The small fraction of spot groups sufficiently long lived to pass through this full evolutionary scenario exhibits little regularity in how much time is spent in the stages A–J described above. A substantial fraction (over half) pass through stages A–F within 10–14 days and then spend a much longer period of 2 or more months decaying to stage J. But some groups do not show this rapid original growth. These may reach Class F only after a month or more and then decay in about the same time. Some groups never reach the most complex states E and F, but all groups end their life at stage J.

The process of spot group decay has received considerable attention. Observations of small unipolar magnetic elements moving preferentially outward from the penumbral edge indicate that in stages G, H, and J magnetic flux is stripped from the spot and carried away into the plage and enhanced network around the active region, and gradually to increasing distance away. However, the process of dissolution has proven difficult to demonstrate, and this question remains a fertile topic for future observation. Large spots that are formed by coalescence of smaller spots generally seem also to divide first into smaller spots, which then decay in situ.

8.1.3 Photometry and Spectra of Umbrae

Our understanding of the thermal structure of sunspots rests mainly on direct measurements of the spot's total radiative output, its temperature structure with optical depth, and its electron pressure. Important advances have been made in these measurements over the past 20 years through development of techniques that reduce the effects of stray light from the penumbra and photosphere into the much fainter umbral spectrum. The temperature and gas pressure structure with geometrical depth, and the fraction of total energy flux carried by radiation can be calculated somewhat less directly using assumptions described in Section 8.2.

The most obvious question concerns the temperature deficit required to explain the dark umbra. The first measurements of this quantity were made almost 50 years ago by allowing the spot's radiation to pass through a small masking aperture onto a thermocouple. After corrections for scattering were perfected, it was found that the total umbral intensity is about 15% of the photospheric value. It follows that the effective temperature of an umbra is about 3700 K, which thus also provides an estimate of the plasma temperature at $\tau = 2/3$ (see Section 2.4.3).

The wavelength dependence of this umbral intensity contrast is of interest. Measurements must be corrected for scattering in the ultraviolet and visible and for the increasing telescope diffraction in the infrared. Care must also be taken to identify the true continuum in the spot's UV and visible spectrum, which is crowded with atomic and molecular lines. The most reliable results show that umbral intensity at disk center is only a few percent of the photospheric value in the UV and rises with wavelength to a relative value of about 0.6 in the IR around 2.5 μm. The umbral T_{eff} determined from weighted integration of this contrast agrees well with the values obtained bolometrically above.

The umbral temperature structure with depth can be obtained from the center-to-limb dependence of either the absolute umbral intensity or its contrast relative to the photosphere. The method is similar to that used to determine the photospheric temperature profile from limb darkening (Section 5.2.2). Early measurements indicated that umbral and photospheric limb darkening were similar. Better data extending into the IR, where scattering is smaller, have shown recently that the umbra/photosphere contrast ratio decreases appreciably near the limb.

Early measurements were also contradictory in their evidence on whether umbral darkness depends on the area, magnetic intensity, or morphology of spots. Better photometric data have confirmed that significant variations of several hundred degrees exist across a given umbra and between different umbrae. However, no good correlation seems to exist between T_{eff} and any of the three factors above.

Photometric measurements over almost two whole 11-year cycles exhibit a surprising monotonic increase of the umbral brightness during the 11 years of a sunspot cycle. The magnitude of the change, which reaches an amplitude of 20% in the IR, is too large to be explained by the migration of spots to lower latitudes over the cycle acting together with their different limb darkening relative to the photosphere. It is also much larger than any uncertainty in photospheric limb darkening over the cycle. The observation indicates that umbrae get progressively brighter (and hotter, by about 400 K) as a solar cycle progresses from its onset at an activity minimum to the next minimum 11 years later. This is a very interesting result that is only beginning to be incorporated into dynamical models of spots and of the cycle itself.

The umbral spectrum is quite different from that of the photosphere. As expected from the lower temperature, relatively high excitation lines of neutrals (such as the Balmer series of H I) and lines of ionized species such as Fe II or Ti II are weakened in umbrae while low excitation lines of neutrals are enhanced. The greater richness of the umbral spectrum is mainly due to the profusion of thousands of molecular lines weak or absent in the photospheric spectrum, and some elements such as lithium are much more easily detected than in the photosphere. Section 3.9 contains an identification of the main molecular species observed in umbrae, but other unidentified molecules also seem to be present.

Estimates of the umbral temperature are obtained from analyses of the curves of growth, intensities, and profiles of the atomic and molecular lines. Values of T range between 3800 K and 4100 K from this method, in reasonable agreement with those obtained from the continuum limb darkening of umbrae.

A direct estimate of the density or gas pressure in an umbra is more difficult to obtain. The appearance of spots near the limb suggests that the density (and thus pressure, since the temperature is lower) might be lower than in the photosphere. Measurements dating back to the observations of A. Wilson in Glasgow in 1769 indicate that as a large spot approaches the limb, the penumbra on the disk-center side appears to become narrower, while that on the limbward side does not shrink in width (Fig. 8-4). Wilson pointed out that this observation might be explained by foreshortening if spots were saucer-shaped depressions in the photosphere, a result referred to as the Wilson effect. Some of the relatively few modern observations that have been made suggest a Wilson depression of about 600 km, but the reliability of the observations has been questioned. Models indicate that if the observations are reliable, the depression might be due at least in part to lower opacity in the umbra although a lower gas pressure at equal geometrical depth is also expected to play a role.

A relatively direct measurement of the electron pressure, P_e, in the umbra is based on curve-of-growth analysis. The shape of the curve of growth for lines of a particular ion (Section 5.5.1) depends upon the width a of the Lorentzian broadening component, which is determined directly by the density through equations (5-17) and (3-40). Comparison of calculated and observed curves of growth using this method yields values about an order of magnitude lower than in the photosphere. However, umbral models show that this is due to much lower ionization rather than to lower gas pressure at equal optical depth.

8.1.4 Mass Motions and Oscillations

Spectral lines observed by placing a spectrograph slit across a large umbra do not show the small-scale wiggly pattern (Fig. 5-7) caused in the photosphere by granular velocities. In fact, study of the spatially resolved velocity field in umbrae (using narrow laboratory discharge tube lines as a wavelength reference to avoid the uncertainties due to use of broad photospheric lines) shows no resolved umbral up- or downflows exceeding 25 m s^{-1} relative to the sun's center of mass. This indicates, as expected, that the intense vertical magnetic field of the umbra greatly reduces the convective flow seen in the photosphere. Nevertheless, measurement of umbral line widths indicates the presence of spatially unresolved motions of about 1 km s^{-1}. Some of this motion is probably associated with the oscillatory wave fields described below.

Fig. 8-4 Illustration of spots showing the Wilson effect. By permission of R. Muller.

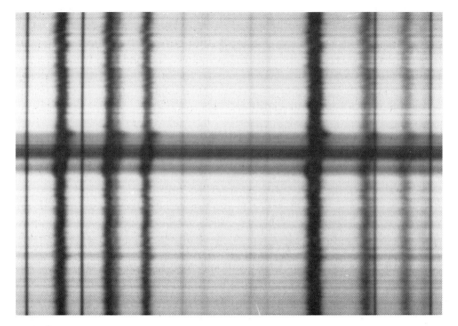

Fig. 8-5 Sunspot spectrum illustrating the Evershed effect. The dark band running horizontally is the spot's dark continuum. The characteristic tilt of the lines caused by penumbral Evershed flow is seen in this very fine spectrum taken at the Observatory del Teide, Tenerife. Courtesy of E. Wiéhr.

The most evident flow in sunspots is observed in the penumbra, as first reported by J. Evershed in 1909. He showed that when a spot is observed near the limb, weak lines exhibit an apparent redshift in the limbward penumbra and a blueshift in the penumbra nearer disk center (Fig. 8-5). In strong lines such as Hα, Na I D, and Mg I b the sense of the Doppler shift is reversed. Evershed showed that the observations implied a mainly radial outflow in the penumbra at about 2 km s^{-1}, although some evidence has also been given for substantial vertical and tangential components. The flow begins near the umbra/penumbra border, achieves maximum speed in the penumbra, and decays within a spot diameter of the outer penumbral boundary.

Early observations indicated that the Evershed outflow was largest in the weakest photospheric lines, decreased progressively in stronger lines, and then reversed sign to increase with line strength as an inflow in chromospheric lines. This dependence upon line strength suggested a height variation. But more recent spectral and imaging observations at higher spatial resolution have shown that the interpretation is more complex, and the line strength dependence does not yet seem to have a widely accepted explanation.

One finding is that the photospheric outflow is observed only in the dark penumbral filaments, where it reaches a velocity up to 6 km s^{-1}. The bright penumbral grains move inward to the umbra at a speed roughly ten times slower. The inward Evershed flow observed in Hα is clearly associated with rapid motions of chromospheric material along dark fibril-like structures that overlay the photospheric penumbra and extend well beyond it. It is most likely that this Hα inflow is just coronal material falling into the umbra along relatively much higher field lines and has little to do with the photospheric Evershed flow. An important question is whether the Evershed flow is steady or whether it represents a reversing flow with a lifetime comparable to the time required to cross the penumbra at the observed speeds.

A clear answer to this question would considerably assist attempts to explain the Evershed flow with mechanisms which include siphon flows of plasma between photospheric regions of unequal pressure connected by the same field lines. A useful test of this explanation might be to compare Evershed flow properties along penumbral field lines connecting two sunspots to those along field lines connecting a low pressure spot to a bright facular region, which is supposed to be at higher pressure.

Besides the Evershed flow that seems to always be present in the penumbra, a marked oscillatory motion occurs in most, if not all, spots. The oscillations in both the umbra and penumbra are clearest in the chromosphere, and we discuss the evidence obtained from line shifts and filtergrams in chromospheric lines such as Ca K and Hα before describing the more complex situation at photospheric levels obtained from studies of weaker lines.

The most easily observed periodic phenomena in spots are the umbral flashes—the recurrent strong brightening (by about a factor 2) of small umbral areas (2–3 arc sec in size) with a period of 2–3 min over a coherence time of up to ten cycles (Fig. 8-6). The brightening in the Ca II lines can be shown to be produced by an actual increase in emission of chromospheric material over the umbra. It is associated with an oscillation of amplitude of 5–10 km s^{-1} in the vertical velocity of the umbral chromospheric levels. The chromospheric oscillation seems to be driven from below since its observation in low chromospheric lines such as Na I D leads that in Ca K by several tens of seconds.

In the chromosphere of the penumbra, running waves have been observed in Hα filtergrams as alternately dark and bright bands of 2000–5000 km wavelength extending over 90° or more around the spot and moving outward from the umbra-penumbra boundary. These waves of 3.5–5 min period are easy to spot in time-lapse films in many, but not all, spots. Spectra of Hα show that the wave produces a vertical velocity oscillation of a few km s^{-1} as it propagates horizontally with a phase speed of 10–20 km s^{-1}. The visibility of these penumbral chromospheric waves in time-lapse films is due to the Doppler displacement of the Hα line by the vertical velocity, which modulates the opacity and thus the visibility of the bright photosphere that is normally obscured below.

The relatively high horizontal phase speed is consistent with the speed expected of compressional waves propagating along magnetic field lines of the penumbral chromosphere. However, the waves are observed to emanate from

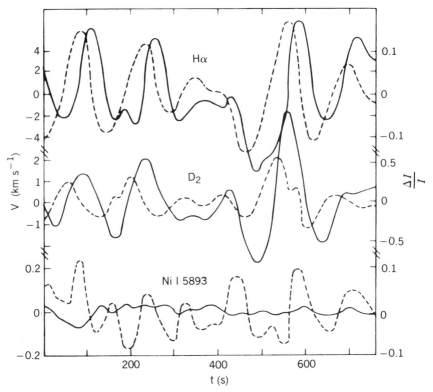

Fig. 8-6 Intensity (solid line) and velocity (dashed) plots of umbral oscillations observed simultaneously at three heights in the atmosphere. Fluctuations high and low in the umbral chromosphere are observed in Hα and in the Na I D_2 line. Photospheric level fluctuations in the umbra are seen in the Ni I line. Negative velocities plotted here are upward on the sun. By permission of F. Kneer.

the umbral edge, and if they were compressional they would follow field lines that are highly inclined to the photosphere, and thus should not be observed crossing the penumbra. A better interpretation seems to be that the basic disturbance is photospheric and travels at an oblique angle to the local photospheric magnetic field lines, so its phase speed is closer to the photospheric Alfvén speed in the penumbra but much less than the chromospheric Alfvén speed. The chromospheric wave is then just the vertical extension of this horizontally propagating photospheric disturbance.

A more complex spectrum of sunspot oscillations is observed at photospheric levels, although the velocities are an order of magnitude lower than in the chromosphere. It appears that the chromospheric oscillations discussed above are the resonant response of these higher levels to a subset of the frequencies observed in the spot at photospheric levels. The 180 s oscillations seen within the umbra have been shown to be associated with the chromospheric umbral oscillation, although the detailed correspondence is only approximate. An oscillation of about 250 s period that begins at the umbra edge and moves across the penumbra seems to be the photospheric driver of the running penumbral waves, as discussed above. However, the evidence for this period is uncertain, and more work on the spectrum of sunspot oscillations is required to verify this result and the general distribution of power in the photospheric oscillation spectrum of spots.

8.2   DYNAMICS OF SPOTS

8.2.1   Thermal Structure of the Umbra

To obtain more complete information on physical conditions in a spot than can be provided by the relatively direct measurements described in Section 8.1.3, it is useful to construct a semi-empirical model in much the same way as one studies photospheric conditions using the iterative procedure described in Section 5.2. Models developed to provide a unified picture of the umbral atmosphere have concentrated so far on conditions in the darkest part of the umbra, where umbral dots and the complex penumbra can be avoided. We describe here the conditions predicted by such a model based mainly on continuum intensities. To within observational errors and uncertainties due to imperfectly known solar abundances, line opacities, and non-LTE effects, the results are in good agreement with other models based on continuum and on lines.

The temperature profile with optical depth in the umbra over the range $10^{-2} < \tau_{0.5} < 1$ is given in Fig. 8-7. The three curves denote the different profiles required to match the umbral brightness at the beginning, middle, and end of the solar cycle, given the increase in umbral intensity over the cycle described earlier. Comparing these with the photospheric curve in Fig. 5-12, we see the umbra is about 2600 K cooler than the photosphere at equal optical depth $\tau_{0.5} = 1$. The temperature drop from the photosphere into the umbra is far greater at a given geometrical depth since we see deeper into the sun in the umbra than in the photosphere. If we accept the Wilson depression of the umbra as 500 to 700 km, then the temperature difference between the umbra at $\tau_{0.5} = 1$ and the convection zone at that same geometrical depth is close to 10,000 K!

Some other interesting physical properties of the umbra are tabulated in Table 8-1. Comparing to Table 5-2, which shows the corresponding values of the photospheric reference model, we see, for instance, that the mass column density is higher above a given optical depth level in the umbra than above the equivalent photospheric level. This means that the umbral opacity per atom is lower, so we would expect to see to a geometrically deeper level below the photosphere in the umbra. This lower opacity helps to explain the Wilson effect, although the smaller density scale height in the cooler umbra must play

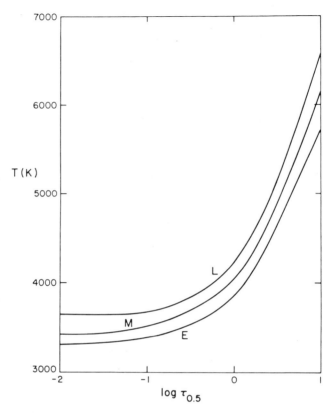

Fig. 8-7 Umbral temperature profile with optical depth. The curves E, M, L refer to models describing umbrae observed during early, middle, and late phases of the solar cycle, as described in the text. From P. Maltby et al., *Astrophys. J.* **306**, 284 (1986).

a role also. It also explains why the umbral density and pressure exceed photospheric values at equal optical depth, since we see to a much deeper level geometrically.

It is interesting that almost 30% of the umbral hydrogen is expected to be in molecular form H_2 at heights somewhat above $\tau_{0.5} = 1$. Also, the electron concentration is only about 0.01% at $\tau_{0.5} = 1$. These electrons are obtained largely from easily ionized metals, rather than from hydrogen, which is neutral to within a few parts per million at this level of the atmosphere. Thus the visible layers of the umbral atmosphere are very close to being neutral, which also implies their electrical conductivity is relatively low.

8.2.2 Why Spots Are Cool

To answer the question "what is a spot?" we need to explain both its relative coolness and its intense, extended, and relatively long-lived magnetic field. In

TABLE 8-1 Model of an Umbral Core

h (km)	m (g cm^{-2})	τ	T (K)	$n_{\rm H}$ (cm^{-3})	$n_{\rm e}$ (cm^{-3})	$P_{\rm total}$ (dyn cm^{-2})	ρ (g cm^{-3})
1662	2.152E − 05	2.150E − 06	6820	4.817E + 11	5.281E + 10	5.897E − 01	1.129E − 12
1027	6.837E − 04	1.620E − 05	5950	1.910E + 13	9.134E + 11	1.873E + 01	4.475E − 11
746	6.443E − 03	3.000E − 05	3700	3.001E + 14	2.165E + 10	1.765E + 02	7.031E − 10
282	4.022E − 01	3.000E − 03	3400	2.210E + 16	1.899E + 11	1.102E + 04	5.178E − 08
95	2.996E + 00	1.000E − 01	3540	1.695E + 17	1.082E + 12	8.209E + 04	3.972E − 07
48	5.557E + 00	3.160E − 01	3700	2.991E + 17	2.202E + 12	1.522E + 05	7.007E − 07
0	9.801E + 00	1.000E + 00	4040	4.627E + 17	5.821E + 12	2.685E + 05	1.084E − 06
−52	1.618E + 01	3.980E + 00	5120	5.733E + 17	3.838E + 13	4.433E + 05	1.343E − 06
−104	2.377E + 01	1.260E + 01	6420	6.683E + 17	1.779E + 14	6.512E + 05	1.566E − 06

Adapted from Maltby et al., *Astrophys. J.* **306**, 284 (1986).

modern theories the low temperature is caused by the intense vertical field, so these two main features are closely linked.

The lowered sunspot temperature implies a local deficit in radiative flux which must be accounted for. The missing energy is quite large, of order 10^{36} ergs from a large spot when integrated over the spot lifetime. The missing flux exceeds the energy budget for the entire solar corona and chromosphere, for instance, although it is unlikely that it has anything to do with the heating of the sun's nonthermal atmosphere.

In principle, the missing heat flux might simply be diverted by convection into the surrounding photosphere and radiated away with no storage implied. This seems to be ruled out since radiometry of the total solar irradiance shows clear dips when spots appear on the disk. The notion that the missing energy can be reradiated by the bright faculae also runs into difficulties. The most likely interpretation is that during the irradiance dips the sun's luminosity is temporarily decreased, so the heat flux that would normally give rise to photospheric radiation is either being temporarily stored as heat or has been converted to another form such as kinetic or magnetic energy.

The most promising explanation of the spot's coolness, and of the fate of the missing energy, seems to lie in the blocking of convection by intense vertical magnetic fields. This explanation was first put forward by Biermann in 1941, and some recent evidence tends to strengthen the argument. The basic idea is that the horizontal motions of overturning convection are inhibited by the magnetic volume force $\mathbf{j} \times \mathbf{B}$ in the presence of a strong vertical magnetic field. In the limit where the field energy density $B^2/8\pi$ is much larger than the kinetic energy density $\frac{1}{2}\rho v^2$ of the convection (Section 4.3.3), the normal overturning should be inhibited, provided the magnetic Reynolds number R_m is large and freezing-in holds. The peak field intensity of 3,000–4,000 G observed in umbrae is certainly large enough to dominate convective motions at the photosphere, and the scales of motion are large enough to ensure $R_m \gg 1$. In this explanation of spot coolness, an equilibrium would be reached in which the convective heat flux blocked below the spot would simply flow around it.

This simple model makes several predictions that can be tested observationally. To obtain some quantitative ideas of how heat flow might proceed around a spot, we can assume that it acts as a thermal insulator or thermal "plug" in a more highly thermally conducting layer—the convection zone. We assume that the heat flux around it is given by the mixing length relation (7-15), which can be expressed as

$$F = K(\nabla T - \nabla T_{ad}) \tag{8-1}$$

where K is the eddy thermal conductivity of the medium due to turbulent gas motions of characteristic velocity u and scale l and ∇T_{ad} is the superadiabatic temperature gradient. It then follows that the heat flow equation in this layer obeys the usual heat diffusion equation given by

$$\rho c_{\mathrm{p}} \frac{\partial T}{\partial t} = \nabla \cdot \mathbf{F} = \nabla \cdot K(\nabla T - \nabla T_{\mathrm{ad}}) \qquad (8\text{-}2)$$

or

$$\frac{\partial T}{\partial t} = \lambda \nabla^2 T \qquad (8\text{-}3)$$

if the thermal diffusivity λ is reasonably constant over the region of interest.

Solutions of this simple relation for eddy heat diffusion around a thermal plug have been helpful in understanding the properties of spots. For instance, steady-state solutions ($\partial T / \partial t = 0$) for a cylindrical plug of depth d show that unless the spot is very shallow it is difficult to explain the sharp edge of the penumbra, since the spot is surrounded by a thermal "shadow" leading to a local temperature depression. Also, it is found that the values of eddy thermal conductivity required to avoid a bright ring around a shallow spot are broadly consistent with values of K estimated from ρ, c_{p}, u, and l in the upper convection zone. Photometric studies of the photosphere just outside the penumbra show that any spatially diffuse bright ring in broad-band continuum must be below the 0.1% brightness enhancement level. Certain older observations reported enhanced brightness of relatively much higher contrast at blue wavelengths. But these observations seem to refer to magnetic faculae whose contrast is well known to increase with decreasing continuum wavelength and in lines (see Section 8.3).

The time-dependent behavior of this model also suggests how the sunspot's missing heat flux might be stored. When an area of the photosphere is prevented from radiating, the local increase of temperature and thermal energy below the obstacle diffuses outward and back inward into the sun. This perturbation heats the photosphere slightly and also represents storage of energy (thermal and potential) in the convection zone outside the spot. After the spot decays, the photospheric luminosity is left somewhat higher than the total heat flux into the bottom of the convection zone due to nuclear burning since the photosphere is slightly hotter and its total radiating area has been restored. This stored heat is radiated away until a steady state between heat flux into and out of the convection zone is again reached. This equilibration time might take as long as several hundred thousand years—the radiative relaxation time of the convection zone.

These results suggest that both the steady-state and time-dependent photometric properties of sunspots can be accounted for by a relatively simple model of a spot as a thermal plug. The missing heat flux is not simply diverted —it is most likely stored as thermal and potential energy of the convection zone, probably for a long time. It appears unlikely that the energy is stored within the spot magnetic field, since even spots that are decaying in true area cause an irradiance decrease. This would not be expected if the stored missing energy were increasing as the volume integral of magnetic energy, $B^2/8\pi$.

Conversion of heat into kinetic or wave energy might help to cool the spot, but so far it has proven difficult to identify how this might occur. For instance, calculations show that the outward Evershed flow might carry sufficient kinetic energy to account for the missing spot flux. However, some spots have no penumbra, which suggests the mechanism could not be general. But the absence of an Evershed outflow in spots lacking penumbrae needs to be checked.

A large enough flux of waves (Alfvén or otherwise) moving upward into the chromosphere would easily be detected, and firm upper limits from Doppler shift observations seem to rule this out. Conversion of blocked convective energy into a downward flux of Alfvén waves has been suggested. This idea is harder to test observationally, but it implies an improbably high efficiency of heat energy to wave energy conversion since less than about 20% of the photospheric flux emerges in the umbra. Models of wave propagation in the umbral atmosphere also indicate that oscillatory power should be trapped in an umbral cavity (Section 11.1.5).

Observations of the penumbra provide some further test of the spot refrigeration mechanism. Here the field lines run more nearly horizontally to the photosphere and the field is somewhat weaker, but convection is still inhibited and carries only about 70% of the photospheric heat flux. Simulations indicate that in a horizontal magnetic field, plasma convection should take the form of long rolls aligned with the field lines. The upwelling material would be brighter than the returning downflow. Although good spectral observations of the fine penumbral filaments are difficult to obtain, there seems to be some evidence for such a correlation between line shifts and the brightness field.

8.2.3 Dynamics of Sunspot Evolution

In the most widely held view of spot evolution, a spot is born when magnetic flux at the photosphere happens to be pushed together and intensified somewhat by supergranule motions. The ram pressure of supergranule flow is sufficient to achieve a field intensity of a few hundred gauss at photospheric levels. Such a field inhibits convection, leading to an instability since the spot location will then cool, causing its gas pressure to decrease. This leads to further contraction and intensification of the field by the horizontal gas pressure gradient. The ultimate field intensity achievable if the spot is effectively evacuated can explain the observed umbral field intensities.

The observation that as much as 20% of the photospheric radiative flux is radiated by an umbra could create difficulties if the umbra is supposed to be monolithic, since overturning convection should be completely suppressed in the umbral field, although oscillatory convection could still transfer energy. An interesting way out is to suppose that the umbra consists not of a single magnetic flux tube, but of a loose cluster of many individual flux tubes. These flux tubes might be anchored deep within the convection zone and thus be held together at the photosphere in part at least by magnetic buoyancy in much the same way as a cluster of balloons. The cluster model would help to explain how convection can penetrate the spot to provide the required heat flow and also produce umbral granulation and perhaps umbral dots. Whether or not such a model can explain the observed umbral magnetic field profile depends on issues such as the relative roles of upwelling convection currents, of the horizontal component of the magnetic "tethering" tensions, and of the gas pressure deficit in the spot, in determining the dynamic equilibrium that holds the flux tube together.

In the cluster model, the spot is taken to be stabilized by the inherently long time scale of slow motions at the flux tube footpoints deep within the convection zone. If a monolithic flux tube is taken to be confined by a deficit of gas pressure in a shallow layer around the photosphere, it will be pinched inward in this layer, while flaring outward both above and below. It has been shown (see Section 11.1.2) that the outward flaring is stable to fluting in a gravitational atmosphere, provided the curvature has the sense expected above the photosphere. Below the photosphere, however, the flute instability must rapidly destroy any tendency to formation of monolithic flux tubes. Given the presence of light bridges spanning umbrae, and the general irregular outline of umbrae, the stabilization of such clearly nonaxisymmetric flux tubes for months poses a problem for monolithic models.

The gradual decay of a spot might be caused by reversal in direction of the supergranule motions that originally acted to form it and hold it together. Such a reversal could be caused by the warming of layers below the spot by the blocked heat. This will increase the gas pressure and tend to counteract or reverse the ram pressure of the supergranule flow. In a cluster model, the eventual decay is explained by the continuous slow motions of the footpoints. Decay of the following end of a flux tube might proceed more rapidly than in the preceding end if it were more inclined to the vertical. The inclination would enable plasma motions to unravel the rope more effectively and thus help to explain the tendency of f-spots to form groups of several smaller umbrae rather than the large unipolar spots more often observed in the preceding half of an active region. The rate of decay of sunspot area A seems to behave as $dA/dt \propto -A^{1/2}$. This is what we might expect if decay proceeds by a passive chipping away at the edge of a spot.

In a different approach to the problem of sunspot evolution, it is proposed that spots are formed at the photosphere from long-lived, strongly twisted flux ropes originating beneath the convection zone. In this model, supergranule motions play no role in assembling the spot or determining its decay. The spot is the result of gradual emergence of a "tree" of magnetic flux whose "trunk" is the strongly twisted helical flux rope. Relative motions between the trunk and the extended branches of flux (observed as the surrounding small-scale fields of like polarity) are determined mainly by magnetic tensions. The twist of the flux ropes is considered to provide their stability, and its time evolution (or unraveling of the flux ropes) is supposed to lead to their decay.

Although intuition might suggest it, calculations do not support the idea that twist stabilizes magnetic flux tubes. Nevertheless, this model deserves attention since in many other respects it explains the observations of the relative motions of flux tubes, for instance, better than the model in which supergranular motions push the field lines around. Certain observations, such as the evidence on brightening of umbrae over the 11-year cycle and also the evidence for a rough energetic balance between the missing radiative flux of spots and the excess radiation of faculae within active regions, have not yet been used to constrain either model.

The possible role of larger-scale (than represented by supergranules and granules) convective eddies in stabilizing spots and determining the location of their emergence deserves close attention. Recent studies of the patterns of active region emergence and morphology during solar cycles 20 and 21 show, for instance, that large-scale magnetic polarity structures such as shown in Fig. 8-15 can emerge up to three solar rotations before any spots begin to form along their neutral lines. This implies that the polarity structure in the background field is more fundamental than the active region or spot group—which is contrary to the Babcock model (see Chapter 11). Also, the spatial scale ($1-5 \times 10^5$ km) and longevity (3–6 solar rotations) of active region complexes both exceed the scale (3×10^4 km) and lifetime ($\sim$ 1 day) of supergranules by large factors and cannot be governed by supergranule evolution. Even the lifetime of the longest-lived p-spots is difficult to understand if supergranular motions are to play the governing role assumed in the most widely used model of sunspot evolution.

Such ideas need to be developed more quantitatively through statistical studies of key parameters such as shear and divergence of photospheric magnetic field and plasma motions at neutral lines. Unfortunately, such studies require long, patient work with extended and often complex data bases, and results at an acceptable level of statistical significance are difficult to achieve.

8.3   FACULAE

8.3.1   Structure and Evolution

Whenever even a small spot forms, it is accompanied by a brightening in the surrounding atmosphere called a facular area that is easily seen in white light near the limb (Fig. 5-1, 8-8). This brightening sometimes precedes the spot's appearance by as much as several days and outlasts it by a significant multiple (2–3) of the spot's lifetime, so faculae are much longer-lived than spots. The brightening in white light near disk center is barely detectable but increases to between 5 and 10% at the limb. In line radiations, faculae (or plages, as they

Fig. 8-8 High-resolution image of faculae near the limb obtained at the Pic du Midi Observatory, France, at a wavelength of 5750 Å with a filter of 100 Å passband. By permission of R. Muller.

-0.4 Å   +0.4 Å

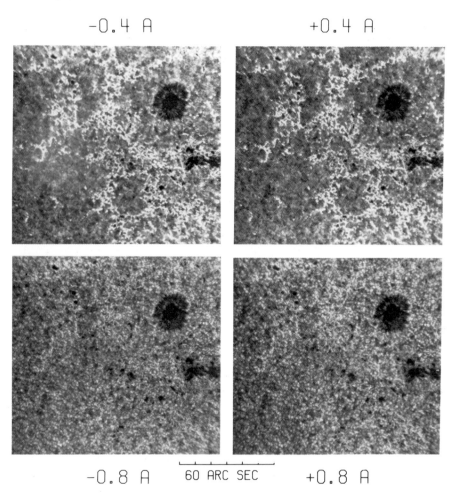

-0.8 Å   60 ARC SEC   +0.8 Å

Fig. 8-9 Narrow-band images of faculae near a spot on the disk at four positions in the wing of the Mg I b-line. By permission of R. Dunn, National Solar Observatory, Sacramento Peak, New Mexico.

are called when their overlying chromosphere is observed in strong Fraunhofer lines) are easily seen even near disk center with relatively high contrast (Fig. 8-9).

As can be seen from Figs. 8-8 and 8-9, the facular areas are irregularly distributed within an active region. The extended patch of brightness seen at low spatial resolution actually consists of many small bright elements packed together in the active regions. Outside the active regions, similar small facular elements are more loosely distributed in a photospheric network that covers the entire solar surface. About 90% of the active region faculae (by area) are found to be associated with spots. They are observed to pass through a phase of at least several days associated with spot appearance in the active region when their structure is compact and their brightness is high. The progressive fragmentation of the bright compact faculae into the active network elements and finally into the quiet network is due to the general dispersal of magnetic flux tubes thought to be caused by supergranular motions. This process is discussed in more detail in Section 8.3.3.

8.3.2 Physical Measurements

Measurements on photospheric faculae tend to be more difficult than those on spots because the individual elements are very small and have low continuum contrast over most of the disk. We have more direct observations of facular conditions in the high photosphere (through continuum observations near the limb and from monochromatic line observations, both of which show faculae at good contrast) than at the layers around $\tau_{0.5} = 1$, where most of the radiation is emitted. On the other hand, facular observations are less affected by scattered light than umbral measurements since facular elements are brighter than their surroundings over most of the spectrum.

Photometric measurements with good spatial resolution (1–2 arc sec) in narrow continuum passbands indicate that the facular contrast increases toward the limb, with values of about 10% seen around $\mu = \cos \theta \sim 0.2$ at 5000 Å. The contrast also varies roughly as inverse wavelength in the visible region. The faculae are approximately 0.1–1% brighter than the photosphere even at disk center when observed in narrow-band continuum passbands in the visible and near-infrared. But the actual excess intensity of individual small facular elements is uncertain since no such photometry in passbands uncontaminated by lines has been carried out yet at spatial resolution better than 2–3 arc seconds. In lines such as Mg I b or the CN bandhead near 3889 Å the contrast of faculae can be considerably higher, even at disk center, but it depends sensitively on exact position in the line, and on the passband used. A recent finding is that faculae tend to be dark, like little sunspots, when observed near disk center at around 1.63 μm in the infrared. The photospheric opacity is lowest around that wavelength, so the deepest facular layers are observed.

A key question is the scale and shape of the elemental facular points observed as brightening in lines and continuum and their relation to the smallest magnetic elements observed in the network and active regions. It has been established that the magnetic elements are cospatial with bright facular structures seen in continuum and weak lines down to a scale of at least 1 arc sec. It is more difficult to show that the field elements are precisely cospatial with the even smaller-scale brightness structures called filigree (Fig. 8-10), which are observed under conditions of excellent seeing, in the far wings of Fraunhofer lines. These filigree structures are as small as 150 km and seem to form a network roughly coincident with magnetic network and magnetic elements in faculae.

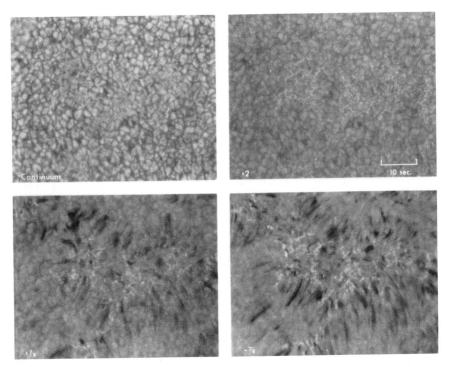

Fig. 8-10   Filigree elements in the magnetic network, imaged in the continuum, at $\pm \frac{7}{8}$ Å, and also at 2 Å, from the center of Hα. By permission of J. Zirker, National Solar Observatory, Sacramento Peak, New Mexico.

It is very difficult to obtain magnetograms at such fine spatial resolution, so the exact relation of the magnetic and brightness fields is elusive at this smallest scale. It is clear, however, that bright magnetic flux elements can have dimensions well below 700–1000 km.

Given the dynamical equilibrium of flux tubes described in Chapter 11, we expect that measurements made in continuum and at different levels of line opacity might show an increasing cross-sectional area of the flux elements with height, as the ambient gas pressure declines rapidly above $\tau_{0.5} = 1$, and the magnetic flux tube expands. Images of the network in chromospheric lines such as Ca K certainly show brightness structures of much larger characteristic scale than the filigree, and this is probably indicative, at least in part, of expansion in the flux tube diameter. But the lack of chromospheric Zeeman measurements precludes a direct comparison between the horizontal extent of the magnetic flux tubes at this level and in the photosphere. At photospheric levels, magnetograms made in various lines show no clear difference in cross-sectional area of the magnetic elements over the accessible height interval of a few hundred kilometers. But the increase calculated at these levels from models is below the resolution limit anyway, given the small diameter of the elemental facular flux tubes.

The observation that faculae are increasingly brighter than the photosphere when observed in lines formed at greater heights above $\tau_{0.5} = 1$ suggests that the temperature gradient in the facular atmosphere is lower than in the photosphere. It is also found that faculae become easily visible in continuum at disk center when images obtained in two continuum passbands widely separated in wavelength (and thus in H$^-$ opacity) are subtracted. Since we see to different photospheric levels in the two passbands, the difference in intensities can be directly related to the temperature gradient around $\tau_{0.5} = 1$ and indicates that the facular temperature gradient around $\tau_{0.5}$ in the flux tube is about 25% lower than in the photosphere at equal optical depth. Similar results have also been obtained from comparison between faculae and photosphere of the equivalent width difference of two Fraunhofer lines formed at different levels in the deep photosphere.

The recent finding that faculae increase the total solar irradiance by a factor that is closely approximated by their projected area and wide-band photometric contrast provides another constraint on their energetics. The enhancement of solar irradiance is far too large to be explained by the excess emission from chromospheric and coronal plage layers overlying the facula. Since this facular contribution increases the total irradiance even when integrated over time scales longer than a solar rotation, it represents a real increase in heat flux passing through the photosphere rather than just an angular redirection of radiative flux as might be suggested by the increasing facular contrast near the limb (see exercises).

The gas motions inside flux tubes are best studied by analyzing the circularly polarized component of facular line radiation since isolation of this component ensures that conditions in intensely magnetized plasma within the small-diameter flux tubes are being studied. No evidence is found for any steady up- or downflows within the flux tubes in such analyses. However, the broadening and asymmetry of the circularly polarized line profile indicates the presence of relatively large (2–4 km s^{-1}) motions that are turbulent or possibly associated with waves or oscillations on time scales shorter than tens of minutes.

8.3.3 Why Faculae Are Bright

A semi-empirical model of the temperature distribution in a facula and overlying chromospheric plage is plotted in Fig. 8-11. It is based on observations in the wings of the Ca K-line, where the opacity variation with increasing distance from the line core corresponds to a height variation from the chromosphere down to $\tau_{0.5} = 1$. Similar models have also been based on the Mg II line wings and on agreement with core strengths of weaker Fraunhofer lines. Only the qualitative features of such models can be regarded as significant, since these observations have much coarser spatial resolution than

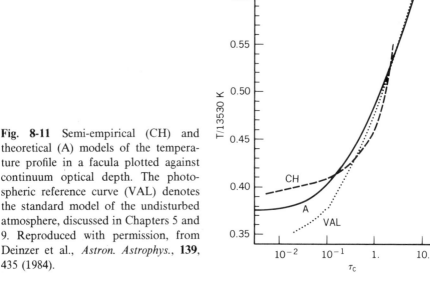

Fig. 8-11 Semi-empirical (CH) and theoretical (A) models of the temperature profile in a facula plotted against continuum optical depth. The photospheric reference curve (VAL) denotes the standard model of the undisturbed atmosphere, discussed in Chapters 5 and 9. Reproduced with permission, from Deinzer et al., *Astron. Astrophys.*, **139**, 435 (1984).

required to resolve the individual facular flux elements. Estimates of the area-filling factor in a facular region are uncertain, but they suggest that the intense flux tubes might occupy roughly 10% of the total area. It follows that the differences between the facular model and photosphere that are indicated in Fig. 8-11 are likely to significantly underestimate the real differences.

The model shows that the facula is hotter than the photosphere at equal optical depth, at least everywhere above the deepest levels accessible to observation in the visible continuum. This is expected, given the direct observation that the facular brightness temperature exceeds that of the photosphere at all wavelengths in the visible and ultraviolet. Since the facular and photospheric temperatures converge around $\tau_{0.5} = 1$ in each atmosphere, it follows that the facular temperature gradient must be less than that of the photosphere, also as indicated in Fig. 8-11. Models such as this also indicate that at equal optical depth the facular atmosphere is denser, and thus at higher gas pressure than the photosphere.

Comparison of relative facular and photospheric physical conditions at equal optical depth is informative if the goal is to understand facula-to-photosphere variations in radiations emitted by the two atmospheres, such as line and continuum brightness and line profiles. However, to achieve a dynamical understanding of the facula, we need to place the two atmospheres on a common geometrical scale. To a first approximation, this can be achieved by shifting the whole facular atmosphere inward by the depth of the facular Wilson depression—the difference in geometrical depth between the levels of

$\tau_{0.5} = 1$ in the facula and photosphere. The size of this shift in faculae cannot be measured directly, but from the model estimates of the difference in facular and photospheric opacities and mass column densities, it appears to be about 200–300 km.

When the two models are placed on a common geometrical depth scale, we see that at all levels, the facular atmosphere is considerably cooler, less dense, and thus at much lower gas pressure than its surroundings. According to the dynamical equilibrium outlined in Chapter 11, the difference in total pressure is made up by the strong facular magnetic field.

The finding that magnetic flux tubes cause local depressions in the photospheric surface of $\tau_{0.5} = 1$ of several hundred kilometers suggests a simple explanation of the excess brightness of photospheric faculae. In both sunspot and facular flux tubes the energy balance seems to be determined by heat transport by radiation and residual convection from below plus net radiation into the flux tube from the side. These energy inputs are balanced against radiative losses into free space from the top of the facular or sunspot atmosphere.

In the spot's large-diameter flux tube, the inhibition of convection along the spot axis is relatively much more important than any excess radiation into the umbra from the hot wall of the Wilson depression, and the result is a relatively cool atmosphere. In the much smaller diameter facular flux tube, the lateral radiation is relatively much more important since it scales as the flux tube radius r, while the amount of convective inhibition scales as r^2. Models of radiative transfer in such flux tubes confirm that radiation from the hot convection zone into the facula (which is cooler at equal geometrical depth) can result in an atmosphere that appears hotter than the photosphere (i.e., at equal *optical* depth), even when viewed along its axis. A profile of temperature with height in such a theoretical facular model is also given in Fig. 8-11.

The observation that the facular continuum contrast rises rapidly near the limb is broadly in agreement with this explanation since we expect to see the facular wall at more nearly normal incidence (and so observe radiation from relatively hot layers) when we view the facula at a large angle from the vertical. This model predicts that the facular contrast should peak near the limb and not continue to rise to the very edge, which seems to be confirmed by the most reliable data.

An alternative explanation is that faculae are heated nonthermally, even at photospheric levels, by large fluxes of waves, by electrical current dissipation, or by systematic upflows of hot plasma from deep layers. However, the recent observations in the infrared at 1.63 μm, where we see some 35 km deeper into the photosphere than $\tau_{0.5} = 1$, would seem to argue against this explanation. Faculae are found to be darker, not brighter, than the photosphere in the 1.63 μm images. It is difficult to understand why nonthermal heating or systematic flows should cease to be effective at the 35 km deeper level seen at $\tau_{1.6} = 1$ if they heat the facula at $\tau_{0.5} = 1$. However, the observation of cool material in the faculae at 1.63 μm is consistent with the expectation that at the deepest observable levels, radiative heating of the facula from the hot convection zone becomes ineffective because the opacity is too high to enable photons to penetrate into the facula from the side. Since facular magnetic fields inhibit convection from below, these deep facular layers should be cool just as in spots.

8.4 OBSERVATIONS OF SOLAR MAGNETISM

8.4.1 The Sunspot Magnetic Field

The Zeeman splitting of Fraunhofer lines in an umbral spectrum is shown in Fig. 3-12. The most accurate measurements, made in certain Fraunhofer lines formed only in umbrae (and thus less susceptible to photospheric and penumbral scattered light contamination) indicate intensities between 1900 and 2500 G for pores and small spots. Umbrae of diameter larger than about 6000 km seem to exhibit increasing magnetic intensity with area, up to about 3000 G. These are measurements of the total field intensity since they do not select any particular polarization component of the line.

The most reliable measurements of magnetic intensity across a sunspot, which now include observations on Zeeman-sensitive infrared lines, indicate that the intensity decreases by about a factor of 2–3 from the center of the umbra to the outer penumbral edge, so that even in the penumbra of a large spot the intensity is about 1500 G.

The orientation of the sunspot field has also been measured from center-to-limb observations. Although the data exhibit large scatter, the field appears to be closely vertical on the axis of the umbra and becomes progressively more inclined until it reaches a horizontal orientation at the outer edge of the penumbra. The vertical field gradient found from comparisons of field strengths in different photospheric and transition region lines is about 0.5 G km^{-1} over a several thousand kilometer height range above $\tau_{0.5} = 1$. Observations with vector magnetographs support these main features derived from studies of the line-of-sight component of **B**. In addition, they provide information on the geometry of a spot's transverse field. Figure 8-12 shows that the field orientation can approximate that expected of a potential field reasonably closely, at least over most of the umbral and penumbral area. This may not hold for all spots.

There is evidence that the field intensity within an umbra can vary by at least a factor of 2 over small scales of a few hundred kilometers. This is found by comparing the magnetic profiles of lines formed at low umbral temperatures with those formed at higher, photospheric temperatures in the tiny brightenings called umbral dots, which show lower field intensities (see Section 8.1.1). The small-scale geometry of the umbral field is also uncertain. Observations of linear polarization in Zeeman-split lines observed even at disk center

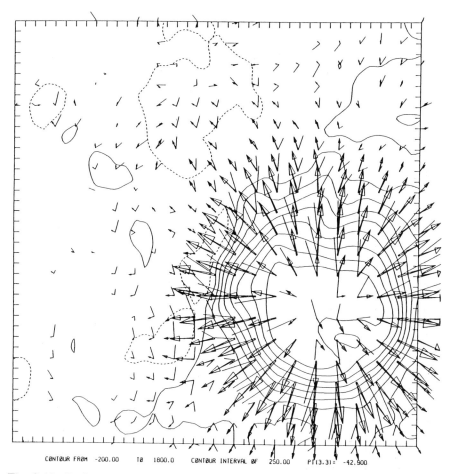

CØNTØUR FRØM -200.00   TØ 1800.0   CØNTØUR INTERVAL ØF   250.00   PT(3,3)= -42.900

Fig. 8-12 Projection of observed field (arrows) and calculated potential field (lines) onto the photospheric surface for a large spot. By permission of M. Hagyard.

indicate a transverse field component that may be associated with small-scale polarity reversals (small dipoles) on the scale of the umbral dots. But the evidence is not yet clear and better observations are required.

The magnetic flux measured in a spot whose diameter lies between 8 and 60×10^3 km ranges between 5×10^{20} and 3×10^{22} Mx. At the outer edge of the penumbra, which is often surrounded by an extended annulus of relatively field-free plasma called a moat, time-lapse magnetograms show small magnetic features moving into and out of the penumbra, at speeds of about 1 km s^{-1}. These (subresolution) magnetic knots seem to provide a net transfer of sunspot polarity magnetic flux across the moat at a rate of order 10^{19} Mx hr^{-1}, sufficient to explain the decay of a spot within a few days.

8.4.2 Photospheric Fields in Faculae and Magnetic Network

Magnetograms of the whole disk, such as Fig. 8-13, and at higher resolution in Fig. 8-14, show a hierarchy of spatial scales in the photospheric magnetic field. Outside sunspots, which are the largest discrete flux elements, the bipolar active regions seen clearly in Fig. 8-13(b) also contain the structured field patterns of faculae. These merge into the general magnetic network that covers the photospheric surface even when no active regions are present, as in Fig. 8-13(a).

A remarkable finding, illustrated particularly well in Fig. 8-14, is the discreteness of the solar magnetic field, which consists of small flux elements clustered together on a wide range of spatial scales and separated by a relatively unmagnetized plasma. The magnetic fluxes measured in discrete elements outside of spots range between roughly 10^{19} Mx for the largest discrete facular elements and 5×10^{17} Mx for the smallest network elements, whose lifetime is only a few tens of minutes. The total magnetic flux in facular

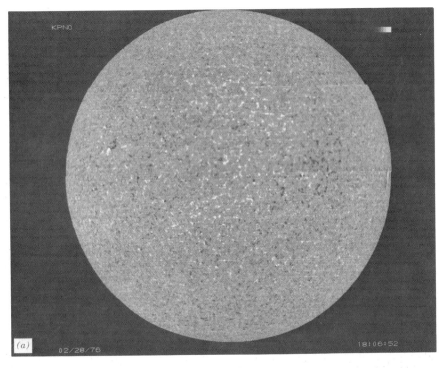

Fig. 8-13 Full-disk photospheric distribution of the longitudinal magnetic field component under (a) quiet and (b) active sun conditions. Black and white indicate opposite magnetic polarities. National Solar Observatory, Kitt Peak, Arizona. By permission of J. Harvey.

elements within an active region is comparable to the spot flux, as might be expected.

The field intensity outside of spots is difficult to measure directly since magnetographs measure flux and some estimate of the scale of the structure is necessary if it does not fill the field of view. However, Zeeman measurements under unusually good seeing conditions, and also in infrared lines such as the Fe I λ 1.5648 μm line, where the wavelength splitting is increased in proportion to λ^2, indicate that field strengths are in the range 1000–2000 G, even in the network elements.

An indirect technique can also be used which depends upon comparing the Zeeman signal measured using two lines of different Lande g-value and does not require that the flux elements be spatially resolved. It relies on the saturation of magnetograph response to strong fields. With some assumptions regarding the symmetry of the flux element, a unique intensity can be determined even though the element may not fill the field of view. When compared with the fluxes measured in these elements, the intensity results

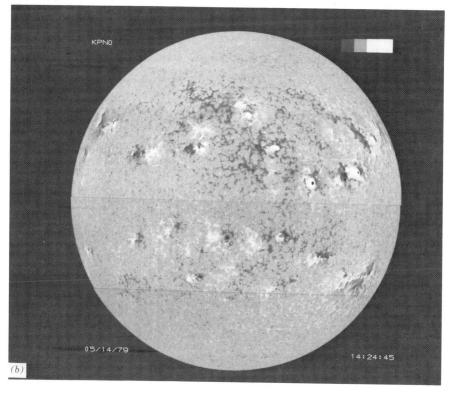

Fig. 8-13 *(Continued)*

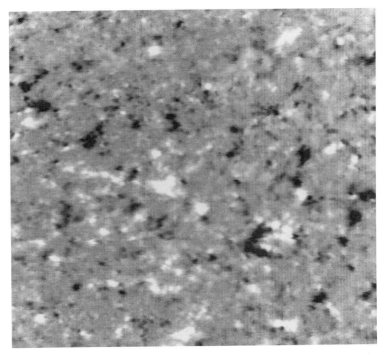

Fig. 8-14   Photospheric longitudinal magnetic fields in the quiet sun, shown at high spatial resolution. The field of view here is approximately 512 × 512 arc sec. National Solar Observatory, Kitt Peak, Arizona. By permission of J. Harvey.

obtained in this way indicate that all the photospheric network fields consist of small-diameter (< 300 km) flux elements of high intensity ($B > 1000$ G) interspersed with relatively nonmagnetic plasma.

Additional information on the scale and intensity of the fields comes from the observation that facular and network flux elements correspond spatially to the bright photospheric structures described in Section 8.3.2. The precision of the correspondence on the smallest scales is not quite clear, but the smallest bright filigree elements seen in Fig. 8-10 have dimensions of less than 300 km. It is found that the number of these bright elements within the field of view is proportional to the magnetic flux measured with magnetograph apertures of increasing size. This result confirms that the filigree and flux tubes are closely related, although it does not determine their scale.

The main difference between fields in sunspots and elsewhere in the photosphere seems to lie in the scale of the flux elements. In an umbra, the field appears relatively continuous if not entirely homogeneous. In faculae, however, studies show that the area-filling factor of flux tubes is only about 10%, and the global-filling factor of photospheric fields including network and active regions is only a few percent. Magnetograms of the highest sensitivity have recently revealed so-called intranetwork fields of mixed polarity within the supergranule cells. Their intensity, of order 10 G when measured with a few arc seconds spatial resolution, may indicate truly weak fields, or perhaps kilogauss flux tubes of 10^{16} Mx flux or less.

The interconnections between the flux elements have been studied in active regions using vector magnetographs. Some results for the vicinity of a sunspot are illustrated in Fig. 8-12. More generally, the transverse component of the photospheric field in an active region exhibits large-scale regularities, and also complicated connection patterns. The field lines of a calculated potential field (carrying no current) rooted in the same photospheric magnetic elements is also plotted in Fig. 8-12. Certain differences between the observed and potential field lines, such as helical structure around spots and a shear in the field lines crossing the neutral line between opposite polarities, are often observed and are indicative of electric currents along magnetic field lines in the photosphere.

Such magnetograph observations can be used to calculate the vector $\nabla \cdot \mathbf{B}$ and thus to estimate the magnitude and direction of the electric currents. It is found that the free energy stored in such current systems (running along magnetic field lines) is typically a few percent of the total energy of the potential component of the active region magnetic field. The currents appear to be generated by twists in the field lines of a flux tube caused by rotational plasma motions at the photosphere or deeper levels.

8.4.3 Large-Scale Structure and Evolution of the Photospheric Field

Magnetic flux emerges at the photosphere partly in the form of bipolar magnetic regions which contain a flux of between approximately 10^{19} and 5×10^{22} Mx. The typical evolution of a large bipolar magnetic region of about 10^{22} Mx is shown in Fig. 8-15. As noted previously, only a small fraction of emerging bipolar regions go on to form such large active regions. To within the 10–20% uncertainties, the emerging flux is always balanced between the opposing polarities, as one expects from $\nabla \cdot \mathbf{B} = 0$. Within a few hours, a small area of intense bipolar field emerges over a scale of about 10^4 km diameter. This bipolar region expands relatively rapidly within the next few days to a scale about ten times larger as it rotates toward the limb [Fig. 8-15(a)]. More than half of the flux in this region is associated with sunspots, the rest with faculae.

Figure 8-15(b) shows how over the next several solar rotations, the magnetic region becomes less compact as the flux elements of both polarities become increasingly fragmented and intermingled with the surrounding fields of the quiet magnetic network. The bipolar configuration also becomes highly sheared by photospheric differential rotation, which is quite pronounced at the 25–30° latitude of this magnetic region. That is, the neutral line between the two polarities rotates by roughly 60° from its orientation on January 12th to March 9th. Three months after its first appearance the large areas of field in

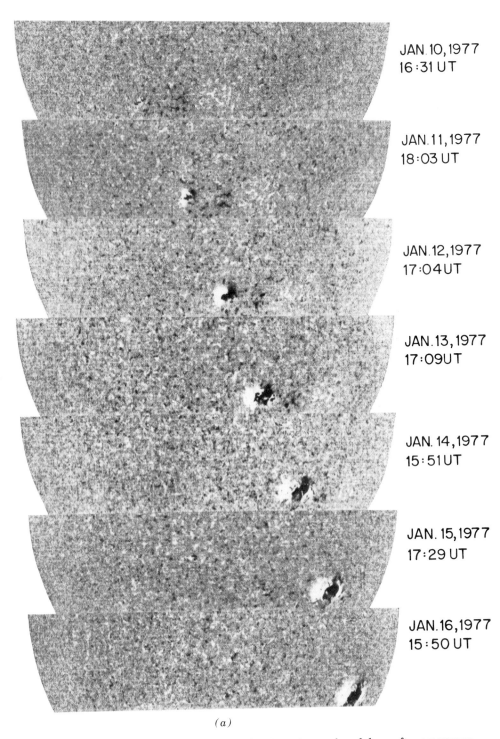

JAN.10,1977
16:31 UT

JAN.11,1977
18:03 UT

JAN.12,1977
17:04 UT

JAN.13,1977
17:09 UT

JAN.14,1977
15:51 UT

JAN.15,1977
17:29 UT

JAN.16,1977
15:50 UT

(a)

Fig. 8-15 Evolution of photospheric fields (a) on time scales of days, after emergence, (b) slow decay on time scales of months. Magnetograms obtained at Kitt Peak Observatory. By permission of N. Sheeley.

280

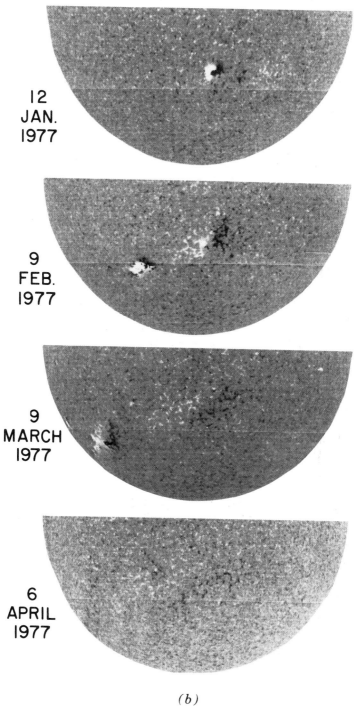

12
JAN.
1977

9
FEB.
1977

9
MARCH
1977

6
APRIL
1977

(b)

Fig. 8-15 *(Continued)*

this magnetic region have decayed into small, subresolution elements indistinguishable from the general network.

The process that leads to this dispersal of fields is usually represented as diffusion caused by the turbulent plasma motions at the photosphere. Detailed studies show that the photospheric flux tubes often move about faster than does the photospheric gas in supergranular flows. This may indicate the influence of deeper motions or of accumulated tensions in the field lines, which, when released, might occasionally propel the photospheric flux elements at relatively high speed. Numerical simulations of the field's global evolution show that the observed behavior can be reproduced quite well over several months and years given the observed latitude and longitude distribution of active region formation. These simulations assume the usual photospheric differential rotation with latitude and an isotropic diffusion coefficient broadly consistent with the lifetime, scale, and velocity field of photospheric supergranules.

Simulations of this kind are able to explain the observation, mentioned in Chapter 7, that the long-lived, sheared magnetic regions found at relatively high latitude tend to rotate rigidly rather than differentially with latitude. The reason seems to be that the angle between the axis of the structure and the equator depends only on the rate of poleward translation of the fields (determined by the diffusion coefficient and by a possible poleward meridional flow) and by the local rotation rate. These parameters, which are roughly fixed, determine the shape of the bands independently of time. Thus the band rotates rigidly, although the long-lived fields that constitute it rotate differentially.

Bipolar magnetic regions differ widely in the maximum size and flux that they achieve, and in their lifetime. These quantities are related—large magnetic regions are distinguishable from their background for three to four solar rotations, as seen in Fig. 8-15(b). Towards the other end of the spectrum, barely resolvable bipolar regions containing a flux of about 10^{19} Mx can be detected using magnetograms. Either Hα or X-ray images help to determine whether chromospheric and coronal field line connections exist between the photospheric magnetic elements. Several hundred are present on the sun at any given time, although their lifetime is only about one third of a day.

To maintain this number on the sun, new dipoles must be appearing at the same rate as they are disappearing. Given their short lifetime, their birth rate (per day) must be about three times their instantaneous number on the sun. From this high rate one finds that more magnetic flux emerges in the very small dipoles than in larger and longer-lived active regions at any phase of the solar cycle except possibly near maximum activity. Recent studies indicate further that substantial flux emergence occurs in the network and intranetwork fields all over the sun. Although this issue is not yet settled, such evidence suggests that the total rate of new flux appearance at the photosphere varies less over the cycle than does the shape of its size spectrum.

Whatever the size distribution of flux emergence turns out to be, there is no doubt that at any phase of the solar cycle most of the photospheric magnetic

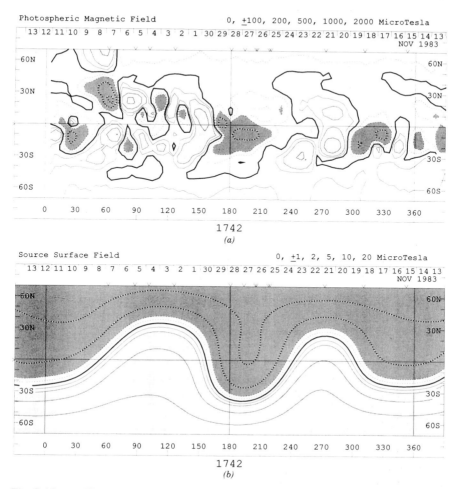

Fig. 8-16 (a) Illustration of the "tennis-ball seam" pattern of the global magnetic field seen in spatially smoothed magnetogram data. The magnetic measurements from which panel (a) is derived were obtained at Wilcox Solar Observatory, Stanford University. By permission of T. Hoeksema.

flux at a given time is in the larger bipolar magnetic regions that last several days or more. The reason is that the larger regions last much longer, and this greater lifetime more than compensates for their lower emergence rate.

Photospheric field observations at low spatial resolution, either with a large magnetograph aperture or by smoothing of higher-resolution data, reveal some interesting large-scale patterns. Figure 8-16 illustrates the patterns formed by the smoothed active and quiet region fields over the whole solar surface (360° in longitude), as observed during ten solar rotations, thus almost a year. As the coarseness of resolution is increased, a wavelike pattern of the field polarity is seen to organize the sun's magnetism to look like the seams on a baseball.

Analysis shows that both the active region and network fields contribute significantly to this pattern, which is due primarily to a large-scale variation in the balance of net magnetic polarity rather than to differences in field strength. The decay of active regions tends to distribute flux so as to reinforce pre-existing large areas of net polarity. This pattern often remains steady over several years, although large-scale changes can be relatively abrupt. So-called giant regular cellular structures have been discerned in the distribution of active regions and Hα filaments, which coincide with these patterns. The observation that active regions are organized into such a large-scale polarity structure is an important clue in predicting their emergence in restricted "active longitudes" and probably in understanding the mechanism of solar magnetism. Such active longitudes are illustrated in Fig. 8-17.

As discussed further in Chapter 11, one of the most important questions in studies of solar magnetism deals with the competing roles played by passive

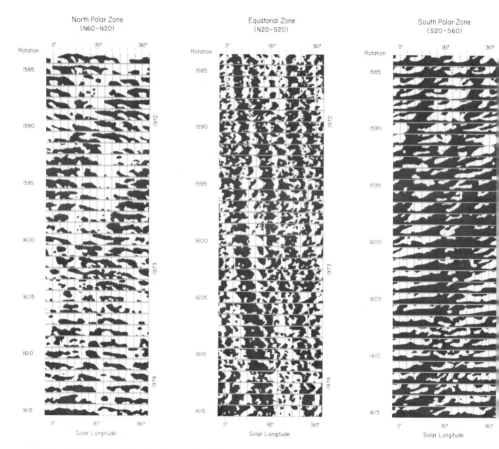

Fig. 8-17 The distribution of active regions illustrating the phenomenon of active longitudes. By permission of P. McIntosh.

diffusion of active region fields versus the organization imposed by deeper-seated large-scale cell patterns. It is not yet clear whether one or both determine the photospheric distribution of fields and their solar cycle evolution. Large-scale polarity structures such as those seen in Figs. 8-16 and 8-17 suggest that a pattern is imposed on the solar field by convective cells much larger than supergranules, although diffusion of emerged active region fields probably determines the field evolution within such cellular structures.

It has also been recognized since the observations of Babcock that the net magnetic polarity observed at the two solar poles is opposite and bears a definite relation during the solar cycle to the sunspot fields at lower latitudes. The observations are relatively difficult since the fields are seen near the limb and their longitudinal component is small and close to the noise threshold of magnetographs. Early evidence for such polar fields given by Hale was based on Zeeman measurements that have since been found to have insufficient sensitivity to reveal such weak mean fields.

At higher resolution the polar fields measured with photoelectric magnetographs are resolved into network, including the polar faculae. The mean polar field measured at latitudes above about 60° is presumably the smoothed resultant of these small-scale structures. Its mean intensity and total flux are uncertain, with values ranging from 4×10^{21} Mx in each polar cap (for a mean intensity of about 1 G) to about 10 times higher. The value of polar flux is of considerable interest since it is believed to determine the magnetic field in the solar wind (Chapter 12).

8.4.4 Global Structure of the Sun's Magnetic Field

Our current understanding of the sun's magnetic field geometry is obtained mainly from the distribution of magnetic polarities observed at the photospheric surface, together with information on how these magnetic elements are connected by field lines observed as loop structures in the chromosphere and corona. Examination of magnetograms such as Fig. 8-13 reveals a distinctive alternation of white and dark polarities in the active regions. In the sun's northern hemisphere the white polarity tends to be located in the "preceding" (westerly) part of the active region, and the dark polarity is located in the "following" part. In the southern hemisphere, the sense is reversed. This law of alternating active region magnetic polarities was stated by G. Hale and S. Nicholson in 1914, based on sunspot magnetic field measurements made at Mt. Wilson.

This Hale-Nicholson polarity rule is remarkably well obeyed, and it carries the most important evidence discovered so far about the sun's magnetic field geometry. According to the generally accepted interpretation usually attributed to H. Babcock, the field at the low and midlatitudes occupied by active regions consists of toroidally wound flux tubes running more nearly east-west than north-south. Where these subsurface flux tubes break through the photosphere in active regions, they form Ω-shaped stitches of flux extending into the chromosphere and corona. Since the total magnetic flux of a large active region, of order 10^{22} Mx, seems to exceed the total flux in the polar regions above 60° latitude, this low-latitude, mainly toroidal flux appears to be the dominant global component of solar magnetism, except possibly at very low activity levels.

The toroidal field pattern at mid and low latitudes is evident not only in the active regions but also in the longitudinal alternation of polarities seen in the quiet regions. As pointed out in Section 8.4.3, the background quiet region fields exhibit a predominance of one polarity over the other within very large scale unipolar magnetic regions whose polarity alternates in longitude around the sun.

A good idea of the global connection pattern of solar magnetic field lines can be obtained from the chromospheric and coronal loop structures described in Chapter 9. From the evidence given there we infer that distant active regions can be connected by huge coronal loop systems following field lines that span hundreds of thousands of kilometers and often cross the equator. The geometry of these magnetic connections varies on the time scales of active region evolution, but in general it is more complex than the purely toroidal geometry suggested by the simplest interpretation of the Hale-Nicholson law.

For instance, a significant fraction of the low-latitude flux tubes do not simply run below the photosphere between active regions and above it within them, as the Babcock model postulates. The fraction of the total flux in coronal flux tubes connecting between active regions is difficult to estimate, but it seems to account for a substantial part of the total active region flux. Moreover, these coronal connections between active regions have a substantial poloidal north-south component, as well as the purely toroidal east-west component.

Comparison with observed coronal structures indicates that to a very good approximation, the field structure in the chromosphere and corona can be represented by a potential $(4\pi\mathbf{j} = \nabla \times \mathbf{B} = 0)$ field extrapolation of the observed distribution of the line-of-sight magnetic field component at the photosphere. This means that the solar field above the photosphere is mainly generated by subphotospheric current systems. Coronal and chromospheric currents running along field lines account for only relatively minor distortions of this potential field.

It is of some interest, particularly in studies of how the solar magnetic field maps into the solar wind and interplanetary medium, to expand the photospheric field into its spherical harmonic components. Results for the years 1959–1974 indicate that the harmonic found to be dominant most often, particularly at high activity levels, is a dipole lying in the equatorial plane. A dipole oriented north-south, the original concept of a solar magnetic field, was found to be prominent only during times of low activity. Higher-order quadrupole and octupole structures were also important at times. The mapping of these fields into the interplanetary medium is discussed in Chapter 12.

ADDITIONAL READING

Observations and Interpretation of Sunspot Phenomena

R. Bray and R. Loughhead, *Sunspots*, Chapman and Hall, London, 1964.

L. Cram and J. Thomas, (Eds.), "The Physics of Sunspots," Sacramento Peak Observatory Publication (1981).

P. Maltby et al., "A New Sunspot Umbral Model," *Astrophys. J.*, **306**, 284 (1986).

R. Moore and D. Rabin, "Sunspots," *Ann. Rev. Astron. Astrophys.*, **23**, 239 (1985).

J. Thomas, "Oscillations in Sunspots," *Austral. J. Phys.* **38**, 811 (1985).

Observations and Interpretations of Faculae

G. Chapman, "Active Regions from the Photosphere to the Chromosphere," in *Solar Active Regions*, F. Orrall (Ed.), Colorado University Press, p. 43, 1981.

R. Muller, "The Fine Structure of the Quiet Sun," *Solar Phys.*, **100**, 237 (1985).

Structure of the Sun's Magnetic Field

N. Sheeley, "The Overall Structure and Evolution of Active Regions," in *Solar Active Regions*, F. Orrall (Ed.), Colorado University Press, p. 17, 1981.

C. Zwaan, "Solar Magnetic Structure and the Solar Activity Cycle," in *The Sun as a Star*, NASA Publication SP-450, S. Jordan (Ed.), p. 163, 1981.

Y. Wang, A. Nash and N. Sheeley, "Magnetic Flux Transport on the Sun," *Science* **245**, 712 (1989).

EXERCISES

1. Calculate the magnetic pressure exerted by a sunspot or facular field, and compare it with the ram pressure of granular and supergranular velocities and with the kinetic gas pressure at the photosphere. Taking into account the depth of the Wilson depression, is it reasonable that the measured umbral fields might be confined by a gas pressure deficit relative to the surrounding gas at equal geometrical depth below the photosphere?

2. If the umbral temperature is only about 4000 K, its light is much redder than photospheric radiation. Plot the spectral distributions of 4000 K and 6400 K black bodies, and calculate the umbral intensity contrast I_u/I_{ph} relative to the photosphere. Compare your results with the observed values of approximately 0.05 at 4000 Å, 0.1 at 5500 Å, 0.15 at 7500 Å, 0.23 at 1 μm, and 0.58 at 2 μm. Comment on errors of modeling and measurement that bear on this comparison.

3. Calculate the phase speed of a sound wave propagating along magnetic field lines in the penumbral chromosphere, taking the gas temperature as 10^4 K. Referring to the inclination of umbral magnetic field lines in Fig. 11-2, discuss the geometry of such a wave and comment on why the basic disturbance responsible for driving chromospheric running waves is more likely to be photospheric. Include a calculation of the Alfvén speed in the photospheric penumbra in your considerations.

4. Assuming the magnetic pressure gradient just balances the gas pressure deficit in a completely evacuated flux tube, calculate the decrease of B with height in the photosphere for a flux tube of facular intensity and diameter. Is it likely that the increase of cross-sectional area could be measured directly at photospheric levels?

5. Calculate the current density **j** required at the atmospheric levels where $(\nabla \times \mathbf{B})/B \sim 1/L$, where $L = 5 \times 10^4$ km is the scale of an active region magnetic field system. Calculate the potential drop along the magnetic field lines, given photospheric resistivities (see Chapter 4). What is the power dissipated by this current system per unit volume and how does it compare with the radiative energy budget at the photosphere?

6. Assuming that a facula is about 0.5% brighter than the photosphere at disk center (i.e., at $\mu = \cos\theta = 1$) at a wavelength of 5000 Å and about 4% and 10% brighter at $\mu = 0.5$ and $\mu = 0.2$ respectively, draw a rough plot of its limb-darkening curve relative to that of the photosphere at the same wavelength. Using this plot determine at what disk positions the facula (i) has greatest absolute intensity and (ii) makes its greatest contribution to the total solar irradiance. Reconcile your findings with the increased visibility of faculae near the limb.

7. Using the angular distribution of facular radiation assumed above, calculate the excess radiative flux of a facula relative to the photosphere at 5000 Å, using equation (2-7). Compare this with the missing heat flux (per unit area) of a spot. If the excess radiation of the faculae in an active region were to balance the missing radiative flux of the spot, what ratio of facular to spot area would be required, and how does it compare to typical measured areas?

9

The Chromosphere
and Corona

During a total solar eclipse the sun's glaring disk is covered by the moon, revealing a pearly white, subtly structured halo extending to a distance of several solar radii beyond the photospheric limb. This extended atmosphere of the sun is called the corona. For a few seconds just before and also just after eclipse totality, a thin reddish annulus is also seen around the rim of the photosphere. Its striking color compared to the white corona prompted early observers to name it the chromosphere (color-sphere).

We owe the discovery of these phenomena to the remarkably exact coincidence in angular diameters of the sun and moon as seen from Earth, and early studies of the chromosphere and corona relied entirely on observations during eclipses. The inventions of the spectrohelioscope, spectroheliograph, and coronagraph described in Chapter 1 gradually made observations of the outer solar atmosphere more accessible. But the art of eclipse observations with temperamental equipment transported to distant sites has played a large part in the lore of coronal and chromospheric research.

Our present recognition that the feeble lights of the chromosphere and corona are emitted by plasmas much hotter than those of the immensely brighter photospheric disk emerged in large part from the identification of the coronal forbidden lines by W. Grotrian in 1939, and more conclusively by B. Edlén in 1942. Additional evidence for very high temperatures came from the large coronal density scale heights noted on eclipse plates, and from the surprising width of the coronal forbidden lines first measured by B. Lyot. The basic problem posed by this rather amazing discovery is to find the mechanism responsible for heating these layers by nonthermal energy transport, such as waves or electric currents. Heating by radiation, convection, or conduction from the cooler photosphere is ruled out by the second law of thermodynamics.

Imaging in chromospheric optical lines and in the EUV and soft X-radiations that reveal coronal structures on the disk has proved particularly useful in studying the dynamics of these layers. The first observations of solar plasmas in the temperature range between chromospheric and coronal values were made in their microwave emissions in the 1950s. But the complex structure of this transition region of temperature was only revealed by EUV telescopes flown on *Skylab* in 1973. Their intricate morphology is largely determined by the structure of magnetic fields extruding from the sun's convection zone.

Investigations of the corona and chromosphere have come to occupy most of solar research in the past two decades. This is due in part to the large volume and high quality of EUV and soft X-ray data obtained from satellites flown since the Orbiting Solar Observatory (OSO) series beginning in the late 1960s, and also from rockets. Valuable insights have been gained about the connections between solar magnetism, plasma motions, and nonthermal heating of the sun's outer atmosphere. Nevertheless, a clear identification of the mechanism(s) that cause the sun's atmospheric temperature to jump to millions of degrees just above the photosphere still eludes us. Also, the propulsion of spicules, the prominent chromospheric plasma jets seen leaping outward from the limb, is not understood. New ideas and observations seem to be required to make progress on these classical problems.

9.1 THE CHROMOSPHERE

9.1.1 Observations of Structures and Motions at the Limb

The limb chromosphere can be observed on photographs taken in broadband light at eclipses when the relatively weak chromospheric line radiations are not obscured by scattering of photospheric light in the Earth's atmosphere. But the contrast of chromospheric structures is greatly enhanced by monochromatic imaging in strong chromospheric lines such as $H\alpha$ or $Ca\,II$ H and K. Even moderately narrow interference filters of a few angstroms passband show the chromosphere at the limb outside of eclipse. Limb imaging is useful for identifying some of the basic chromospheric structures and for determining their heights. Spectra taken at the limb, particularly during eclipses, also afford the best opportunity to study the chromospheric radiations without the confusion of the many photospheric lines present on disk spectra.

The limb observed with a narrow-band filter tuned to wavelengths around $H\alpha$ is shown in Fig. 9-1. The chromosphere is usually defined to begin in the temperature rise just above the temperature minimum region. This occurs about 300–400 km above the level of $\tau_{0.5} = 1$, thus just above the photospheric limb seen in the topmost panel of Fig. 9-1. The elongated, tilted structures known as spicules are evident in the bottom panels of Fig. 9-1, taken with the

Fig. 9-1 Spicules photographed at the limb through a filter tuned in wavelength through the Hα line from the red wing (top) to the line center (bottom). The red wing filtergrams tend to show spicule plasma receding from the observer. National Solar Observatory, Sacramento Peak, New Mexico.

filter tuned to the center of the Hα line. The largest can be measured to heights of about 15 arc sec above the limb. More typically they extend over roughly 10 arc sec, with thicknesses of between 1 and 2 arc sec. It is difficult to determine from the limb observations whether the appearance of diffuse chromospheric emission also seen in Fig. 9-1 arises from interspicular material, or whether it is simply the superposition of many optically thin spicules along the line of sight.

The lifetimes and motions of spicules can be studied with a combination of time-lapse photography and spectra. They seem to last typically between 5 and 10 min when followed in time-lapse Hα films. The largest and longest lived occur near the poles, and their size may be connected to the different atmospheric structure associated with the polar coronal holes discussed below. Exceptional macrospicules, which are best observed in EUV lines, are found to extend 40,000 km beyond the limb.

Spicule motions are measured in two different ways. The simplest procedure is to plot their extension above the limb against time. This rate of elongation has been estimated in many studies carried out in Hα and gives a speed transverse to the line of sight of roughly 25 km s^{-1}. Doppler shifts of the spicular plasma are more difficult to measure. They yield a line of sight velocity of about 5 km s^{-1}. The two measurements can be reconciled if we

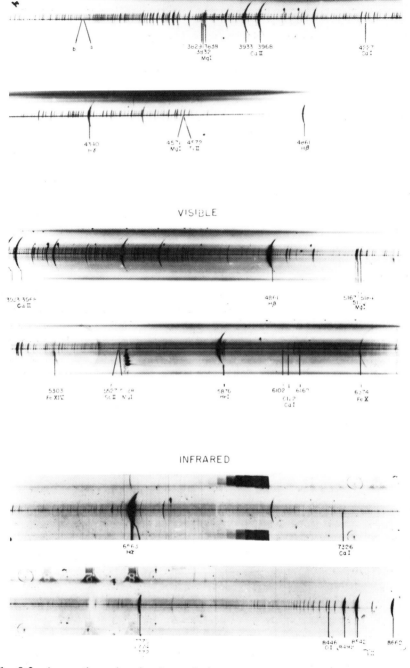

Fig. 9-2 A negative print showing a flash spectrum of the chromosphere from the ultraviolet below 3500 Å to about 8700 Å in the infrared. Obtained at Khartoum, Sudan, during the 1952 eclipse by an expedition from the High Altitude Observatory, Boulder, Colorado. By permission of G. Athay.

take the Hα expansion velocity of 25 km s^{-1} as the true plasma flow speed along the axes of spicules whose average tilt from the radial (and thus also out of the sky plane) is estimated to be about 20°.

Studies of chromospheric limb spectra provide some of the most direct information about chromospheric chemical composition, temperatures, and pressures. They are the only unambiguous source of information on the height dependence of these properties. Figure 9-2 shows a good chromospheric "flash" spectrum taken in the few seconds around eclipse totality when the radiations of the chromosphere appear most brilliantly.

The lowest chromospheric layers below about 1000 km, which produce this spectrum, emit a rich variety of atomic lines of heavy elements. They also emit strongly in the lines of hydrogen, in both neutral and ionized helium, and in the bands of CN. The low Balmer and Paschen lines of hydrogen, the Ca II H- and K-lines, and some of the singly ionized metal lines extend to between 5 and 10 times this height, where they are resolved into spicules. Attempts to explain the relative intensities of lines in this spectrum only began to meet with success when the techniques of non-LTE radiative transfer were brought to bear upon this problem in the 1930s by D. Menzel and later by R. Thomas, G. Athay, and others.

The fundamental point that thermodynamic equilibrium (and thus a balance between collisional and radiative processes in the plasma), does not hold even approximately in the tenuous chromospheric plasma, came to be generally accepted by the 1950s. The techniques for interpretation of the detailed shapes of the optically thick chromospheric lines such as Hα or Ca K are now relatively well understood, but the computations are complex, and the conclusions tend to be dependent upon the uncertain geometry and velocity structure of spicules and other magnetic structures described below. This complexity has made for relatively slow progress in understanding the dynamics of the chromosphere, and even of the information it holds on the solar abundance of helium.

9.1.2 Observations on the Disk

Observations on the disk in the strongest Fraunhofer lines provide additional information on the structures seen at the limb. Narrow passbands of 0.5 Å or less, usually obtained with photoheliographs using birefringent filters, or with spectroheliographs, are required to take advantage of the highest opacity in the cores of Hα, Hβ, and Ca H and K and thus achieve adequate contrast to study the disk structure. Figure 9-3 shows full disk photoheliograms obtained in Hα and Ca K for comparison with the underlying structures seen in white light. The sunspots so evident in white-light photographs are harder to find in the chromospheric radiations. But the faculae, which are barely detectable in white light except near the limb, are seen to be overlaid by bright chromospheric areas called plages (French for beaches) which are easily observed near spots at any position on the disk.

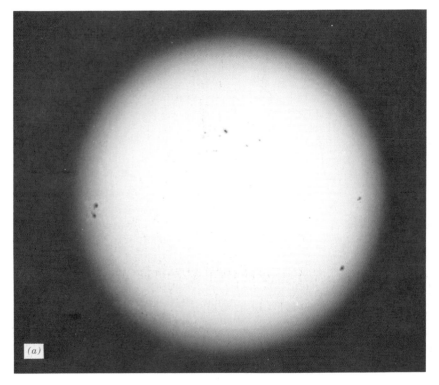

Fig. 9-3   The solar disk imaged sequentially in (a) white light, (b) Hα and (c) Ca K. National Solar Observatory, Sacramento Peak, New Mexico.

The most prominent feature of the quiet chromosphere outside of active regions is the network of bright mottles that extends over the whole disk. Comparison of the Hα filtergram shown in Fig. 9-4 with the accompanying photospheric magnetogram shows that these bright mottles are roughly cospatial with the magnetic network discussed in Section 8.4. As discussed later, the physical mechanism responsible for the bright mottles and for the more extended plages seems to be similar. This chromospheric network defined by the Hα and Ca K mottles was first discovered in early spectroheliograms by Hale, Deslandres, and others, more than 60 years before the magnetic fields outside sunspots were properly observed in the 1950s.

The appearance of this network depends on the line used to observe it, as seen in Fig. 9-3. It also varies between filtergrams obtained at different positions within the line profile. As illustrated in Fig. 9-5, the network is still seen in the wings of Hα, but it is marked now by dark, more elongated structures. These fan out from the network in a bushlike arrangement. The three-dimensional geometry of the dark mottles seen near the limb is evident. Detailed studies indicate that the dark, elongated mottles seen on the disk in the Hα wings are the same spicules that we observe in profile at the limb,

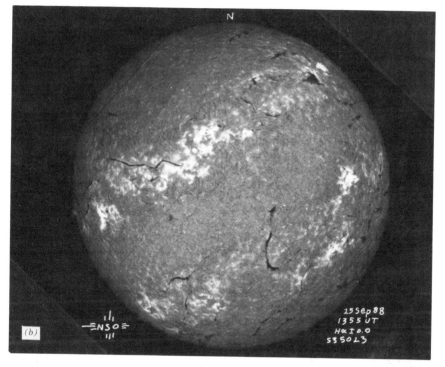

Fig. 9-3   *(Continued)*

although this is difficult to prove since it is hard to trace these very fine structures across the limb on a single exposure. It is encouraging that the lifetimes of individual dark mottles are found to be similar to the 5–10 min measured for spicules.

If we accept that dark Hα mottles are spicules seen in absorption on the disk, it appears from photoheliograms such as Fig. 9-5 that the upper chromosphere consists of little else besides spicules clustered in the network. Since these spicules cover only about 1–2% of the solar surface, it appears that the chromospheric volume is largely empty, with occasional jets of dense spicular material protruding from below. Chromospheric material is observed in spectra taken of the much larger fraction of the disk covered by the supergranule cells, but it seems to be confined to a relatively thin layer lying within a few arc seconds above the photospheric limb. The implication is that over the supergranule cells, the plasma temperature might rise to coronal values over an extremely short distance above the limb. This is consistent with reports of coronal lines observed down to only a few arc seconds above the limb during eclipse totality.

In active regions, the increased number density of magnetic flux tubes gives rise to the extended bright chromospheric plages which consist of many bright

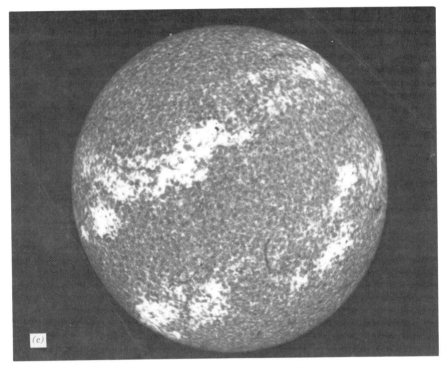

Fig. 9-3 *(Continued)*

mottles packed close together. As shown schematically in Fig. 9-6, the plages occupy areas of intense radial magnetic fields, just as do the bright mottles of the network. Extending out of the plages we see dark structures similar to the dark mottles discussed above in the quiet network, but more elongated. These are referred to as fibrils.

Figure 9-6 illustrates how fibrils are similar to the spicules or dark mottles, except that they are bent over and extend along the solar surface at a height of about 4000 km rather than protruding radially outwards. Their length of typically 15×10^3 km and thickness of about 2000 km somewhat exceed spicular dimensions, but their lifetimes, and the proper motions along their axes of between 20 and 30 km s^{-1} are both similar.

As illustrated in Fig. 9-6, the roughly radial geometry of spicules is consistent with their occurrence in magnetically quiet regions. Here chromospheric fields tend to be vertical, extending well outward into the corona before returning to the photosphere in relatively distant areas of opposite magnetic polarity. Near active regions the field is known to connect more locally to opposite polarity, often over distances of 10,000 km or less. Here the chromospheric field lines run more horizontally and give rise to fibrils.

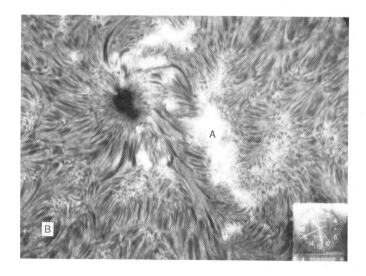

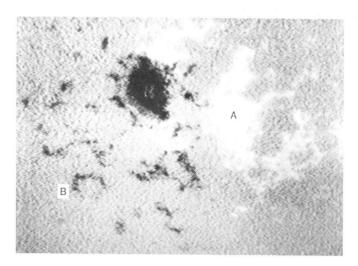

Fig. 9-4 A nearly simultaneous Hα photoheliogram (top) and photospheric magnetogram (bottom) showing the correspondence of chromospheric structures and magnetic field. The letters A and B denote examples of plage and network (bright mottles in Hα), respectively. Photoheliogram and magnetogram obtained at Big Bear Solar Observatory, California Institute of Technology.

Fig. 9-5 Photoheliogram in the wing of the Hα line, of an active region and network near the limb. Arrows point to examples of dark mottles. National Solar Observatory, Sacramento Peak, New Mexico.

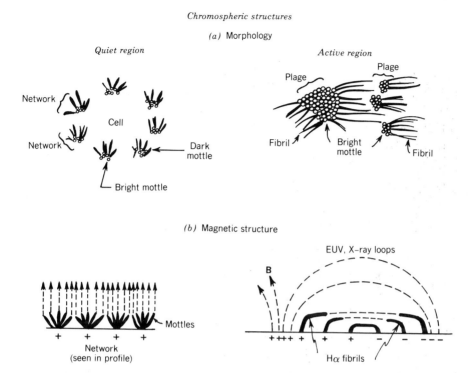

Chromospheric structures

(a) Morphology

Quiet region   *Active region*

(b) Magnetic structure

Fig. 9-6 Schematic diagram showing chromospheric magnetic structures observed in Hα.

Quite often the fibrils seen issuing from one plage can be traced to their termination in another plage of opposite polarity, as illustrated in Fig. 9-6. These connections can sometimes be followed over great distances, even extending between active regions in different hemispheres. In Hα we see only the portions of such long flux tubes near the two opposite polarity footpoints. The intervening plasma is usually too hot to be visible as most of the flux tube length extends upward into the corona. One expects to see more of the extended connections between active regions in hotter or more opaque plasma radiations such as the Lyman α line, in the EUV resonance lines of heavy ions, and in X-rays. Figure 9-11 shows some good examples of such far-ranging coronal loops.

The pattern of fibrils around an active region is often reminiscent of iron filings aligned around a bar magnet. This suggests magnetic ordering, although the field in chromospheric structures is difficult to measure directly, mainly because the strong chromospheric lines in which the spicules, fibrils, and mottles are visible have small Lande g-factors, making magnetograph measurements difficult. Consequently, our knowledge of chromospheric magnetic

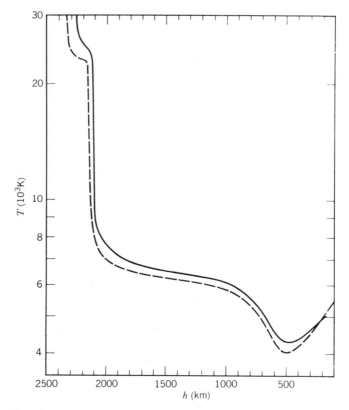

Fig. 9-7 Plot of temperatures versus height in the chromosphere for the cells (dashed) and network (solid). Adapted with permission from J. Vernazza, E. Avrett, and R. Loeser, *Astrophys. J. Suppl.*, **45**, 635 (1981).

fields is based almost entirely on comparisons between Hα fibril structures and photospheric fields extrapolated into chromospheric layers.

Such extrapolations have been discussed in Chapter 8. Comparisons between calculated fields in an active region and the chromospheric Hα fibril patterns observed at about the same time indicate that the main features of the fibril geometry can be reproduced with a potential field ($\nabla \times \mathbf{B} = 0$) calculation, but better fits can be achieved when substantial currents are assumed to flow along magnetic field lines.

Such currents do not violate the force-free condition in the chromosphere since $\mathbf{j} \times \mathbf{B} = 0$, but their presence does imply substantial free energy in the field within this layer since they require $\nabla \times \mathbf{B} \neq 0$ and thus a twisting of the field lines not present in a potential (current-free) field. This free energy is available for release in atmospheric heating and in transient eruptions such as flares and filament activations.

9.1.3 Physical Conditions

The complex morphology of the chromosphere described in the previous section makes it difficult to construct a model of temperature and density that can be related to the magnetic structures such as mottles, fibrils, and spicules. Observations on the disk show that even in the quiet sun, the network exhibits considerable contrast at all chromospheric heights. This indicates that a satisfactory model, even of the lowest chromospheric layers, must have at least two components to take into account the most evident difference between bright network mottles and the darker cell centers.

The two-component model presented here for the low chromosphere is based on EUV observations of the continuum emission between 400 and 1400 Å made from *Skylab*. The spatial resolution of the EUV spectrometer used to measure these continuum intensities near disk center was 5 arc sec. This is sufficient to distinguish the bright network mottles from the darker cells, but not enough by far to resolve the individual network magnetic flux tubes. This means that the "network" component presented in the model indicates the kinds of changes in the atmosphere that can produce higher chromospheric intensities, but their magnitude is likely to be greatly underestimated.

The iterative technique used to construct this model is similar to that described in Chapter 5 for modeling the photosphere. Hydrostatic equilibrium is assumed, which is defensible in the lower chromospheric layers since large-scale velocities observed there are typically $1–2$ km s^{-1} and thus well below the sound speed. A significant fraction of the "pressure" is provided, however, by small-scale turbulent velocities of larger amplitude that do not produce line shifts, although they are observed as line broadening.

The model is one-dimensional, so that each of the two components is assumed to have large horizontal extent compared to the depth of the layer. This assumption is reasonable in the cells, whose dimensions of about 30,000 km far exceed the depth of at least the low chromospheric layers, as we shall see. It is not obviously good in the network, where the horizontal and vertical scales of elements seem to be comparable.

The chromospheric temperature profiles with height in the cells and network are plotted in Fig. 9-7. For both components, the model indicates a rapid temperature rise from minimum values around 4500 K located some 500 km above $\tau_{0.5} = 1$ to a broad temperature plateau of between 6×10^3 K and 7×10^3 K, about 500 km higher. This plateau extends out to about 2000 km and is followed by a very rapid rise toward coronal values within a few hundred kilometers. More detailed models of the atmospheric structure in the transition region and corona, which take into account the diverging magnetic field geometry at these levels, are discussed in Section 9.2.4.

The height distributions of temperature, density, pressure, and other physical variables for the network and cells are given in Table 9-1. As noted above, the "cell" distribution may describe that atmosphere quite well. Thus, over most of the sun's surface, the low chromosphere can be visualized as a roughly

TABLE 9-1   Chromospheric Physical Conditions in Cells and Network.

h (km)	m (g cm^{-2})	T (K)	n_H (cm^{-3})	n_e (cm^{-3})	P_{total} (dyn cm^{-2})
			cell model		
2245	3.942E $-$ 06	23000	1.383E $+$ 10	1.436E $+$ 10	1.080E $-$ 01
2135	4.328E $-$ 06	18000	1.981E $+$ 10	1.876E $+$ 10	1.186E $-$ 01
2093	4.722E $-$ 06	7830	4.906E $+$ 10	2.694E $+$ 10	1.294E $-$ 01
1915	7.932E $-$ 06	6715	1.115E $+$ 11	2.992E $+$ 10	2.173E $-$ 01
1280	1.392E $-$ 04	6090	3.408E $+$ 12	6.022E $+$ 10	3.815E $+$ 00
705	7.364E $-$ 03	4890	2.667E $+$ 14	5.945E $+$ 10	2.018E $+$ 02
505	5.096E $-$ 02	4020	2.271E $+$ 15	2.557E $+$ 11	1.396E $+$ 03
250	6.159E $-$ 01	4755	2.322E $+$ 16	2.645E $+$ 12	1.687E $+$ 04
			network model		
2245	6.868E $-$ 06	50000	1.152E $+$ 10	1.273E $+$ 10	1.882E $-$ 01
2135	7.442E $-$ 06	24000	2.537E $+$ 10	2.625E $+$ 10	2.039E $-$ 01
2093	7.725E $-$ 06	10800	5.743E $+$ 10	4.453E $+$ 10	2.116E $-$ 01
1910	1.253E $-$ 05	7140	1.671E $+$ 11	4.912E $+$ 10	3.434E $-$ 01
1280	1.936E $-$ 04	6320	4.588E $+$ 12	8.883E $+$ 10	5.305E $+$ 00
705	8.856E $-$ 03	5100	3.077E $+$ 14	9.045E $+$ 10	2.426E $+$ 02
505	5.713E $-$ 02	4270	2.398E $+$ 15	2.900E $+$ 11	1.565E $+$ 03
250	6.167E $-$ 01	4775	2.315E $+$ 16	2.668E $+$ 12	1.690E $+$ 04

Symbols: h is the geometrical height above $\tau_{0.5} = 1$; m is the mass of plasma in a column of 1 cm^2 cross-section above h; T is the temperature; n_H and n_e are the hydrogen and electron number densities; and P is the total pressure, including a contribution from plasma turbulence. Adapted from J. Vernazza et al., *Astrophys. J. Suppl.* **45**, 635 (1981).

spherical shell of roughly 1500-km thickness and about $6-7 \times 10^3$ K tempera-ture. The particle density in this shell decreases rapidly from about 10^{15} cm^{-3} near the temperature minimum to about 10^{11} cm^{-3} near 2000 km.

Figure 9-8 illustrates the approximate heights of formation of various important chromospheric radiations, in this model. The model accounts very well for the intensity behavior of the EUV continua of He I, H I, C I, and Si I between 400 Å and 1400 Å as expected, since it is iterated to produce agreement with the observed fluxes in these radiations. It also provides reasonably good fits to the rather complicated profile shape of the strongest chromospheric emission lines such as Ca II H and K, Mg II H and K, and of the strongest Lyman and Balmer lines.

As seen in Fig. 9-9 the Ca II resonance lines observed in absorption on the disk show broad wings, an increase of intensity toward the line center, and a sharper dip at line center. Since the opacity in such a line increases toward its center, the shape can be broadly interpreted as a map of brightness tempera-ture with depth in the solar atmosphere. In the far wings, where opacity is least, we see to the local H$^-$ continuum level somewhat below $\tau_{0.5} = 1$. Closer

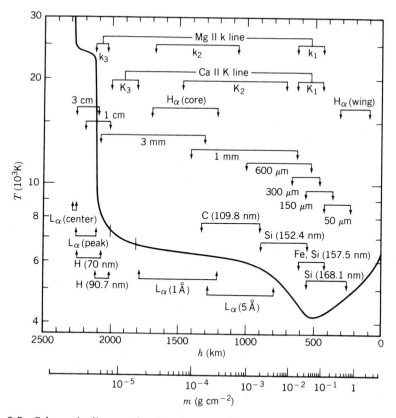

Fig. 9-8 Schematic diagram showing the approximate regions of formation (in height and in mass column density) of various chromospheric radiations. The temperature plateau at 20,000 K does not appear in more recent calculations. From J. Vernazza, E. Avrett, and R. Loeser, *Astrophys. J. Suppl.*, **45**, 635 (1981).

to the line center, as the opacity increases, we see higher in the photosphere, where the temperature decreases and the intensity drops. At the point of first minimum, called K_1, we are observing the temperature minimum region. The increase of T_b with opacity even closer to line center, marked K_2, is caused by the chromospheric temperature rise. Finally, the K_3 dip or central reversal arises because near the line core the opacity is so high that we see layers of low density whose emission falls well below that of a plasma of the same temperature in LTE.

Physical conditions in the highly dynamical spicule and fibril structures are more difficult to determine. Models are based mainly on their chromospheric radiations in the visible spectrum, particularly in the strongest lines of Ca II, He I, and H I. They indicate spicule temperatures and electron densities of around 9000 K and 10^{11} cm^{-3} at 2000 km and about 16×10^3 K and 4×10^{10} cm^{-3} at projected heights of $8-10 \times 10^3$ km above the limb. The total plasma

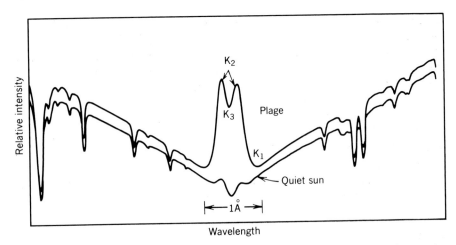

Fig. 9-9 Profiles of the Ca K-line of the quiet sun (lower curve) and of a plage region (upper curve). By permission of O. White.

densities are likely to be considerably higher, given the estimated hydrogen ionization of only about 10% in such structures. Even less is known about the physical conditions in the fibrils, but studies of their EUV continuum absorption indicate that their densities are at least as high as those seen in spicules, and their temperatures are comparable or lower.

The basic shape plotted in Fig. 9-7 for the chromospheric temperature profile in the quiet sun and also for plages and sunspots can probably be understood in terms of the behavior of radiative losses in the sun's atmosphere. The location of the temperature-minimum region at about the same geometrical height in the three magnetically different atmospheres seems to be a consequence of the strong temperature dependence of photospheric emissivity, which approximates that of a black body. This acts as a thermostat, tending to keep the temperature fairly constant through these layers.

An increase in the heating function per unit volume, Q, would then act mainly to move this temperature minimum inward from a layer of optical depth τ_0 to a larger optical depth τ_1, obeying the relation

$$Q = \Delta\tau\, B(T), \qquad (9\text{-}1)$$

where $\Delta\tau = \tau_1 - \tau_0$. Given the small scale height of τ around the temperature-minimum region, an order of magnitude increase in Q requires only about a 100–150 km decrease in height. But even such small differences in the geometrical height scales between the atmospheres of a plage or sunspot and the quiet sun can have dramatic effects on the dynamics of magnetic flux tubes, as discussed in Chapter 8.

The temperature plateau between 6000 and 7000 K in these models probably arises from the steep dependence of chromospheric line radiation on electron density since the electron supply is determined by the collisional ionization of hydrogen at temperatures above about 5500 K. In hotter solar plasma the electron supply is strongly temperature dependent, so that even a small temperature increase produces sharply higher radiative losses. This thermostat ceases to operate when hydrogen becomes mainly ionized at temperatures close to 10,000 K.

The explanation of the chromospheric temperature profile given above in terms of the behavior of plasma radiative losses suggests that we cannot expect to learn much about the specific mechanism that heats the chromosphere by pondering the shape of the curves in Fig. 9-7. To make progress on this interesting question we must calculate the magnitude of the radiative losses with height, as discussed in the next section.

9.1.4 Energy Balance

A first step is to estimate the rate of energy losses from the chromosphere and thus the required power input. Conductive fluxes back to the photosphere (and also into the chromosphere from the corona) can be neglected throughout the bulk of the chromosphere because the thermal conductivity at these temperatures is too low.

The radiative losses can be computed from a model chromosphere. A model is required (rather than just summing up the observed radiations) because many of the strong lines and continua arise in part from photospheric layers, and their contribution must be separated. Also, some chromospheric radiation is merely photospheric light scattered by the chromospheric plasma and does not represent true emission from these plasmas. Finally, the chromosphere is also heated by radiations both from the photosphere below and the corona above, and their contribution must be subtracted.

Such computations indicate that the largest contributions to the overall chromospheric radiative losses of roughly 5×10^6 ergs cm^{-2} s^{-1} are from the Ca II lines, including the infrared triplet lines at $\lambda\lambda$ 8542, 8498, and 8662 Å, followed by the Balmer lines, the Mg II resonance lines, Lyman α, and chromospheric radiations in the H$^-$ ion. The computations indicate that the net radiative losses (i.e., the divergence of the radiative flux) peak at a value of about 10^{-1} ergs cm^{-3} s^{-1} around a height of 800 km. They decrease steadily by a factor of about 50 out to about 2000 km, with a sharp peak of about 5×10^{-2} ergs cm^{-3} s^{-1} around 2500 km. In a steady state these net losses must be balanced locally by some form of heating, so the shape of this loss curve might give some clue as to the nature of the heating process. But detailed calculations of radiative losses from the chromosphere are of limited value unless the other, possibly dominant, contributions to the energy balance of these plasmas (discussed in Section 9.1.6) can be evaluated with comparable accuracy.

To identify the chromospheric heating mechanism it would be helpful to know the lowest atmospheric level at which mechanical heating is required. In principle, this could be determined by finding the level at which the observed atmosphere becomes hotter than a model of a theoretical radiative-convective atmosphere in which no mechanical heating is included. Figure 5-12 shows such a theoretical atmosphere for reference. But the uncertainties both in the radiative-convective atmosphere and in the observed atmosphere are too large to determine whether significant nonthermal heating is present already at the temperature minimum, or perhaps even in the upper photospheric layers. At levels above about $\tau_{0.5} \sim 10^{-4}$, however, the temperatures rise above the values of the radiative-convective model by a significant amount, and mechanical heating must be present.

9.1.5   Chromospheric Heating

The heating mechanism responsible for the chromospheric temperature rise is still uncertain, but the most attractive prospect is provided by shock wave dissipation. The basic idea was put forward by L. Biermann in 1946 and by M. Schwarzschild in 1948 that acoustic waves generated by photospheric turbulence would propagate outward, steepen into shocks, and dissipate their energy to heat the chromosphere and corona. Subsequent calculations and observational tests indicate that such shock heating can account for at least the initial chromospheric temperature rise, and they have narrowed the range of wave types and frequencies likely to be responsible. However, they also suggest that the corona may not be heated in this way, and even the dominant influence of shock heating in the chromosphere is difficult to prove.

Calculations of the power spectrum of acoustic flux in the photosphere using aerodynamic theory suggest that peak vertical fluxes of order 10^8 ergs cm^{-2} s^{-1} might be expected between wave periods of 10–100 s. The actual values are very uncertain due to the effects of compressibility and the very strong dependence on the power spectrum of photospheric turbulent velocities. But acoustic fluxes of the magnitude required to heat the solar atmosphere above the temperature minimum seem quite easy to achieve.

The wave energy fluxes propagating outward from the photosphere can be calculated assuming such an acoustic noise spectrum. It is found that these simple waves dissipate most of their energy in the upper photosphere, where radiative losses during the compression and heating phase are greatest. Nevertheless, sufficient power propagates outward to form shocks and contribute significant heating beyond the temperature minimum.

The results of these calculations are uncertain because the shock theory used is an approximation and the damping by radiation is difficult to calculate accurately. Moreover, the rate of steepening of the shock depends sensitively on the initial acoustic power available and on the geometry of the magnetic field that acts to guide the wave motion. However, the results indicate that the required power can probably be supplied at least to the chromospheric layers up to about 2500 km.

At the same time, the calculations indicate that the dissipation of these shocks by the strong chromospheric radiative losses makes it difficult for sufficient energy to reach even the upper chromospheric layers. The longest-period waves are expected to dissipate most slowly [see formula (4-86)], but observations of Doppler shifts in EUV lines do not indicate sufficient power at these long wave periods to balance losses from the plasmas above $T > 10^5$ K. In fact they may rule out these waves as heating for any but the lowest chromospheric layers. Direct observations of the high-frequency waves of period below 50 s are more difficult, and better measurements of the power at these frequencies will be required to determine whether longitudinal waves can provide the heating higher up.

Comparison of the chromospheric radiative loss profile with the calculated shock heating function shows that both curves drop rapidly with height, roughly following an exponential falloff over the height range 1000–2000 km. Thus the loss function behaves as we might expect if wave dissipation heated the chromosphere. However, the sharp subsequent increase in the calculated losses to a maximum around 2200 km does not agree with wave heating. As we discuss below, it is unlikely that the losses in these upper chromospheric regions are balanced by such viscous dissipation of waves or even by Joule dissipation of electric currents. Rather, it seems that the heating is provided here by a combination of thermal conduction from the corona and advection of heat by downflowing coronal material.

9.1.6 Dynamics of Spicules and Fibrils

The observations of spicules and fibrils indicate that some dynamical process is able to propel large masses of relatively dense and cool chromospheric material along magnetic field lines high into the corona. The heights attained greatly exceed not only the hydrostatic pressure scale height at chromospheric temperatures, but also the heights expected if spicular material were to follow a ballistic trajectory after initial acceleration to the observed relatively low speeds. Many attempts have been made to identify the force that is able to push the spicule so high without ever heating it over 10^4 K or achieving speeds much over 25 km s^{-1}. So far the efforts cannot claim success. Similar problems must be faced to understand the dynamics of fibrils, which seem to be closely related except for their magnetic geometry.

The dynamics of spicules and fibrils may well have broad implications for the mass and energy balance of the whole outer atmosphere of the sun, including the solar wind. The global mass flux of spicular material averaged over the total solar surface is of order 10^{-9} g cm^{-2} s^{-1} and thus about two orders of magnitude larger than the solar wind mass flux. The global kinetic energy flux carried by the spicule mass motions is about two orders of magnitude less than the total energy requirement of even the corona. But the power required to lift them against solar gravity is comparable to the remainder of the total energy budget of the whole atmosphere above the temperature minimum. An even larger power expenditure may be required for their rapid heating if it can be shown that spicules disappear in Hα because their temperature rises to well beyond chromospheric values.

A useful clue is the observation that the upward flux of spicular material is closely related, both in magnitude and spatial location, to strong downflows of plasma in the 10^5–10^6 K range. As discussed in Section 9.2, the downward advection of enthalpy in these transition region flows is at least comparable in magnitude to thermal conduction from the corona back to the chromosphere. To the extent that the spicules may drive a forced circulation of material between the chromosphere and corona, they may determine the coronal density and temperature and thus also to some degree the mass and energy flux in the solar wind.

Several basic mechanisms have been put forward to drive spicules. Besides the need to provide a steady acceleration against gravity, the main observations the mechanism must explain are why these structures occur only in regions of intense vertical magnetic field, why they are so general at these sites, and their lifetimes.

In one attractive model the spicular plasma is driven upward by a compressive simple wave or shock moving along magnetic field lines from the low chromosphere into the corona. Several variants of this model have been proposed, but in general they associate the generation of the initial wave with granular motions at photospheric levels. This explains the general agreement in lifetime of granules and spicule structures.

The main weakness lies in the requirement that the wave must be capable of lifting plasma of densities corresponding to the low chromosphere, to a height of about 10,000 km into the corona. As explained in Section 4.4.4, even strong shocks cannot easily produce spicular densities by compression of the tenuous general medium found above 5000 km, even if mass conservation could somehow be satisfied. Thus the material must come from dense, low levels. The problem then is to explain how the dense material gets so high. This might be achieved by a single kick from a strong shock, but the velocities required in a ballistic trajectory would be far in excess of those observed, and the powerful associated heating would produce a coronal, not a chromospheric structure. So far, calculations based on such models have been limited to the case of adiabatic disturbances. In practice, the strong radiative losses in these chromospheric plasmas rapidly attenuate the amplitude of wave motions propagating in these layers.

One possible solution is to consider spicule driving by a relatively long-period wave to which power is continuously supplied from behind by a piston pushing upward over 5–10 min. The amplitude of the wave, which will be limited by radiative losses, need not be larger than required to maintain the 25 km s^{-1} spicule speed, but the momentum is supplied continuously throughout the spicule lifetime. This basic idea was put forward already by D. Osterbrock in 1961, but unfortunately, subsequent work on the topic has focused more on relatively high frequency waves.

Another mechanism put forward relies on magnetic reconnection to provide the spicular acceleration. In an X-type reconnection geometry (see Section 11.1.6), outward flows are expected from the reconnecting region. The spicule might be identified with the flow generated by the local Joule heating in such a reconnecting region. It might also be driven by a melon-seed acceleration effect on the plasma, which is compressed by the squeezing together of the magnetic field lines. Although these magnetic mechanisms can probably produce structures similar to spicules and fibrils, they require the additional assumption that merging of opposite polarities is a very general process occurring continuously in all network and at all plage edges.

A third class of mechanism associates spicules with thermal and dynamical instabilities at the corona-chromosphere interface. For instance, it has been suggested that since the conductive heat flux back into the chromosphere from the corona seems too large to be radiated away, it may instead produce powerful local heating. The rapid chromospheric pressure increase, particularly in the network where conductive flux is channeled down by the converging magnetic field lines, might then produce a dynamical instability. This could be similar to the Rayleigh-Taylor instability described in Section 11.1.2, and might cause spicule-like jets to be expelled upward.

9.2 THE CORONA AND TRANSITION REGION

9.2.1 Spectrum and Radiation Mechanisms

The white coronal light seen during an eclipse actually consists of several distinct spectral contributions. Near the sun, out to about $2R_\odot$ from disk center, the main contribution is from photospheric light scattered off coronal electrons. The high speeds of the scattering electrons smear out the Fraunhofer lines, making them essentially undetectable, except the H- and K-lines in careful photometry. This featureless continuum, which is of closely similar color to the photosphere, is called the K (for the German "kontinuerlich") corona. Its intensity a few arc minutes beyond the limb is about 10^{-6} of the disk-center intensity and falls by roughly another factor of 100 between 1 and $2 R_\odot$. It is polarized by the electron scattering, with the electric vector parallel to the limb.

Beyond approximately $2R_\odot$ the contribution of the Fraunhofer, or F, corona becomes increasingly important. This faint light also originates in photospheric rather than coronal emission, but is now diffracted by the much slower moving dust of the interplanetary medium. The Fraunhofer lines are clearly visible in the spectrum of this near-sun enhanced zodiacal light.

Unlike the visible coronal continuum, the coronal line spectrum at visible and ultraviolet wavelengths originates by atomic radiations from the hot plasma. As discussed in Chapter 3, the lines detected in the visible region are all forbidden transitions between low-lying fine structure states of heavy, multiply ionized atoms, such as iron. The best known are the λ 5303 green line of Fe xiv, the λ 6374 red line of Fe x, and the λ 5694 yellow line of Ca xv. The difficulty of explaining these strong unknown lines, which were first observed in 1869, posed one of the major puzzles of solar research for 70 years. Their successful identification by Grotrian and by Edlén was one of the major breakthroughs of solar research since the necessity for high temperatures required to produce these multiply ionized species proved that the corona was two orders of magnitude hotter than the photosphere.

The resonance lines of the most abundant coronal ions are found in the ultraviolet and X-ray regions. Some solar spectra at these wavelengths, presented in Figs. 3-3 and 3-4, show the contributions of continua and lines down to the shortest wavelength X-ray line, observed at 1.9 Å. Both the forbidden and permitted coronal line emissions are observable only in the relatively low regions of the corona, since their intensity depends on the square of the density as shown in Section 3.8. Beyond about 10^5 km above the limb, only observations of the F-corona and meter wave radio measurements provide information on coronal structure.

The quiet sun's meter wave emissions have brightness temperatures of approximately 10^6 K and thus arise from coronal plasmas. Emissions of roughly 50 cm to 1 m wavelengths arise from transition region plasmas at temperatures between about 10^5 and 10^6 K. Over most of the disk, the radiation arises from thermal free-free emission. In regions of intense magnetic field, gyro-radiation from thermal and also relativistic electrons plays a role.

9.2.2 Structures of the Corona and Transition Region

White-light photographs during eclipses show coronal structure down to the smallest resolvable spatial scales around 1 arc sec. Examples of two such photographs, made near solar activity minimum and maximum, are shown in Fig. 9-10. Near activity minimum the corona appears symmetrically elongated about the equator as seen in Fig. 9-10(b) because active regions are found at low latitudes. The corona seen in Fig. 9-10(a) shows the effects of higher latitude active regions present at a more active phase of the solar cycle. The most prominent structures are the large streamers which extend to between 1.5 and 2 $R_\odot$ with their axes often tilted at a noticeable angle from the radial direction. The so-called helmet-type streamers typically consist of relatively bright coronal arches lying roughly in the sky plane with low coronal intensity in the center near their bases. A large prominence can usually be seen in Hα within this dark region. The streamer tapers off above the arches into extended, roughly parallel fine structures extending well beyond the helmet structure itself. The arch systems appear to delineate the magnetic field lines connecting between different latitudes.

Other large streamers seen in Fig. 9-10(a) are associated with bright active regions at their bases, rather than a void and a prominence. Such active streamers have shorter lifetimes of a few weeks instead of months. It is likely

Fig. 9-10   White-light photographs of (a) the active corona observed in 1980 in India by an expedition of the High Altitude Observatory and (b) the quiet corona observed in 1954 in Wisconsin. Plate (a) by permission of D. Sime, and plate (b) by permission of J. Bahng.

that, as the underlying active region evolves into an extended bipolar region, the character of its coronal fields and structures evolves from the active region streamer to a helmet type.

Near the poles the coronal intensity is generally depressed, particularly near activity minimum. Shorter high-latitude streamers, called polar plumes, border this region now known to correspond to more or less permanent coronal holes whose properties are discussed below. The polar plumes deviate noticeably from the radial direction and tend to angle down towards lower latitudes as expected if they follow the lines of force of a global solar magnetic field.

The higher-density plasma structures of the lower corona can be observed at the limb with coronagraphs equipped with narrow-band filters passing the green and red coronal emission lines. Closed-loop structures are seen in active regions. Vertical structures, suggesting more open field lines, are evident in the more quiet limb regions.

The appearance of the corona on the disk is best revealed in broad-band soft X-ray photographs such as shown in Fig. 9-11. The most striking features

(b)

Fig. 9-10   *(Continued)*

are the very bright active region condensations and loops, the fainter and more extended loop systems often extending over the equator or between active regions, and the dark coronal holes. Another remarkable feature are the X-ray bright points, found to be the sites of recently emerged, intense bipolar fields in generally quiet regions. Similar structures are seen in monochromatic images in EUV resonance lines of ions such as Mg x or Si xII that are abundant at coronal temperatures.

The transition region plasmas in the temperature range between about 10^4 and 10^6 K can be observed best in the strong EUV resonance lines of multiply ionized heavy atoms such as Ne VII, O VI, and O IV. In Fig. 9-12 we see some of the more sharply defined loops and condensations defined in these cooler radiations. The brightest features are sunspots, whose looplike and plumelike structures are illustrated in more detail in Fig. 9-13. Another notable difference in the transition region radiations illustrated in Fig. 9-12 is the gradual disappearance of the quiet magnetic network as we observe plasmas of progressively higher temperature. As shown in Figs. 9-11 and 9-12, the network is not seen in coronal radiations.

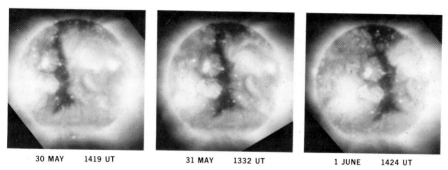

| 30 MAY | 1419 UT | 31 MAY | 1332 UT | 1 JUNE | 1424 UT |

Fig. 9-11 The corona on three consecutive days in 1973 imaged in soft X-rays by the grazing incidence telescope flown on *Skylab* by American Science and Engineering, Inc. By permission of L. Golub.

Network structures in the transition region temperature regime can now also be observed from the ground using microwave interferometers, as at Westerbork in the Netherlands and at the Very Large Array (VLA) in the United States. Although the nominal spatial resolution is about 1 arc sec at a few cm wavelength, the observations do not show quite as fine structures as the EUV and X-ray data, probably because the small structures change during the several hours of observation required to synthesize the required field of view.

The significant decrease in coronal green-line emission relative to quiet sun values over extended areas now called coronal holes was first noted from limb observations by M. Waldmeier in 1957. EUV observations from the *OSO 4* and *OSO 6* satellites first revealed their structure on the disk and showed how common they were. In soft X-rays the emission from a hole is about two to three times lower than in adjacent coronal structures, which are usually systems of high, roughly parallel loops. Holes are also seen prominently in coronal EUV lines, such as Mg x λ 625 and Fe xv λ 284, and in coronal meter wave emissions.

The contrast of the dark holes becomes progressively less in cooler transition region emissions in the EUV and centimeter waves, and they actually appear bright in centimeter and millimeter wave emissions of $\lambda < 2$ cm, for reasons discussed below. They are, however, relatively easy to see as dark structures in He I and He II emissions, such as the λ 304 and λ 584 resonance lines, and in the λ 10,830 line. This visibility in the helium lines seems to be due to the important role of coronal EUV radiation in populating the upper levels of He I and He II ions present in the chromosphere. Thus the weakening of coronal emissions in the hole tends to decrease the helium emission also.

Around activity minimum, holes can cover up to 20% of the sun's surface, and those at the poles are essentially permanent. The lower-latitude holes are

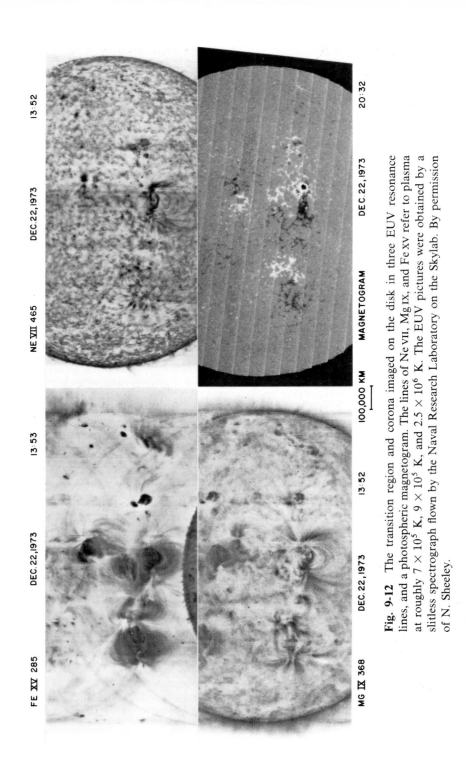

Fig. 9-12 The transition region and corona imaged on the disk in three EUV resonance lines, and a photospheric magnetogram. The lines of Ne VII, Mg IX, and Fe XV refer to plasma at roughly 7×10^5 K, 9×10^5 K, and 2.5×10^6 K. The EUV pictures were obtained by a slitless spectrograph flown by the Naval Research Laboratory on the Skylab. By permission of N. Sheeley.

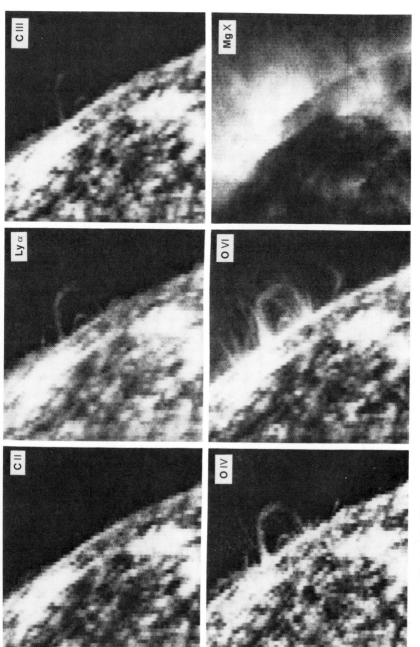

Fig. 9-13 EUV emissions from loop structures associated with a spot group at the limb. The lines of C II, H I, C III, O IV, O VI, and Mg X are emitted by plasmas at temperatures of roughly 2×10^4, 1.5×10^4, 7×10^4, 1×10^5, 3×10^5, and 1.5×10^6 K. These observations were obtained with the Harvard EUV spectrometer flown on *Skylab*.

connected at one time or another to one or both polar holes, and they last for as long as ten solar rotations. Holes form near the centers of large unipolar magnetic regions, and comparisons with calculated magnetic field lines indicate that they are regions of open, perhaps diverging, magnetic fields, although their formation seems to also require the nearby presence of newly emerged closed active region fields.

The development and decay of the polar holes during the solar cycle is closely related to the magnitude and extent of the sun's polar magnetic fields, which are difficult to study by direct Zeeman observations. The implications of important recent research on this topic are discussed in Chapter 11.

9.2.3   Magnetic Fields and Plasma Motions

Coronal magnetic fields are too weak and the Doppler widths of the lines are too large for direct Zeeman measurements to be made reliably, although this may change with the prospect of using infrared coronal lines since the Zeeman effect increases as λ^2. Some interesting techniques have recently been developed using microwave spectral observations of electron gyro radiation around spots. Although the results are more model-dependent than direct Zeeman measurements, they give promise of new tests of the field strength and geometry, at least in active regions. Some Zeeman effect measurements have also been made over sunspots in the transition region radiations of ions such as CIV formed at $T \sim 10^5$ K. They indicate fields of intensity roughly 1 kG over umbral areas whose photospheric fields are about 2 kG.

Most of our understanding of magnetic fields in the corona and transition region is derived from comparisons of the geometry of coronal structures, with field lines calculated from photospheric magnetograms. A comparison of this kind using white-light coronal structures is shown in Fig. 9-14. The detailed correspondence obtained is far from perfect. But it does indicate that closed coronal magnetic field lines correspond to areas where we observe loops in X-rays and EUV on the disk or bright white-light streamers at the limb. Open field lines are usually calculated in regions where coronal holes appear, both in the active latitudes, and at the poles.

The widths of coronal lines place a firm upper limit of about 25 km s^{-1} on velocities in the coronal plasma occurring on spatial scales below the resolution limit of about 1 arc second. This is in accord with the relatively small proper motions of coronal features such as loops and streamers, which are typically between 1 and 10 km s^{-1}, although occasional transient displacements at up to 50 km s^{-1} occur. Spectra of active regions also indicate coronal downflows of 20–30 km s^{-1} over plages, with some evidence for systematic coronal upflows over spots. Recent observations indicate that, in coronal holes, EUV lines tend to be blueshifted relative to the rest of the disk, suggesting a general outflow increasing with temperature to about 12 km s^{-1} at around 1.5×10^6 K.

In the outer corona, observations of the Lyman α intensity near the limb provide an important new technique to study the outflow velocity in the region

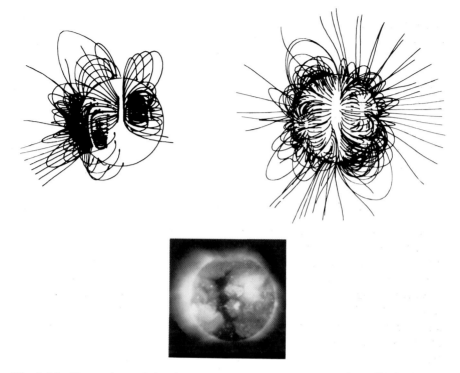

Fig. 9-14 Comparison of the observed X-ray corona and calculated magnetic fields. X-ray image from the American Science and Engineering telescope on *Skylab*. By permission of L. Golub.

where solar wind is likely to be formed. At the heights of interest, around $4R_\odot$, densities are too small for significant coronal emission of Lyman α, so the observed radiation at λ 1216 is produced by resonance scattering of chromospheric Lyman α off whatever H I is locally present in the corona.

The intensity of this Lyman α radiation is determined both by the chromospheric intensity and by the systematic velocity of the coronal H I relative to the chromosphere. Scattering is largest when this velocity is zero. So-called Doppler dimming occurs when the relative velocity increases. From observations of both the scattered Lyman α intensity and the local coronal white-light intensity at least an upper limit on the velocity of the coronal flow can be determined. Outflow speeds of 65 and 100 km s^{-1} have been found by this technique in the quiet sun and in coronal holes, respectively.

Wave motions have been sought in the transition region and corona over the range of periods between several hundred seconds and a few tens of seconds characteristic of waves observed in the photosphere and chromosphere. Fluctuations in velocity and intensity are observed in the $T \sim 10^5$ K plasma, and these might contain a substantial amount of power in upward propagating waves. However, the variations appear to be mainly aperiodic,

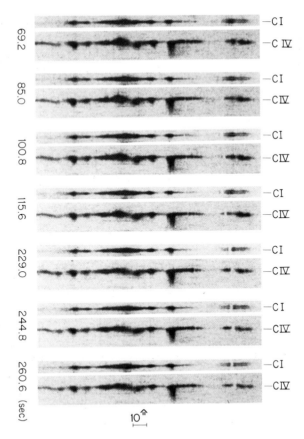

Fig. 9-15   Time sequence of spectra in the C I and C IV lines showing highly blueshifted (i.e., downward) plasma jets changing over time scales of a few minutes. Data obtained during a flight of the HRTS rocket spectrograph by the Naval Research Laboratory. Reproduced by permission of G. Brueckner.

probably due to disruptions in the wave trains caused by the time taken to reach these heights, which is comparable to the lifetime of chromospheric and transition region features.

The transition region velocity field is remarkable in several respects. First, a systematic redshift of transition region lines near disk center relative to the limb is observed in the network. This indicates a general downflow at speeds up to 15 km s^{-1} in the network plasmas of temperature around 10^5 K. The global mass flux in this downflow is comparable to the mass upflow rate in spicules, so it is likely that the downflow represents the return to the chromosphere of spicular plasma heated to well beyond chromospheric temperatures. In active regions, downflow speeds of 70–100 km s^{-1} are commonly observed in the "coronal rain" of plasma cooling from 10^6 to 10^4 K, draining along field lines into sunspots.

The explosive events found in the network when observations of transition region line shapes are made at high spatial and time resolution are another interesting phenomenon. Some of these events are shown in Fig. 9-15. Blue shifted velocities of up to 400–500 km s^{-1} are deduced from the line shapes. The structures seem to be magnetically confined jets or exploding loops of relatively high-density plasma and they appear to occur frequently enough to contribute to the coronal mass and energy balance.

9.2.4 Physical Conditions in Closed and Open Magnetic Structures

Comparison of EUV and X-ray pictures with calculated coronal fields shows that three kinds of structures can be usefully distinguished. These are (a) magnetically closed loop structures of active regions, (b) the closed loops of quiet regions, whose footpoint separation is larger so the coronal magnetic intensity is lower, and (c) the open magnetic structures of coronal holes, where the field lines extend out into the solar wind and the mean field intensity is lowest. Analysis of the plasma conditions in these structures reveals differences in their energy balance which seem to be caused by the differences in magnetic geometry.

The large bright loops observed in active regions are most easily studied because their plasma structure both along and across the magnetic field can be spatially resolved with EUV, soft X-ray, and optical telescopes. Their stability is remarkable. Changes in a loop's appearance are observed to occur over the plasma's radiative cooling time scale of a few minutes. But the magnetic structure can last for a few days and even weeks. This far exceeds the few minutes that would be required for an Alfvén wave to traverse the loop, setting the time scale for its destruction by an MHD instability.

The density and temperature structure of the coronal plasma in these loops can be derived from photos taken through filters transmitting various soft X-radiations. The X-ray contrast between a loop and its surroundings is typically a factor of 10 or more. Comparison with white-light pictures during eclipses reveals the same loops in the low corona off the limb as seen in the X-rays. Since the white-light enhancement arises from electron scattering which is independent of temperature in the fully ionized coronal plasma, the loops must be considerably enhanced in density over their surroundings.

More detailed information on loop physical conditions is obtained most directly from studies of EUV line intensities. The intensity of a line formed over a temperature range T_1 to T_2 in a plasma of homogeneous density N_e is given by integration of equation (3-26) along the line of sight to yield

$$I = \frac{hc}{4\pi} C_{mn} A \int_{T_1}^{T_2} G(T) N_e^2 \, dT. \qquad (9\text{-}2)$$

Here A is the abundance of the element relative to hydrogen and C_{mn} is the collision cross-section from the ground state to level n. $G(T)$ is a temperature-dependent function which gives here the ground-state population density of the ion producing the line.

When a loop is observed transverse to its long axis, the temperature gradient along the line of sight (i.e., across the magnetic field lines) is found to be sufficiently high that the temperature regime T_1 to T_2 over which the line is formed can often be taken to occur over a shorter distance than appreciable variations in N_e^2. A mean value of $G(T)$ can then be taken outside the integral, and equation (9-2) can be written in terms of the spatial coordinate as

$$I = \frac{hc}{4\pi} C_{mn} A \langle G(T) \rangle \int_{x_1}^{x_2} N_e^2 \, dx, \tag{9-3}$$

where $x_2 - x_1$ is the distance between the T_2 and T_1 isotherms. The integral

$$\int_{x_1}^{x_2} N_e^2 \, dx \tag{9-4}$$

is an important quantity called the emission measure of the plasma volume since it determines the plasma emissivity by any binary collisional process.

Root mean square estimates of the loop electron densities can be derived using relation (9-3) to interpret the observed EUV loop intensities. However, the values between 5×10^9 and 5×10^{10} cm^{-3} derived in this way are systematically between 10 and 100 times lower than those obtained by techniques which measure the true local electron density.

More accurate densities of coronal and transition region plasmas can be obtained from the intensity ratios of forbidden and permitted emission lines, preferably emitted by a given ion. Some examples are the λ 5694 and λ 5445 lines of the coronal Ca xv ion and the lines $\lambda\lambda$ 977, 1176 of the transition region ion C iii. The density sensitivity of the ratios arises because the probability that an ion excited to the upper state of the forbidden transition will decay radiatively (rather than through de-excitation by an electron collision) decreases with increasing plasma density. Since the probability for radiative decay of the permitted transition is much higher, the likelihood of a collisional de-excitation of this upper level is small, so the intensity ratio of a forbidden to permitted line decreases with rising electron density.

The much higher densities of order 10^{11}–10^{12} cm^{-3} typically found from line ratios are supported by the observed pressure broadening of the highest lines of the Balmer and Paschen series of hydrogen, although this comparison can only be made for the plasmas at temperatures below about 2×10^4 K, where the hydrogen lines are formed. The difference between the true densities measured by these techniques and the rms values obtained from the emission measure indicates that the loop plasma is highly filamented. Even when a loop looks homogeneous in an EUV or X-ray observation made at a spatial resolution of a few arc seconds, only a small fraction, between 1 and 10% of its volume is actually occupied by plasma at that temperature. Direct evidence for filamentation along the field lines can be seen in the highest-resolution Hα photoheliograms, such as Fig. 10-15.

The temperature structure of large loops that emit brightly in coronal and transition region radiations and are often associated with sunspots is particularly easy to study. EUV pictures in transition region radiations (such as Fig. 9-13) reveal roughly coaxial loops whose width increases systematically with the temperature of the emitting plasma. Decreased emission is often observed in coronal EUV lines at the location where the transition region emission is strong. These observations indicate that spot loop temperatures rise from a cool core at $T \sim 10^4$ K occupying a sizable fraction of the loop volume to coronal values of $T > 10^6$ K over a distance of about 10^4 km.

Cool material at chromospheric temperatures high in coronal loops is often observed even in structures whose footpoints are not in sunspot umbrae. Similar temperature gradients transverse to the field exist in most loops. However the temperatures of the hottest and coolest plasmas, their densities, and their detailed distribution can vary substantially.

There is little evidence for systematic height gradients in either plasma temperature or density within a loop. The hydrostatic density scale height at the coronal temperature exceeds the vertical scale of most loops observed in X-rays or coronal EUV radiations, so we do not expect to see height gradients in these emissions. The cooler loop plasmas are generally observed to be in a state of rapid downward motion, and their density is often measured to be as high near the loop apex as near its base.

In the quiet sun, loops are less distinct. Conditions in the quiet transition region and low corona are best studied from disk observations of EUV line intensities in the network and analyzed using equation (9-3) to derive emission measures. In the model presented here the magnetic field in these quiet regions is calculated from its photospheric sources in the network flux tubes; their cross section increases as the force-free field expands into the chromosphere and low corona. The model is also constrained by the assumption that the energy balance is governed by the relation

$$\nabla \cdot \mathbf{F}_c + \nabla \cdot \mathbf{F}_R + \nabla \cdot \mathbf{F}_M = 0, \qquad (9\text{-}5)$$

where $\mathbf{F}_c$, $\mathbf{F}_R$, and $\mathbf{F}_M$ are respectively the fluxes of conductive and radiative heat, and of mechanical heating. Thermal conduction is assumed to follow the magnetic field lines from the hotter corona since cross-field conduction in these tenuous plasmas can be neglected.

The quantity obtained most directly from such a model is the emission measure distribution with temperature, which is plotted in Fig. 9-16. The large maximum seen in the emission measure around $1–5 \times 10^6$ K indicates that most of the coronal emissivity (not necessarily most of the plasma) is found in that temperature regime, which might be called "the temperature of the corona."

The steeply decreasing emission measure between about 10^5 and 10^6 K is interesting in view of the maximum in the radiative loss function around 10^5 K (Fig. 4-2) of a plasma of solar chemical composition. This inverse relation between the radiative power and the observed emission measure suggests that the transition region temperature structure might be determined by the radia-

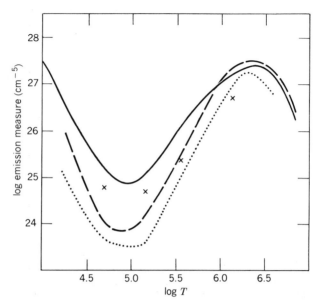

Fig. 9-16   Distribution of emission measure in two bright coronal loops (solid line and long dashes). Emission measure curves for the quiet sun (short dashes) and active region material outside the loops (crosses) are also shown. From J. Raymond and P. Foukal, *Astrophys. J.*, **253**, 323 (1982).

tive properties of solar plasmas, just as the structure of the temperature-minimum region and chromosphere seem to be determined by the thermostatic effects of plasma radiations at lower temperatures.

Comparison of the emission measure profiles given in Fig. 9-16 for various solar structures shows also that plasmas of all temperatures are observed in pretty much any quiet or active solar structure. Thus temperatures as high as 4×10^6 K indicated by, e.g., the EUV emissions of Si XII, S XIV, and Fe XVIII, are observed both in the quiet and active corona, although in active regions they are brighter, implying that the emission measure in the hottest plasma is higher. Except during flares, there is little evidence that active region plasmas attain higher temperatures than quiet regions—but they certainly contain more, or at least denser, material at coronal temperatures in the range $1–5 \times 10^6$ K.

When the transition region and coronal emission measures are modeled assuming hydrostatic equilibrium, the profile of temperature and density with height that is found is shown in Fig. 9-17. We see that the relatively weak transition region emission measures shown in Fig. 9-16 imply a very rapid temperature increase from chromospheric to coronal values over a scale of a few hundred kilometers. It is questionable whether such a thin transition layer actually occurs in the quiet network or anywhere else on the sun. The rapid downflows of transition region material suggest that the plasma of temperature

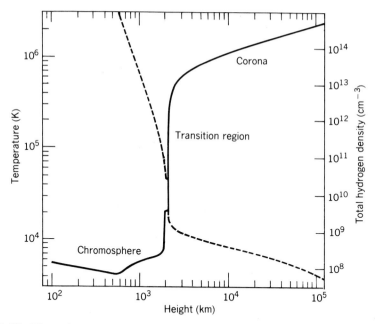

Fig. 9-17   Plots of temperature (solid line) and density (dashed) from a model of the quiet network. Reproduced with permission from A. Gabriel, *Phil. Trans. Royal Soc. London*, **281**, 339 (1976).

between about 2×10^4 and 5×10^5 K is more likely to be distributed over a considerable range in height. Models that attempt to take this motion into account contain an additional advective term $\nabla \cdot \mathbf{F}_{adv}$ in the energy balance equation (9-5) representing the enthalpy flux of the falling, cooling plasma.

Models of the transition zone underlying coronal holes indicate that its structure is quite different from that of the quiet sun. Its temperature gradient seems to be about 5–10 times lower and its density several times less than in normal quiet regions. According to equation (9-4), a lower temperature gradient increases the spatial extent $x_2 - x_1$ over which a given line is emitted. This tends to compensate for the lower electron density and explains why the transition region radiations are of similar intensity inside and outside coronal holes. Such a difference in structure may also explain the bright microwave emission of holes, although it is less clear how this occurs.

Models of the conditions in coronal holes to distances of about $5R_{\odot}$ indicate that the density in the hole is about 2–3 times lower and the temperature is about 5×10^5 K below quiet sun values. The coronal hole magnetic geometry seems to be not only open, with field lines extending into the solar wind, but actually diverging, so that at least out to about $2R_{\odot}$ the hole's cross-sectional area increases with height as much as 10 times faster than R^2. This result is found from the calculated field lines in coronal hole areas. It can also be seen directly from the appearance of the X-ray and K-corona near the poles, as seen in Figs. 9-10 and 9-14. As discussed below and in Chapter 12, this rapid divergence may play a role in the acceleration of solar wind in holes.

9.2.5 Heating and Dynamics of Coronal Loops and Holes

The disk observations in EUV and soft X-radiations described above show the close connection between patterns of coronal emission and magnetic structures such as active regions and holes. This association makes it clear that the magnetic field plays an important role in determining the wide range of physical conditions found in these coronal structures. However, a clear answer cannot yet be provided to the related questions "what mechanism produces the high coronal temperature?" and "what processes determine the mass of coronal plasmas in the sun?"

The total losses from the solar corona by radiation, thermal conduction, and advection are approximately 3×10^{28} ergs, or about 5×10^{5} ergs cm^{-2} s^{-1}. The heating rate required to balance these losses is therefore only about 1% of that required to heat the chromosphere and merely a part in 10^{5} of the sun's total energy output. A fair answer to the question of why the corona is so hot might then be "because it is so tenuous and radiates so little," since if the energy losses from a plasma are made small enough, even a tiny energy input can achieve high temperatures.

The problem of coronal heating is made more interesting by the observation that the power requirements of different coronal structures vary widely. For the brightest loops they are an order of magnitude larger than can easily be supplied by at least some of the heating mechanisms that might be expected to play a role.

The most likely sources of nonthermal energy input to the corona are the disturbances generated by the ceaseless agitation of the sun's convective layers. These disturbances are observed at the photosphere in the form of granulation, supergranulation, magnetic flux tubes, and wave motions. They are all ultimately driven by solar convection and carry far more free energy than is required to heat both the corona and the chromosphere. The problem of coronal heating is then to identify which form of abundant free energy (a) propagates upward from the photosphere, (b) is transmitted (rather than reflected or absorbed) in the chromosphere and transition region, and (c) is dissipated in the corona (rather than passing outward into the solar wind).

The problem would be more tractable if there were good reasons to believe that one mechanism heats all coronal structures. The requirements placed on the heating in holes and bright active region loops are quite different, and such simultaneous constraints might help to narrow the choices. Unfortunately, there is no reason except a desire for simplicity to support that view. The two mechanisms presently favored involve the dissipation of the mechanical and electromagnetic energy of MHD waves and the Joule dissipation of dc currents running along coronal magnetic field lines. These two mechanisms are actually related, but in their original form as proposed over 40 years ago they were seen as quite distinct.

The compressional waves emitted by granules carry mechanical energy outward from the photosphere. As discussed earlier, this energy is dissipated into heat by viscosity in shocks that form as the waves steepen in the chromosphere. Coronal electric currents were considered to be generated by the twisting of coronal flux tubes caused by the "freezing" of their photospheric footpoints to turbulent convective motions. The electromagnetic free energy of these currents could be dissipated by Joule heating if the coronal plasma resistivity were appreciable and the currents could be restricted to sufficiently small volumes so that self-induction would not slow their decay.

More recent findings have formed the present view of these basic mechanisms in several ways. For one, the coronal magnetic fields must play an important role in the energy balance of that medium. We know that both thermal conduction and plasma motion are constrained to follow coronal field lines, so it might be mainly the anisotropy of energy transport that produces the elongated loops. However, compressional waves emitted from the photosphere into a coronal medium filled by largely horizontal magnetic fields will tend to couple to the additional oscillatory modes provided by this field and produce an MHD wave. Such an MHD disturbance can be usefully visualized as some combination of a longitudinal (sound) wave and of a transverse Alfvén wave (Section 4.4.5).

This inevitable coupling makes it likely that the heating rate related to the MHD wave dissipation is also determined by the field geometry. In the situation that is simplest to visualize, one can invoke the enhanced piston action of granules acting on flux tubes in the network to produce a relatively intense upward flux of purely compressional waves. These are guided along the essentially rigid field lines of low-β plasma into coronal loops whose footpoints are anchored in the network.

Many other wave-heating scenarios have been proposed, including waves travelling along the surfaces defined by intense flux tubes. Alternatively, the dissipation of magnetic field-aligned electric currents would provide a heating function that is naturally associated with the geometry of coronal magnetic fields. The choice between MHD waves or of dc electric currents as the dominant coronal heating mechanism depends on how well each can supply and dissipate the required energy flux in various observed coronal structures.

Both of the mechanisms can explain how the coronal heating function might be enhanced along certain field lines and thus help to explain the existence of loops. Currents of sufficient magnitude to provide the coronal heating are observed to flow at coronal heights while the evidence for a sufficient wave flux is less clear. From equation (4-76), we see that for Alfvén waves, the flux is given by

$$F = \rho v^2 v_A,$$
(9-6)

where v_A is the Alfvén speed. In the brightest loops, energy balance requires $F = 5 \times 10^6$ ergs cm^{-2} s^{-1}, so with a particle density close to $n_e = 10^{10}$ cm^{-3} plasma velocities $v \sim 10$ km s^{-1} are required. Typical rms velocities obtained from forbidden line widths are somewhat larger than this, although the relation between the rms velocity found from a line width and the value of v that can be accommodated in equation (9-6) requires consideration of how the line is formed in a moving atmosphere.

Searches for waves in the period range around 5 min in the transition region have not revealed periodic motions of sufficient amplitude required to propagate the coronal heating flux. But such negative tests are not entirely convincing since the "waves" responsible for coronal heating could in principle range in period from the hours associated with slow propagation of twists along flux tubes, down to seconds.

The mechanisms of energy dissipation by waves and currents are not easy to study since they depend on details of the coronal plasma structure on very small spatial scales. Sound waves (or magnetosonic waves if magnetic flux tubes act as wave guides) seem to dissipate too easily by viscous damping in shock waves. The difficulty there lies in understanding how such compressive waves could penetrate past the low chromosphere. Alfvén waves are much harder to dissipate. Their compressional effects are only of second order in amplitude and should be small except for unusually violent disturbances. The most attractive prospect for heating most loops seems to be hybrid MHD waves of reasonable amplitude and period.

The basic difficulty in electrical heating lies in achieving sufficient dissipation over significant volumes so that the Joule dissipation rate j^2/σ becomes appreciable. The standard electrical conductivity given by equation (4-99) is far too high (see exercises). Plasma turbulence can increase this resistivity when a large compression is applied to a small volume, or when excessively large currents are generated in the tenuous plasma, and electron speeds exceeding the electron thermal velocity are required to carry the current in formula (4-98) and thus in (4-99). A reasonable mechanism might be the ion-acoustic instability, which generates longitudinal ion sound waves (see Section 4.4.7), whose electrostatic fields can produce several orders of magnitude higher scattering of the charge-carrying electrons than that expected from electron-ion collisions alone.

The conditions most likely to favor electrical heating probably exist in very restricted regions of the plasma where pressures and current densities might be high. But it is difficult to transport energy by thermal conduction across field lines from such small volumes to larger regions of loops which are required to produce the observed emission measures in X-rays and EUV radiations. Steady heating of the corona by this process would then seem to require that such regions of transient energy release be ubiquitous. This is possible, but observations of sufficient spatial resolution are not available to decide this point. This difficulty of power distribution is avoided in wave heating, where the dissipation is more evenly distributed over large volumes.

Both of these general mechanisms probably play a role in coronal heating. Observations suggest that electrical dissipation may be most important in low-lying loops associated with recently emerged fields, where field gradients and time rates of change are found to be largest. Wave heating is more attractive for the larger loops and also for coronal holes.

In the coronal holes the momentum carried by Alfvén waves [formula (4-77)] could help to explain the rapid acceleration of the high-speed wind streams that emanate from these regions. Radiative losses from the dark coronal holes are relatively low, but the enthalpy flux of the outflowing plasma is of order 5×10^5 ergs cm^{-2} s^{-1} and thus comparable to radiative losses from closed coronal loop regions. As discussed in Chapter 12, the requirement is to explain how the relatively high velocities are produced without the large mass fluxes that would result if thermal pressure gradients provided the driving force. An intense flux of Alfvén waves pushing on the plasma in the region around $2-5R_\odot$ without heating it could provide the required momentum.

Whatever process heats the corona, the thermal conduction back down along field lines will tend to increase the chromospheric temperature. Locally increased coronal heating will then increase the pressure scale height at the footpoint of a loop, and increase its density. Excessive density should eventually lead to thermal instability (Section 11.1.3), which might explain the presence of cool filaments of plasma within the loop structure. This cool dense material near the tops of loops will tend to fall under gravity because of its Rayleigh-Taylor instability (Section 11.1.2) when supported over a column of much more tenuous coronal plasma. The persistent downflows of transition region material found in active region loops, and also in the network, may well be related to such thermal and dynamical instability of the coronal loop plasmas.

This basic model of mass and energy balance is attractively simple but it does not incorporate the spicule contribution to the mass balance, which may be dominant. The possible role played by the explosive events found in the network (Section 9.2.3) also deserves consideration. A satisfactory model must explain all the phenomena of comparable energy and mass flux observed in these atmospheric layers.

ADDITIONAL READING

The Chromosphere

G. Athay, *The Solar Chromosphere and Corona*, D. Reidel, 1976.

S. Jordan, "Chromospheric Heating" in "The Sun as a Star," NASA publication SP-450, S. Jordan (Ed.), p. 301 (1981).

G. Athay, "The Chromosphere and Transition Region-Current Status and Future Directions of Models," *Solar Phys.*, **100**, 257 (1985).

J. Vernazza, E. Avrett, and R. Loeser, "Structure of the Chromosphere," *Astrophys. J. Suppl.*, **45**, 635 (1981).

The Transition Region and Corona

J. Zirker, "The Solar Corona and the Solar Wind" in "The Sun as a Star," NASA Publication SP-450, S. Jordan (Ed.), p. 135 (1981).

G. Withbroe and R. Noyes, "Mass and Energy Flow in the Solar Chromosphere and Corona," *Ann. Rev. Astron. Astrophys.*, **15**, 363 (1977).

J. Zirker, "Progress in Coronal Physics," *Solar Phys.*, **100**, 281 (1985).

G. Vaiana and R. Rosner, "Recent Advances in Coronal Physics," *Ann. Rev. Astron. Astrophys.*, **16**, 393 (1978).

D. Wentzel, "Coronal Heating," in "The Sun as a Star," NASA Publication SP-450, S. Jordan (Ed.), p. 331 (1981).

EXERCISES

1. Calculate the expected height of the chromosphere if this layer were in hydrostatic equilibrium. Also calculate the height reached by a spicule if it followed a ballistic trajectory with an initial velocity of the observed magnitude. Comment on the implications of your results for the dynamics of the chromosphere.

2. Calculate the shock strength required to balance chromospheric radiative losses around 2000 km using the formula (4-86) for shock periods between 50 and 300 s. Also calculate the plasma velocity behind such a shock and compare it to chromospheric observations.

3. What is the maximum compression that can be achieved behind an adiabatic shock (see Section 4.4.4). Is this sufficient to form spicules out of material at coronal densities? How can spicular densities be explained?

4. Calculate the typical speeds of coronal electrons. Explain quantitatively (see Section 3.9.1) why Fraunhofer lines are essentially invisible in the K-corona.

5. Calculate the rms electron density that would be measured in a magnetic loop consisting of elongated filaments filling 10% of the loop volume and containing an isothermal plasma of electron density 10^{12} cm^{-3}. Assume that all the loop plasma is confined to those filaments. Comment on the usefulness of emission measures and other density diagnostics in inhomogeneous solar plasmas.

6. How is the temperature of the corona determined? Is there a unique coronal temperature? Discuss understanding of temperature differences between quiet and active coronal regions.

7. Calculate the Joule heating rate in the general coronal volume of active regions using the formula (4-99) for plasma electrical conductivity. To estimate the current density, use the relation $\nabla \times \mathbf{B} = 4\pi\mathbf{j}$ or $j \sim B/4\pi L$, where L is a characteristic scale over which B is observed to vary substantially. Compare this heating rate to the coronal power requirement. How localized would this current need to be for the speed of electrons carrying the current to exceed the electron sound speed?

8. Explain why the sun's centimeter emission is observed to be limb brightened, and why this brightening disappears in meter waves.

10

Prominences and Flares

Several pinkish clouds extending to 50,000 km or more above the photosphere can often be discerned at the limb during eclipse totality. These are called prominences, and their intricate structure is easily observed out of eclipse through a small telescope equipped with monochromatic filters passing strong chromospheric lines such as Hα or Ca K. When these same structures are observed in absorption on the disk, they appear dark and are called filaments.

Prominences have been the objects of close study since the photographic investigations carried out during the 1860 eclipse, which showed that they were solar, rather than meteorological, phenomena. Analysis of their spectrum yielded the first identification of helium and still provides probably the best estimate of that element's abundance in stars like the sun. Time-lapse films of the fascinating forms and motions of prominence material played a key role in the early development of magnetohydrodynamics. The remarkably stable equilibrium between magnetic and gravitational forces achieved by cool prominence material perched high in the corona still challenges our understanding of plasma confinement in solar magnetic fields.

Recent results indicate that disruptions of this MHD equilibrium are closely associated with ejection of large volumes of solar plasma into the interplanetary medium as coronal transients. The process of filament disruption seems also to be linked to charged particle acceleration in the very rapid and violent releases of energy often observed in active regions, called flares. Very rarely, flare emissions are bright enough to be visible on the disk in white light, as in the case of the famous first sightings of a flare by Carrington and Hodgson in 1859. More usually, a flare can only be detected by imaging in chromospheric and coronal radiations, whose relative enhancement is far greater than in the photospheric continuum.

The power of flares seems to be supplied by remarkably quick dissipation of stored coronal magnetic field energy. The primary energy release at coronal levels heats the local plasma in the affected loop structures in a few seconds and then raises the temperature of underlying chromospheric layers. The largest events cause appreciable heating even of the photosphere.

Flares often eject large masses of coronal and chromospheric material into interplanetary space, and the powerful wave that propagates ahead of this material can bulldoze through pre-existing magnetic structures, causing major rearrangement of the heliosphere to well beyond the Earth's orbit.

A substantial fraction of the energy released in at least some flares goes into acceleration of nonthermal particles whose energies range up to several GeV. Particle collisions in the flare's highly energetic thermal and nonthermal plasmas lead to nuclear reactions and production of both γ-rays and neutrons, both of which have now been directly detected near the Earth.

Large solar flares are easily the most powerful transient phenomena in the solar system, and their effects have a significant impact on terrestrial systems. The dramatic brightness increase and majestic eruption of plasmas in large events also make flares the most exciting solar phenomena to observe and study. Their rich spectrum of intense radiations over the full wavelength range from the γ-ray region to kilometer radio waves presents a unique opportunity to learn how cataclysmic magnetic energy release may proceed in space plasmas observed on various scales throughout the universe from the Earth's magnetosphere to active galaxies. For all these reasons the still-elusive mechanism responsible for flare energy release has become the holy grail of solar research, and the emphasis on flare research has for many years outstripped the effort invested in study of other solar phenomena.

10.1 PROMINENCES AND FILAMENTS

10.1.1 Observations and Physical Conditions

Some examples of prominences at the limb are shown on the Hα photoheliogram in Fig. 10-1. It is useful to distinguish three main types of structure in this picture. The most common are the high, bladelike quiescent prominences, which are usually found well away from active regions. An example of their intricate structure is seen at higher spatial resolution in Fig. 10-2. Around active regions one often observes matter condensing and falling, particularly into sunspots. Some of this "coronal rain" is also seen in Fig. 10-1. Most prominent in Fig. 10-1 is a bright radial structure, called a spray. Sprays and surge eruptions are associated with flares. Along with loop prominences that often form after flares (Fig. 10-15) they are included in the broad class of "active" prominences.

Figures 10-3 and 10-4 illustrate some common forms of filaments seen in Hα on the disk. These include the relatively long, narrow active region filaments that are usually too low to appear as prominences at the limb, as well

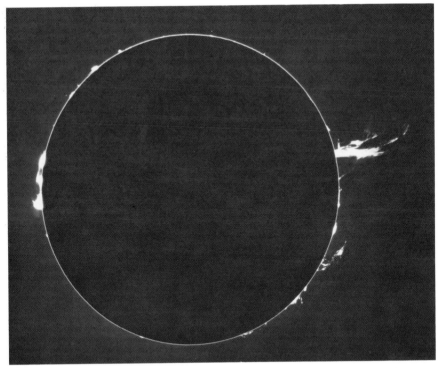

Fig. 10-1   Spray (3 o'clock), active (4 o'clock) and quiescent (9 o'clock) prominences photographed in the Hα line with a coronagraph at Haleakala Observatory, Maui. By permission of M. McCabe.

as quiescent filaments exhibiting the hedgerow structure seen in the prominence shown in Fig. 10-2.

Filaments can be very long lived. Quiescents typically last for two to three solar rotations. However, both quiescent and active region filaments sometimes disappear for a few days or more. This sudden disappearance within a few hours is called a disparition brusque. When viewed on time-lapse films, the sudden lifting and fading of such an immense structure is an impressive sight. Sometimes a prominence reforms in the original location a few hours or days later.

Filaments are found preferentially in two latitude belts—in a strip at high latitude (called the polar crown) and in the active midlatitudes. They form within extended volumes of low coronal emission called filament cavities. These are situated along neutral lines of the longitudinal magnetic field (i.e., in regions of horizontal field). However, not all neutral lines are marked by filaments, and usually only a small fraction of the cavity length is occupied by a filament.

Direct measurements of the prominence magnetic field using the Zeeman and Hanle effects indicate fields of 5–10 G in quiescent prominences. The field

Fig. 10-2 A quiescent "hedgerow" prominence photographed in the Hα line. Sacramento Peak Observatory photograph.

in the much lower active region filaments has not yet been observed directly, except during eruptions, when intensities as high as 200 G have been reported. The Hanle effect is observed in lines such as Na I D_3 whose origin in the tenuous prominence plasma is mainly by resonance scattering of photospheric light. This scattering produces a linear polarization of the scattered radiation, parallel to the limb. The Zeeman splitting of the atomic levels by the prominence magnetic field can cause a partial depolarization of the line radiation and also a rotation of its linear polarization plane. The degree of residual measured polarization left after this Hanle depolarization effect can be related to the magnetic field intensity and direction using models of line radiative transfer in the prominence plasma.

Results from this technique indicate a field direction in prominences that is essentially horizontal to the solar surface, but making an angle of typically 25 degrees to the long axis of the prominence. In addition, it is found that in some prominences the direction of this sheared field is opposite to that expected from the distribution of adjacent photospheric fields, as discussed below.

The spectrum of quiescent prominences (see Fig. 10-5) is similar to that of the chromosphere, and a temperature of about 7000 K or less is found from analysis of, e.g., the widths of optically thin lines. This plasma is much cooler than the surrounding corona at the same height, but its density (at least its rms value) seems to be much higher. Typical electron densities found from the

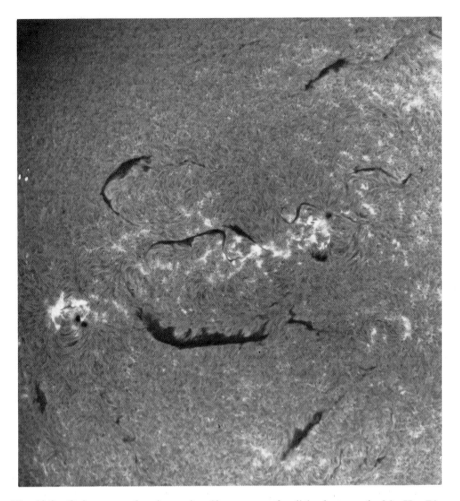

Fig. 10-3 Quiescent and active region filaments on the disk photographed in Hα. Big Bear Solar Observatory photograph.

Stark broadening of the high Balmer lines lie between 10^{10} and 10^{11} cm^{-3}. The plasma densities might be of order 10^{12} cm^{-3} if the hydrogen ionization in this cool plasma is below 10%, as the best data indicate, so the gas pressure might be similar to coronal values. The darkness of filaments persists in the EUV and microwave radiations of the transition region. It is not clear whether this darkness is caused mainly by a lower plasma pressure at these wavelengths or by a thinner transition zone than exists over the quiet chromosphere.

Even quiescent filaments and prominences contain plasmas moving at high speeds. Time-lapse movies of large prominences such as that in Fig. 10-2 show consistently downflowing motions at about 5 km s^{-1} of the knots and condensations that make up the spatial fine structure. But attempts to find

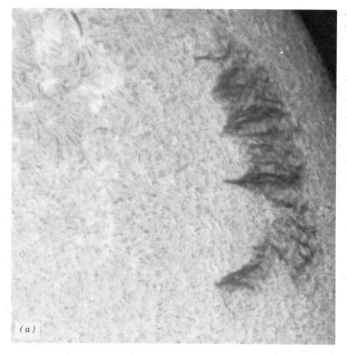

Fig. 10-4 Examples of (a), (b), (c) quiescent filaments and (d) an active region filament, imaged in Hα. Note the thinness of the filaments, illustrated in (c). Big Bear Solar Observatory photographs.

systematic downflows from Doppler shifts have yielded ambiguous results. In active region filaments, material is seen flowing along the filament axis, generally into a sunspot located at one end of the structure. Studies of line widths in quiescent prominences indicate substantial turbulence near their tops, whereas motions near the prominence bottom are smaller and the structure is also more sharply defined.

10.1.2 Dynamics

The long lifetime of quiescent prominences draws attention to the remarkable stability of their MHD equilibrium. The factors that cause a quiescent prominence or active region filament to undergo a disparition brusque are also important since filament eruptions appear to produce many of the coronal transients and associated shock waves in the interplanetary medium.

The total mass in a few large prominences is comparable to that in the rest of the coronal volume. Moreover, the prominence plasma seems to be draining downward at a rate sufficient to deplete the corona within a few hours. These two observational results imply a substantial and more or less continual

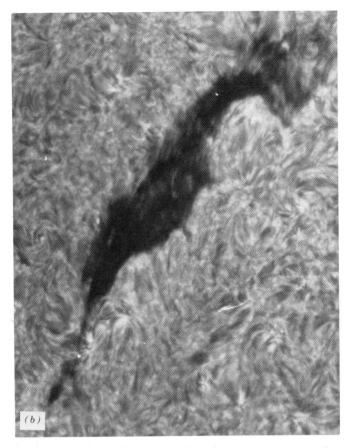

Fig. 10-4   (*Continued*)

circulation of material that can readily condense into the prominence state. This circulation must play a significant role in the corona's mass balance.

The spontaneous cooling and condensation of thermally unstable coronal plasma probably plays a role in prominence formation, as first pointed out by K. Kiepenheuer in the 1940s. From the discussion of Section 11.1.3 the observed time and length scales of filament formation seem too short to be explained entirely in terms of thermal instability in plasmas of a mean coronal density around $10^9 \, \text{cm}^{-3}$. But as pointed out earlier, true plasma densities in the inhomogeneous corona may be ten times higher. Turbulence may also substantially reduce plasma thermal conductivity, so that this difficulty may not be fundamental. Prominences might then be visualized as static structures in which spontaneously condensing plasma is suspended by a magnetic basket of horizontal fields. However, the rapid draining of prominence plasmas

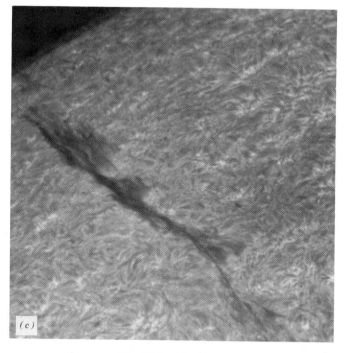

Fig. 10-4   (*Continued*)

suggests that even in the long-lived quiescents a more efficient mass supply is required, perhaps from the chromosphere below.

One possibility is a siphon mechanism similar to that described in Section 11.1.4. The gas pressure gradient required to draw material up against gravity could be generated by slightly different conditions between two photospheric footpoints of a loop. Alternatively it might be produced by condensation and draining down of plasma cooling near the loop apex and flowing down one loop leg, thus drawing plasma up from the other footpoint. In either case, one might visualize a progressive sagging of the field lines acting to increase the hold time of the plasma near the loop apex, as discussed below.

The classic issue of how prominence material is supported against gravity may be clarified by the interesting recent observations of filament field geometry. One of the two models that have been put forward is shown in Fig. 10-6(a). As suggested originally in 1957 by R. Kippenhahn and A. Schlüter, a horizontal field is bowed downward by the prominence mass. A Lorentz force acting to balance gravity is generated by a current running transverse to the field lines, along the filament axis. The magnetic field associated with this current alters the original magnetic field to the bowed shape. The filament is supported by the tension in the bowed field lines, whose footpoints are

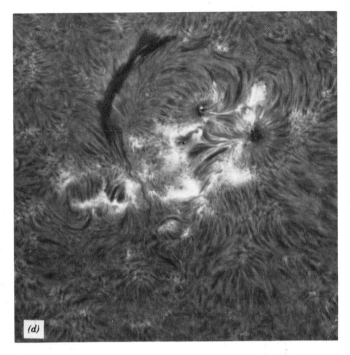

Fig. 10-4   (*Continued*)

anchored into photospheric plasmas. A clear prediction of this model is that the magnetic vector should thread the filament in the direction expected of field lines joining the two polarities observed in the photosphere to either side of the filament.

A different model put forward in 1974 by M. Kuperus and M. Raadu associates filaments with material condensing within a current sheet, as illustrated in Fig. 10-6(b). This is plausible since reconnection at a magnetic neutral line by the so-called tearing mode instability can lead to formation of current filaments with axes aligned along the neutral line. The support against gravity of the dense filament plasma is provided here by the vertical gradient in field-line tension. A net force acts outward because the field lines fan out from their photospheric footpoints. As seen in Fig. 10-6(b) in this model, the (closed) field lines thread the prominence both in the direction expected from straight connection between polarities observed to either side of the filament and also in the antidirection.

It is interesting that the high quiescent prominences, such as those of the polar crown, exhibit a magnetic field structure consistent with the Kuperus-Raadu model. The magnetic field structure of lower quiescent prominences whose height is below about 3×10^4 km is consistent with the Kippenhahn-Schlüter geometry. These latter prominences occur at the active latitudes,

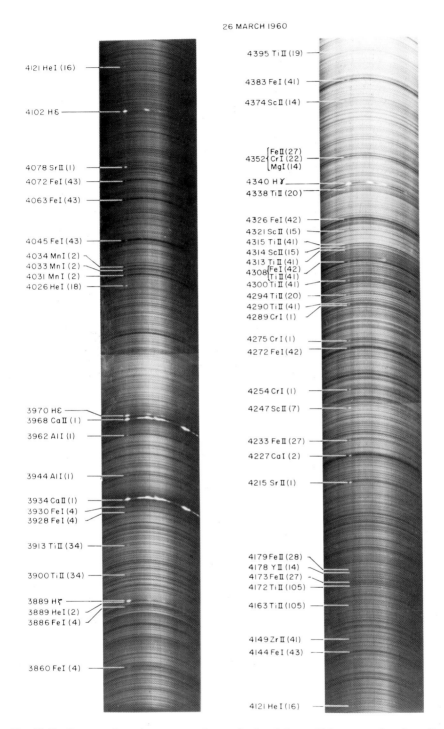

Fig. 10-5 Spectra of a quiescent prominence in the violet and blue spectral regions. By permission of E. Tandberg-Hanssen.

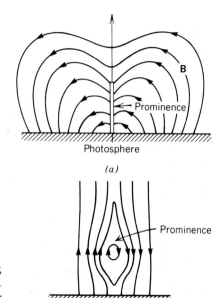

Fig. 10-6   Schematic diagrams illustrating (a) the K-S magnetic geometry, (b) the K-R geometry. Adapted from E. Priest, *Solar Magnetohydrodynamics*, D. Reidel, 1982.

suggesting that these quiescents might be supported in the largely horizontal fields associated with neutral lines in the quiet sun. Between active regions, neutral lines indicating low-lying current sheets (i.e. *not* associated with field lines connecting across the neutral line observed in the photosphere) might promote the formation of midlatitude filaments.

Although magnetohydrostatic support of prominences is likely to be important, other processes may also play a role. Radiation pressure can be shown to be negligible, but the role of outward pressure exerted by Alfvén waves deserves closer study. The momentum of the waves would tend to be absorbed in condensed plasmas, where the Alfvén speed is reduced, just as a beach absorbs the momentum of waves breaking due to friction in the shallow water. In many respects dynamic support by Alfvén waves describes observations of complex prominence shapes better than do the magnetic support models.

Several processes have been identified to explain how a filament can disappear rapidly. Some disparitions brusques seem to be caused by a rapid draining of the filament into the chromosphere. Presumably this drainage is caused by a change in the magnetic geometry that supported it or perhaps by a change of the wave pressure. In other cases, the cool material ceases to be visible in Hα because it is rapidly heated to transition region and coronal temperatures.

In so-called coronal mass ejections, outward ejection of the filament leads to a coronal transient in the interplanetary medium (see Section 12.2.3). The transient travels at uniform or even increasing speed at distances as far out as $3-5$ $R_\odot$. The density is enhanced around the apex of the transient's looplike geometry (see Fig. 12-11) and reduced behind it. The geometry of this loop, such as its fixed footpoint separation and the dependence of its radius of curvature and thickness upon its radius, may also offer some insights into the possible acceleration mechanisms discussed below.

Some coronal transients associated with flares may be driven by plasma pressure gradients produced during rapid heating of the chromosphere and corona. Models of such magnetically channeled blast waves can produce good agreement with many of the observations, provided the initial magnetic configuration is open or at least of the streamer type. But filament eruptions during flares often precede the impulsive heating of the flare plasma, and they proceed at a uniform rate during the flare itself so they appear to be caused by release of magnetic stresses.

One simple model associates the acceleration with an imbalance between the axial and azimuthal fields of a loop (Fig. 10-7). A loop's curvature packs the azimuthal field lines together more densely at the inner edge, so the azimuthal field (associated with a twist in the loop) will be more intense at the inner edge of the loop apex than at the outer. The resulting gradient of magnetic pressure $B^2/8\pi$ will tend to expand the loop outward against the tensions of the axial field lines. In equilibrium, to produce a constant velocity, the difference of these two magnetic forces must balance the downward acceleration of gravity on the contained plasma.

The model assumes the uniform speed that is indicated by the observations (rather than, e.g., a ballistic trajectory). It then predicts that the loop thickness increases linearly with R, the distance from the center of the sun, which is

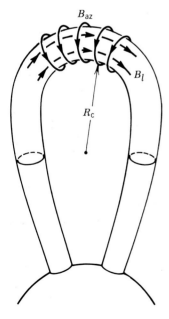

Fig. 10-7 Schematic diagram illustrating the imbalance of magnetic forces in a twisted loop. Adapted from E. Priest, *Solar Magnetohydrodynamics*, D. Reidel, 1982.

roughly consistent with observations. On the other hand it predicts an increase of the loop's radius of curvature R_c that is significantly too small for a given distance R.

The model associates the eruption of filaments with an instability caused by an excess of twist in coronal fields, similar to the kink that occurs in an elastic band that is twisted too far. Magnetic twist is known to be present and can be roughly measured in active regions from the shear of the overall field configuration (Chapter 9). This is an attractive feature, but other models of the magnetic stress imbalance have been put forward, and detailed comparisons with data will be required to identify the fundamental driver.

10.2 FLARES

10.2.1 Observations and Physical Conditions

The sun's global output in soft X-rays occasionally shows pronounced, transient increases of total flux which exhibit a rapid rise on time scales of minutes and a more gradual decay, usually over hours (Fig. 10-8). Imaging in soft

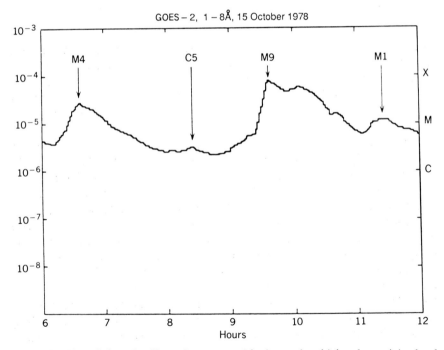

Fig. 10-8 Plot of the solar X-ray flux over a 6-hr interval at high solar activity level, showing several flares. The designations C5, M1, M4, and M9 indicate flares of increasing magnitude in soft X-rays. Adapted from Z. Svestka, in *Solar Flare MHD*, E. Priest (Ed.), Gordon and Breach, London, 1981.

JUNE 16, 1973

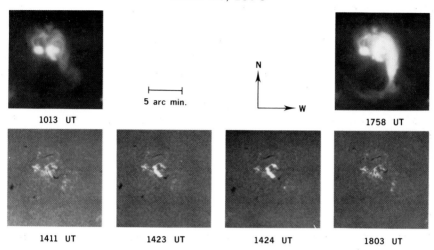

1013 UT   5 arc min.   N → W   1758 UT

1411 UT   1423 UT   1424 UT   1803 UT

Fig. 10-9   A flare imaged on the disk in Hα (bottom) and in soft X-rays (top) by a grazing incidence X-ray telescope on *Skylab*. Note the spectacular X-ray brightening of a large magnetic loop at 1758 UT. By permission of L. Golub.

X-rays shows that the increase is most often caused by a dramatic local brightening of the corona within an active region, called a flare (Fig. 10-9). In Hα photoheliograms the brightening extends into the chromosphere where the intensity rises rapidly in parts of the pre-existing plage. The very brightest flares can even be seen in the photospheric white light, although this is rare (Fig. 10-10).

The basic time behavior of flar   emissions over a wide range of wavelengths is shown in Fig. 10-11, and comparison of these curves provides some useful clues to how the energy release proceeds. An initial "preflare" brightening lasting generally a few minutes (but up to tens of hours in some cases) can usually be detected in EUV and soft X-ray radiations. The highest-energy flare emissions all reach their peak intensity during the dramatic flash phase. Impulsive spikes of emission are often seen in microwaves, EUV, and hard X-rays during this phase. The individual spikes have time scales of seconds or less, and the entire impulsive phase generally only lasts for a few minutes. Hα and soft X-rays may continue to increase for 10–20 min after the flash phase. The subsequent gradual decrease of flare radiations can take several hours. A large flare significantly increases the solar output in the EUV and X-ray regions (Fig. 10-12), although its effect in the visible and infrared, and on total output, is below the 0.01% level.

The detailed morphology of flares as revealed by imaging in hard and soft X-rays, in the EUV, and in Hα is quite complicated. But a division of flares into two basic structural types seems to be most useful at the present time. In

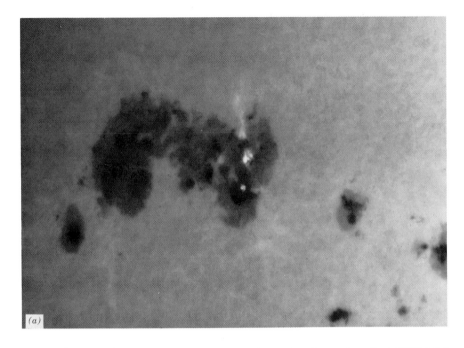

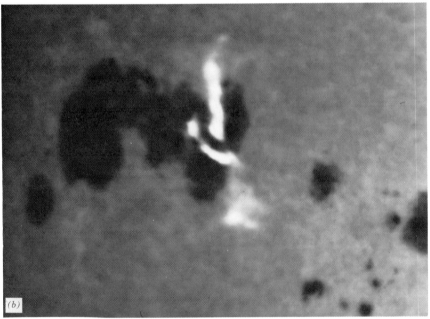

Fig. 10-10 Appearance of the white-light flare of April 25, 1984 at (a) 6200 Å and (b) 3610 Å. Note the enhancement in the ultraviolet. Sacramento Peak Observatory photos by permission of D. Neidig.

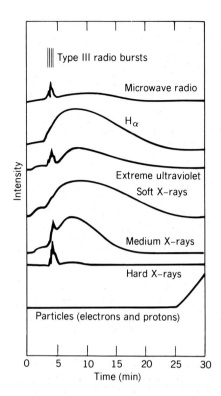

Intensity

Type III radio bursts

Microwave radio

H$_\alpha$

Extreme ultraviolet

Soft X-rays

Medium X-rays

Hard X-rays

Particles (electrons and protons)

0   5   10   15   20   25   30

Time (min)

Fig. 10-11 Time evolution of flare emissions. By permission of S. Kane.

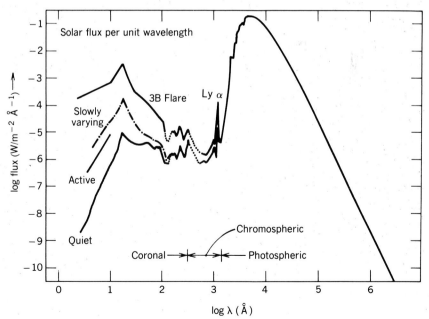

Solar flux per unit wavelength

log flux (W/m^{-2} Å^{-1}) →

3B Flare

Ly α

Slowly
varying

Active

Quiet

Chromospheric

Coronal →|← →|← Photospheric

log λ (Å)

Fig. 10-12 Enhancement of the solar spectrum at short wavelengths by a large flare. Adapted from E. Smith and D. Gottlieb. *Space Sci. Rev.*, **16**, 771 (1974).

the so-called compact flare, a brightening of coronal radiations occurs in a pre-existing loop or arch system, and little structural change is observed in these loops as a result of the flare. Motions, if any, are confined to small surges of chromospheric material accelerated to velocities well below the escape velocity from the chromosphere. Large flares of this kind can involve the entire volume of loop structures in a fully developed active region. The smallest compact flares, on the other hand, can be enhancements of tiny X-ray and coronal EUV bright points in the quiet sun, with no detectable chromospheric signature.

Large flares, however, typically show intricate structure (Fig. 10-13). During an important type of flare, called a two-ribbon event, the activation of the solar atmosphere is quite dramatic (Fig. 10-14). In such an event, changes in the large-scale arrangement of the active region can typically be observed well before the flash phase. The active region filament that often marks a magnetic neutral line of the active region begins to move about noticeably and usually lifts off in a disparition brusque.

The flash phase of this kind of event usually sets in after the filament disruption. In Hα this phase is marked by the brightening of two narrow ribbons, one to each side of the neutral line, which give this kind of flare its name. Observations in soft X-rays or in the EUV show that these Hα ribbons are the footpoints of an arcade of bright coronal loops straddling the neutral line.

Similar, somewhat lower loops can be observed in emission in Hα during flares that occur at the limb (Fig. 10-15) and sometimes as dark loops in Hα absorption after disk flares. Traditionally, such loop systems have been referred to as postflare loops. However, more recent results indicate that the evolution of these loops coincides with successive or continuous primary inputs of energy during a two-ribbon flare, rather than simply a passive decay of the original flash phase.

The largest two-ribbon flares can involve energies of 10^{32} ergs or more. This output is distributed between the thermal energy of plasmas observed mainly as radiations in soft X-rays, EUV, and strong visible lines; the nonthermal energy of charged particles accelerated up to the GeV range; and the kinetic energy of plasmas and associated shock waves propagating out from the sun. In both compact and two-ribbon flares, the partition of the total energy between these three components varies greatly, especially for the smaller events. Some produce energetic streams of nonthermal electrons or eject a coronal transient, while in others the power output appears only as the relatively static heating of magnetically confined coronal plasmas.

The importance or size of a flare can depend on the emissions used to measure it. A commonly used index is the Hα area. As shown in Table 10-1, chromospheric brightenings exceeding 3×10^8 km^2 are divided into flare classes 1–4 with increasing area. Smaller brightenings are called subflares. Most of the smaller flares and subflares are of the compact variety, while the largest events are of the two-ribbon kind.

Fig. 10-13 Appearance of a large flare in Hα at high spatial resolution. Pic du Midi Observatory filtergram. By permission of R. Muller.

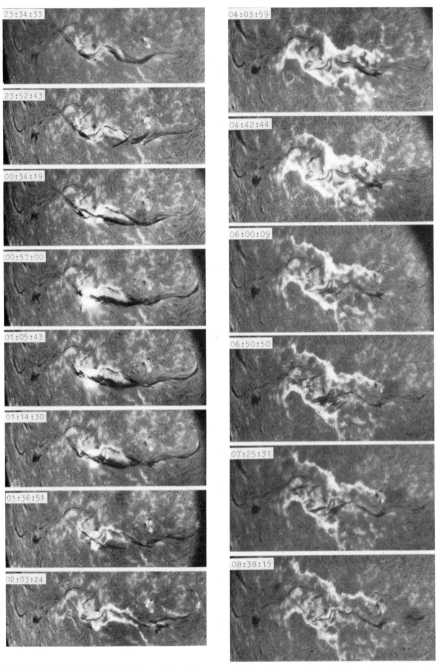

Fig. 10-14 Detailed development of a two-ribbon flare over a 9-h period. Note the complex magnetic fields seen already before the onset (as indicated by the structure of overlying Hα filaments), and the formation of dark postflare loops at 06:50:50. Filtergrams obtained by H. Morishita at Norikura Solar Observatory. By permission of E. Hiei.

Fig. 10-15 Postflare loops seen at high spatial resolution at the limb in Hα. The cross-sectional diameter of the loop (near the top) is about 2.8×10^4 km, and its height above the photosphere is roughly 100,000 km. Big Bear Solar Observatory filtergram.

The power output of large flares far exceeds the thermal or kinetic energy content of the low-β plasmas in the corona, where the primary energy release seems to occur. The most suitable energy source seems to be the coronal magnetic field. Only a fraction of the total magnetic energy of an active region is available to energize a flare. This free energy of the field is that fraction resident in coronal electric current systems (flowing along the magnetic field lines) or alternatively visualized as the twist of the field lines associated with these coronal loop systems.

It would be nice to check this by observing changes in the magnetic geometry during a flare and calculating the power released in such relaxation for comparison to the total flare energy. In practice, the interpretation of such

TABLE 10-1 Flare Classification by Its Area Observed in Hα Images.

Flare Importance	Flare Area (10^{-6} visible hemisphere)
1	100–250
2	250–600
3	600–1200
4	> 1200

observations remains ambiguous for several reasons. As discussed in Chapter 9, magnetic field structure in the low corona can still only be inferred from the geometry of plasma loops observed in soft X-rays, EUV radiations, and coronal forbidden lines. When changes in the geometry of such loops occur during a flare it is never quite certain to what extent the field lines have re-arranged themselves or whether different pre-existing fields at various levels of the corona are successively illuminated by the flare heating. Another problem is that active region fields are continuously changing even in the absence of flares, so it is difficult to decide whether a change observed during a flare is physically related.

Despite these difficulties, studies of the relation between flares and magnetic fields have revealed some useful points. One finding is that larger flares occur more often in active regions whose magnetic field is relatively complex. This complexity is usually the result of emergence of small new bipoles into a pre-existing, more developed active region. The relatively high coronal field gradients extrapolated from photospheric magnetograms in such active regions probably contribute to the release of flare energy, as discussed below.

The free energy content of an active region magnetic field is usually judged by comparing its loop geometry in soft X-rays or Hα with the geometry extrapolated from photospheric magnetograms assuming a purely potential, current-free field. As discussed in Chapters 8 and 9, the observations that loops often cross the active region neutral line at less than a perpendicular angle, and of pinwheel-like structures in Hα around spots, both indicate the presence of magnetic field-aligned currents.

For instance, during two-ribbon flares on the disk, the successively higher loops observed as the event progresses align themselves more nearly perpendicular to the neutral line. This behavior is consistent with relaxation of an initially sheared field, but does not prove it because of the ambiguities described above. In general, sufficient coronal magnetic free energy seems to be available to energize flares, but the direct evidence for its release during the flare is weaker.

The thermodynamic structure of the flaring atmosphere is complicated, and a comparative discussion of how a flare looks in each of its hundreds of strong emissions can easily fill a whole book. Fortunately, interpretation of these data has progressed to the point where the key points with most bearing on the energy release mechanism and the subsequent fate of the released energy are beginning to emerge.

The exact site of the flare's primary energy release is not yet certain, but images in soft and hard X-rays give the most direct information on the distribution of the most energetic plasmas and on their physical conditions. In soft X-rays the flaring plasma occupies a closed loop, or a system of parallel closed loops called an arcade. The loops delineate magnetic flux tubes that cross the active region neutral line and connect to opposite-polarity photospheric footpoints on either side.

Flash phase images of flare loops taken through different X-ray filters, and also monochromatic images taken in different EUV and XUV lines of known temperature of formation, indicate that at least in two-ribbon flares, the hottest thermal plasmas are located near the top of the loops. The temperatures range up to 10–20×10^6 K. The densities found from emission measures and from EUV line intensity ratios can be as high as 10^{13} cm^{-3} in the brightest kernels of flare emission.

The impulsive spikes of the flash phase can generally be measured in radiations requiring excitation energies of up to several hundred keV and occasionally in excess of 1 MeV. The ion energies giving rise to these hard X-rays imply enormous temperatures of about 10^9 K if the emitting plasma is thermal (and probably confined to many small and short-lived volumes). The data can also be taken to indicate the presence of relativistic particles, probably consisting of directed electron beams.

Imaging in hard X-rays is difficult, and spatial resolution is so far limited to about 10 arc sec. Results indicate that the hard X-rays can originate in bright points at the footpoints of the soft X-ray and Hα flare loops in some flares. But in other flares they seem to also originate in more diffuse regions located 50,000 km or higher in the corona.

It is tempting to suppose that the most energetic radiation coincides with the source of the flare energy release. However, the hard X-rays are often not observed until many minutes after the flash phase onset in softer radiations, so it is unlikely that they hold the most direct key to the site of the primary release. In some cases they may arise from the impact of the electron beams on lower layers of the flare loops, while in the higher events they might be more closely associated with progressive stages of magnetic field line reconnection.

Many flares produce microwave bursts whose time behavior is generally very similar to the impulsive hard X-ray bursts and often tracks it in detail, at least in the earlier part of the burst. Imaging of the microwaves with high spatial resolution is relatively difficult. Interferometric studies suggest that microwaves are most intensely emitted in those loop footpoints and condensations which are also brightest in Hα and soft X-rays. The emission is produced by gyrosynchrotron radiation, probably by the same relativistic electron beams that produce the hard X-rays.

The behavior of visible, UV, and EUV emissions formed at temperatures below about 10^6 K carries information on how the flare energy is deposited at lower atmospheric layers during the flash phase and also on how the coronal flare volume cools during the gradual decay phase. The behavior of the impulsive EUV radiations closely parallels that of the hard X-rays and microwaves, suggesting that the EUV arises from downward propagation of the same heating that also produces the Hα ribbons.

Doppler shift measurements of visible and EUV lines provide the best data on flare plasma motions. The often-observed asymmetries of the profiles of Hα and other chromospheric lines during the flash phase indicate first an upward and then, later, a downward streaming of up to 100 km s^{-1} in the cooler plasmas. Even higher turbulent or directed velocities have been reported in the hottest flare plasmas from broadening and asymmetries in resonance lines of ions as energetic as Fe XXIII and Fe XXIV.

Observations in meter waves provide direct information on radiations from electron streams, plasma, and shock waves issuing from the flare region. Several different kinds of short-lived meter wave emissions, referred to as bursts, are observed in association with flares. Their distribution in radio frequency and duration is shown in Fig. 10-16. The most common are the Type III bursts, which occur for only a few minutes at the very onset of many flares and extend over frequencies from a few tens of kHz to hundreds of MHz. Type III bursts are caused by streams of electrons accelerated outward from active regions, which excite plasma oscillation at the progressively lower plasma frequency (see Section 4.4.7) of the successive coronal layers they pass through. The highest frequencies recorded for Type III's correspond to oscillations in the low corona, the lowest to those of interplanetary space near 1 AU.

The connection between Type III's and electron streams has been directly tested by tracking radio bursts out from the sun and then detecting the associated electrons by interplanetary probes and Earth-orbiting satellites. The active region phenomenon responsible for acceleration of the Type III electrons is less certain. Type III's seem to be generated by activated filaments that are often associated with flaring, but also occur in nonflaring active regions.

A different kind of impulsive broad-band continuum emission observed particularly during two-ribbon flares is the Type IV burst. These extend in frequency from microwaves down to about a hundred kHz, but the best-studied component is the so-called moving Type IV observed at frequencies between roughly 10 and 100 MHz. Type IV emission is interpreted to be gyrosynchrotron radiation from energetic electrons magnetically trapped in a plasma cloud accelerated outward from the flare. The emission typically lasts for tens of minutes as the source moves outward to a few solar radii at a speed of from a few hundred to about 1500 km s^{-1}. It is generally considered that moving Type IV bursts are the radio signature of the coronal transients produced by two-ribbon flares and by erupting filaments undergoing disparition brusque.

The third major type of meter wave burst is the Type II, which produces intense radiation at frequencies below a cutoff around 100 MHz. The emission consists of two discrete frequency bands which drift to lower frequency over a time of about 5–10 min. The two bands are interpreted as fundamental and second harmonic radiation from plasma oscillations due to a disturbance moving outward in the corona. The frequency drift rate corresponds to outward velocities of between 500 and 5000 km s^{-1} through progressively lower density layers to beyond $3R_\odot$.

Finally, the Type I or noise storm emissions are very long-lived (up to days) continuum events observed at between 50–400 MHz from active regions. Their production is still not well understood, but their very high brightness tempera-

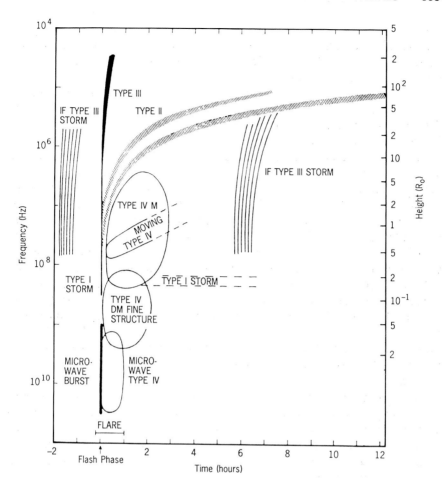

Fig. 10-16 The radio spectrum of a large flare. The low-frequency Type III and Type I storms preceding and following the flare are not necessarily part of the flare. Only one Type III burst has been drawn, although a group of about ten occurs at the flash phase. Only the envelopes of the Type IV burst have been drawn; usually only parts of them are filled. The height scale on the right-hand side corresponds to the plasma level of the frequency scale on the left-hand side. Adapted from H. Rosenberg, *Phil. Trans. Roy. Soc. London A*, **281**, 461 (1976).

tures and close to 100% polarization are best understood in terms of plasma oscillations excited by energetic electrons.

Even the most common Type II velocities around 1000 km s^{-1} are much higher than the local sound or Alfvén speeds, so the disturbance must be an interplanetary shock wave. The large spatial extent of the Type II source measured by interferometers indicates the shock is a spherical disturbance

Fig. 10-17 A big surge photographed in Hα extending to more than 230,000 km above the limb. Filtergram from Hida Observatory, Japan, by permission of Y. Funakoshi.

expanding out from the active region. The very close statistical association of Type II's with the flash phase of two-ribbon flares and with Type IV bursts supports this interpretation.

Besides the filament eruptions and associated coronal transients discussed above and in Section 12.2.3, flares often also produce violent smaller scale mass motions called surges and sprays (Fig. 10-17). In the former, plasma is typically accelerated outward at 50–200 km s^{-1} for a few tens of minutes in a relatively narrow jet extending as high as 1–2×10^5 km, but generally much less. It then falls back, usually along the same curved path. In sprays, plasma is accelerated within minutes to beyond the escape velocity and fans out into a considerable solid angle.

Flares also accelerate particles nonthermally to relativistic energies. The bursts of hard X-rays and microwaves are produced by energetic electrons and the Type III and IV meter waves are produced by electrons of relativistic energies. Large two-ribbon flares seem to be responsible for the acceleration of protons and heavier nuclei to energies of 10 MeV or more, and these relatively rare events are referred to as proton flares. Even less frequent are the cosmic ray events, which can produce protons of over 500 MeV detected through their production of cosmic ray air showers in the Earth's upper atmosphere and thus through their effect on neutron monitors at ground level.

Unlike the energetic electrons, which radiate powerfully by the synchrotron mechanism and through bremsstrahlung in the corona and lower solar atmosphere, energetic protons and nuclei leave much less of a direct radiative signature. On the other hand, the fact that they dissipate their energy less makes them more abundant near 1 AU.

Recent γ-ray observations have demonstrated that nuclear reactions between protons and α-particles occur during flares. The 2.2-MeV γ-ray line, whose emission by capture of energetic neutrons onto hydrogen to form deuterium was predicted by L. Biermann in 1951, has now been detected in several dozen solar flares. Several other γ-ray lines, such as the 0.51-MeV line due to positron annihilation, and also γ-ray continuum have also been detected. Primary neutrons directly produced in the sun (rather than indirectly by cosmic ray spallations in air showers) now seem to have been measured also. They are the result of energetic proton impacts on heavier nuclei.

10.2.2 Energy Release and Dynamics

The flare observations outlined above pose many interesting questions. The most fundamental are those dealing with the source of a flare's enormous energy, the mechanism of its amazingly rapid release, and the transformation of that great power into heat and into plasma and charged particle kinetic energy. Study of these huge solar explosions has always attracted a great deal of theoretical attention, and the trail of the elusive energy source appears to be getting warmer as observations become available of the light and particles most intimately connected with the site of primary energy release.

To within the uncertainties in our present understanding of turbulent plasmas, several competing explanations of the overall flare phenomenon appear to be at least internally consistent in their physics and also sufficiently adaptable to explain the key observations. In this sense, the explanation of solar flares has lost the status of a fundamental enigma. But the task of determining which (or any) of the competing mechanisms specifically explain the observations of compact and two-ribbon flares, of how such mechanisms might scale to flares observed on other stars, and to what extent a flare and its terrestrial effects can be predicted are still very interesting issues.

As pointed out above, it is generally believed that the energy released in a flare is derived from the coronal electromagnetic field. The most simple alternative explanation is that the energy release might occur by rapid adiabatic compression of the high plasma by the confining magnetic field, leading to rapid heating. However, the final coronal field intensities required to achieve this are in the kilogauss range. This discrepancy with observations is one of the difficulties with adiabatic heating, leading to the view that flares represent the dissipation of magnetic energy.

The energy liberated in a large flare, of order 10^{32} ergs, is formidable even compared to the total magnetic energy content of substantial coronal volumes (see exercises). For instance, it significantly exceeds the entire magnetic energy of a large active region loop, even if such a loop's energy were to be completely annihilated during a flare, which is not observed to be the case. In practice, the free energy content of the field associated with currents running along magnetic field lines is less than the loop's total magnetic energy, which is mainly associated with subphotospheric current systems that cannot be tapped in a flare. This implies either that the stored free energy of most of an entire active region is liberated or that the magnetic free energy is continuously replenished from the photosphere during the flare.

It is not clear whether the flare energy is derived from field-aligned currents existing in a coronal magnetic configuration whose equilibrium is disturbed or whether it is provided to the flaring volume by an external driver. If it is the former, then the configuration must be stable to at least small perturbations, otherwise it would not be able to build up the required free energy density. The trigger would then be some disturbance whose amplitude exceeded the critical value that could be sustained.

An external driver could take the form of magnetic flux thrust upward from the convection zone and continuously pushed into pre-existing coronal field lines. It could also be seen as the effect of gas motions at subphotospheric levels acting to twist the footpoints of flux tubes, which then transmit these torsional stresses as Alfvén waves propagating into the corona.

The observation that at least the young, most flare-producing, active regions contain significant field-aligned currents indicates that considerable free energy is always available for release in flares. On the other hand, the frequent association of flares with the emergence of new flux into a pre-existing active region suggests that external driving may play a role also, although its main significance might be in producing a trigger.

The most fundamental question is how so much electromagnetic energy can be released so quickly. In the discussion of plasma self-induction given in Section 4.3.2, it was shown that the time scale for magnetic dissipation in a volume of characteristic dimension L and electrical conductivity σ is of order $4\pi\sigma L^2$ s. Given the minimum volume required for storage of a large flare's energy, it is easy to show that this time scale is about 10,000 years, rather than the few minutes in which a substantial fraction of a flare's total energy release occurs.

This time scale for current dissipation can be decreased somewhat if the standard electrical conductivity [equation (4-98)] is decreased by taking into account plasma turbulence. This turbulence generates electrostatic wave fields which promote additional scattering of the electrons, so decreasing their mean free path and thus the plasma conductivity. However, the decrease in the energy release time scale is still insufficient to explain flare rise times, unless the dimensions of the flare plasma are also drastically restricted. The field gradients that govern the diffusion time scale through equation (4-69) must be high at least in one dimension.

Two basic geometries have been put forward to reconcile the requirements for high local magnetic field gradients (to obtain rapid energy release) and sufficient volume (to obtain enough energy to produce a flare). In one kind of idealized configuration, called a neutral sheet, antiparallel frozen-in magnetic fields are pressed together at a surface (at which $B = 0$, hence the name) until the magnetic gradients transverse to the surface become sufficiently high for reconnection to occur (Section 11.1.6). Energy can then be released at the required rate. A current $\mathbf{j}$ must flow transverse to the field lines and within this surface since $\nabla \times \mathbf{B} \neq 0$, so such surfaces are often called current sheets. Although the thickness of the reconnecting region may be very small, perhaps a kilometer or less, its extent in the other two dimensions can be many tens of thousands of kilometers so the power released by the dissipation of the current can be large.

A different approach is adopted in models where the required current dissipation is achieved by reducing the plasma's ability to conduct electrically along field lines. For instance, the density of field-aligned currents flowing along a coronal loop can become too high to be carried by thermal electrons in the tenuous plasma. In such circumstances, a number of plasma effects can set in which share the common property of being able to support a relatively large local potential drop and thus being able to dissipate current energy at a relatively high rate.

The details of how these processes proceed in flares are far from clear, but laboratory experiments suggest conditions leading to the production of plasma turbulence and an increase of resistance to anomalously high values. Under other circumstances the plasma can polarize locally, forming a space-charge region like a parallel-plate capacitor, called a "double-layer," that can sustain a large potential drop. In yet other cases, MHD instabilities called tearing modes can cause filamentation of the current into narrowly confined channels where the resistivity is relatively low.

This second class of flare models deals with dissipation of currents $j_{\parallel}$ flowing along $\mathbf{B}$, rather than transverse to it, as in current sheets. The geometry usually assumed is a simple loop, which is in best agreement with the observations. The onset of the highly dissipative energy release requires some mechanism to dramatically increase either the current density $j_{\parallel}$ or decrease the local density of charge carriers along the loop's axis. This trigger may be a large-scale MHD instability or perhaps a plasma pressure disturbance.

Both types of mechanism can probably be adjusted to yield the dissipation rates required, so with reasonable external forces invoked to provide power at the required rate, a flare can be explained. The detailed plasma physics of how reconnection proceeds in solar current sheets and what processes can increase resistivity in solar loops, is sufficiently uncertain that it is difficult to identify strict observational tests. Observations of structure are scarcely possible since in current sheets, and probably also in the double layers, the thickness of the volume probably responsible for primary energy release is far below present resolution. The direction of the electric field in the flaring region could provide a direct distinction, since it should be perpendicular to $\mathbf{B}$ in the reconnection models, and parallel to $\mathbf{B}$ in the current limitation scenario. Remote sensing of $\mathbf{E}$ in these plasmas may be achievable if their emission measure can be detected.

The acceleration of large plasma volumes in sprays and disparitions brusques of filaments is closely associated with two-ribbon flares. However, it is not yet clear whether this association constrains the mechanism of flare energy release or whether it is perhaps an independent MHD instability triggered by the same disturbances leading to the flare. Recognition of the close relationship between two-ribbon flares, filament eruption, and particle acceleration has been one of the most interesting directions of flare research in recent years.

Acceleration during the impulsive flare phase accounts for electrons of a few tens to few hundreds of electron volts required to explain the bursts of hard X-rays and microwaves. The source of the accelerating electric fields (which must be oriented parallel to $\mathbf{B}$) is not yet clear. Turbulence and waves may be responsible. The observations near 1 AU of much higher energy nuclei, up to the GeV range, seem to require a second step of acceleration, probably produced by shock waves associated with the ejection of coronal transients.

Both types of flare mechanisms outlined above seem capable of accelerating the substantial particle fluxes observed at energies well in excess of the mean thermal energy. However, if the impulsive bursts of hard X-rays and microwaves are assumed to be produced by a beam of nonthermal electrons decelerated in a dense target like the lower chromosphere, the number of electrons of around 25 keV required to explain the observed radiative fluxes is extremely large, in excess of 10^{37} particles. The dc fields in a double layer might come close to producing such an electron flux, but their existence has yet to be demonstrated.

The shape of the hard X-ray spectrum, which is usually represented by a power law, can also be reproduced by the superposition of several exponential spectra, each produced in a hot thermal plasma of different temperature. If the X-rays are produced by collisions of very hot electrons with ions in a magnetically confined thermal plasma, the number of electrons required to produce the observed hard X-rays and microwaves is a few orders of magnitude lower than the 10^{37} particles required above. Such fluxes might be produced by a wider range of flare energy release mechanisms, possibly including neutral sheets.

ADDITIONAL READING

Prominences and Filaments

T. Hirayama, "Modern Observations of Solar Prominences," *Solar Phys.*, **100**, 415 (1985).

A. Poland (Ed.), "Coronal and Prominence Plasmas," NASA Conference Publication 2442 (1985).

E. Tandberg-Hanssen, *Solar Prominences*, D. Reidel, 1974.

Flares

J. Brown, D. Smith and D. Spicer, "Solar Flare Observations, Interpretation and Theory," in "The Sun as a Star," NASA Publication SP-450, S. Jordan (Ed.), Chapters 7 and 18 (1981).

H. Hudson and K. Kai (Eds.), "Recent Advances in the Understanding of Solar Flares," *Solar Phys.*, **113** (1987).

E. Priest (Ed.), *Solar Flare MHD*, Gordon and Breech, London, 1981.

Z. Svestka, *Solar Flares*, D. Reidel, 1976.

G. Trottet and M. Pick (Eds.), "Particle Acceleration and Trapping in Flares," *Solar Phys.*, **111** (1987).

EXERCISES

1. From the mean speed of thermal electrons and protons and the expression (4-97) for the electron-ion collision frequency, determine whether the plasma dynamics of a prominence is reasonably calculated in the continuum approximation or whether individual particle behavior is likely to be important.

2. Draw the two basic magnetic configurations envisioned for prominence support, and explain the source of the magnetic forces that act against

gravity in each case. Try to draw the current systems that must flow in both cases to provide the Lorentz force required for the support. Compare each geometry with the magnetic field observations of quiescent and active prominences and comment on the comparison.

3. Estimate the upward momentum flux $\rho v \cdot \nabla v$ required to balance the pull of gravity on a prominence [see equation (4-25)], and use equation (4-77) to calculate the Alfvén wave velocity required to supply this flux under ideal conditions. Is this consistent with values of coronal density and magnetic field at the heights of quiescent prominences? Can such a mechanism account for the prominence acceleration during a disparition brusque?

4. The equation of motion of an erupting loop propelled by an imbalance between axial and azimuthal fields is

$$\frac{B_{az}^2}{8\pi R_c} - \frac{B_l^2}{8\pi R_c} = \rho \frac{GM}{R^2}.$$

Evaluate the terms in this equation, and determine how large a fraction the loop's azimuthal field must be of the axial field for such a driving mechanism to work. Estimate the pitch of the loop's twisted field from the ratio B_{az}/B_a, and compare this with the pitch of fields around a sunspot, as revealed by the vortex structure of Hα fibrils, as seen in Fig. 9-4.

5. Explain why impulsive spikes in flare emission are often seen in microwaves, EUV, and hard X-rays, but not in soft X-rays.

6. Explain the mechanisms of the different kinds of meter wave bursts, and relate their occurrence to a flare's time evolution. Use equation (4-91) to estimate the plasma density in the outer corona using the observed frequency behavior of Type II bursts.

7. Estimate the thickness of the reconnecting region required to produce flare energy release at the observed rate. Also, estimate the velocity at which energy must be carried into this region to satisfy the magnitude of the flare's power requirement. How does this velocity compare to the local Alfvén speed if the neutral sheet area is of order 10^8 km^2?

11

Dynamics of the
Solar Magnetic Field

The observed confinement of the photospheric field to discrete sunspot, facular, and network flux tubes in an otherwise nonmagnetic plasma is surprising, and we discuss in this chapter the mechanisms that might intensify the field locally and determine the plasma properties within the flux tubes. We also discuss the processes that seem to bring the field up from the convection zone into the visible atmosphere, determine its evolution over the surface, and drive the flows and wave motions that enable energy to travel outward along field lines into the corona and solar wind.

The oscillation of the sun's magnetic field polarity and structure over the 22-year magnetic cycle gives us our main clues to the processes that have sustained stellar magnetism for billions of years. We discuss here the main ideas put forward to describe the dynamics of this oscillation and its effects on solar structure. Understanding the 22-year magnetic cycle and particularly the associated 11-year cycle in activity is also of increasing practical interest as predictions of the sun's variable influences upon "space weather" near the Earth assume greater importance.

11.1 DYNAMICS OF SOLAR MAGNETIC FLUX TUBES

11.1.1 Dynamical Equilibrium and Geometry

In the absence of other forces, the magnetic repulsion of the field lines within a flux tube would lead to their dispersal on a time scale comparable to the acoustic transit time across such structures. This is under a minute for the network and facular flux tubes and well under an hour for spots. Early attempts to understand the intensification of network fields led to the suggestion that they were maintained by the ram pressure of the converging horizontal supergranular flows. Later observations of kilogauss intensities at photospheric levels showed that the magnetic pressure in the network is too high to be maintained by the advection of momentum at a rate $\rho v \cdot \nabla v$ in a flow of photospheric density with v about 0.5 km s^{-1}.

The most likely explanation of the intensification process has emerged from numerical simulations of convection in a medium that is initially uniformly magnetized. Under plausible solar conditions of the magnetic Reynolds number and Rayleigh number (see Chapter 4) convection tends to push field lines to the boundaries between the overturning cells. This tendency to exclude the field from the convecting plasma creates a two-component medium consisting of intensified and evacuated flux tubes at the cell boundaries and a relatively nonmagnetic cell interior of higher gas pressure. The basic geometry is illustrated in Fig. 5-10. Details of this explanation depend on the uncertain transport coefficients of solar plasmas, but the model provides a reasonable framework for understanding why photospheric magnetic fields are observed to be bunched in discrete flux tubes rather than filling the photospheric surface. A similar process may fragment the solar field within the convection zone. However, its effectiveness has been questioned since the ability to radiate powerfully into space from the photosphere seems to play an important role in the ability to compress photospheric plasmas and intensify their fields.

This model of field intensification suggests that a photospheric flux tube in a spot or facula is held together by a balance between the outward pressure of the field and the inward force caused by a lower gas pressure inside the flux tube. The picture is plausible since we saw in Section 8.2.1 that in an umbra the observed temperature and probably also the gas pressure are much lower than in the photosphere at equal geometrical depth.

We assume further that on all spatial scales of interest the magnetic Reynolds number greatly exceeds unity, so that the magnetic field and plasma are frozen together. We also assume, mainly on the grounds of convenience, that this flux tube is axisymmetric and stands vertically, although a vertical orientation is also suggested by arguments for the buoyancy of flux tubes given in Section 11.1.2.

In the most simplified model, illustrated in Fig. 11-1, the field in the interior of the flux tube is taken to be force free, so that no horizontal forces act upon it. The magnetic volume force $j \times B$ holding the flux tube together is confined to a thin, current-carrying sheath at the flux tube-photosphere interface. This model is too simple for spots, where the field intensity seems to exhibit irregular variations across the umbra. But it provides a convenient point of departure for a discussion of the much more slender facular flux tubes whose structure cannot be resolved with existing observations.

The most general configuration taken up by such a force-free field cannot be derived analytically, and this complicates dynamical studies. A numerically computed configuration for a sunspot flux tube is shown in Fig. 11-2. Comparison with the observed sunspot profiles of $|B|$ shows reasonable agreement

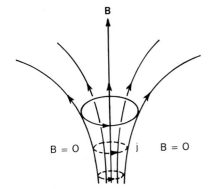

Fig. 11-1 Schematic diagram of a flux tube, illustrating the current sheet of current density j bounding a potential field **B**.

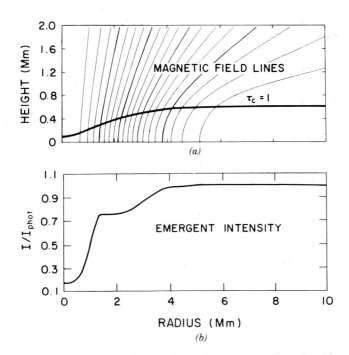

Fig. 11-2 Computed magnetic field line shapes in a sunspot flux tube. The geometric height of the $\tau = 1$ surface is given by the solid line in the top panel. The lower panel shows the intensity distribution across the spot. By permission of V. Pizzo.

although detailed comparison is difficult because a fully consistent calculation is not yet possible.

A simplified, analytically tractable model can be constructed by assuming the flux tube is very slender. In practice it can be shown that this requires a radius less than about four times the photospheric pressure scale height, thus a few hundred kilometers. For such a flux tube the horizontal equilibrium is given by

$$P_e - P_i = B^2/8\pi \tag{11-1}$$

where P_e and P_i are the external and internal gas pressures and B is the field intensity taken to be uniform across the flux tube.

The vertical force balance is maintained by the usual hydrostatic relation

$$\frac{dP}{dr} = -\rho g. \tag{11-2}$$

A general idea of the flux tube radius and shape can be obtained from simple consideration of the behavior of gas and magnetic pressure at different layers of the solar atmosphere. As pointed out in Chapter 4, the behavior of magnetized plasmas at high magnetic Reynolds number R_m is largely determined by the plasma's magnetic pressure relative to its thermal pressure, P. This ratio is denoted by the parameter β, defined as

$$\beta = 8\pi P/B^2. \tag{11-3}$$

If β is taken as the magnetic intensity within a flux tube and P is the external gas pressure, the discussion of flux tube equilibrium given above indicates that $\beta \sim 1$ near the photosphere. Above the photosphere we generally find $\beta < 1$ since the gas pressure decreases exponentially, while the magnetic field of a flux tube falls off more slowly according to a power law in radius whose exponent depends on the local distribution of photospheric fields. When $\beta < 1$, the field expands to fill all space, so the field lines confined to flux tubes at the photosphere rapidly diverge with height and fill the entire chromospheric and coronal volume.

The computed configuration of quiet region network fields is shown in Fig. 11-3. We notice that the boundary above which the flux tubes are expected to have merged forms a magnetic "canopy" whose height is estimated to be roughly 1500 km above $\tau_{0.5} = 1$. The real field configuration at this level, as seen in the pattern of Hα fibrils (Chapter 9), is more complicated since the model assumes the field lines extend to infinity into the solar wind while coronal and chromospheric fields contain tensions along field lines caused by connections between flux tube footpoints anchored in the photosphere at various distances.

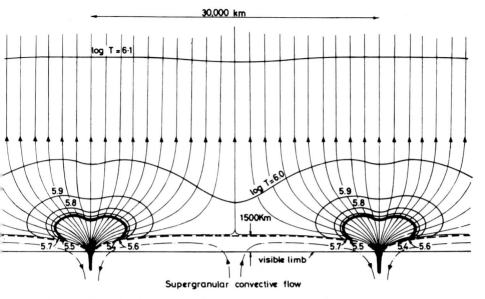

Fig. 11-3   The structure of a magnetohydrostatic network model. Temperature contours between $\log T = 6.1$ and $\log T = 5.4$ are marked. Reproduced with permission from A. Gabriel, *Phil. Trans. Roy. Soc. London*, **281**, 339 (1976).

Below the photosphere, it is usually assumed that the gas pressure increases inward more rapidly than the magnetic pressure, so $\beta \gg 1$. Whatever the actual β-value deep in the convection zone, the magnetic stresses of even the 2–3 kG umbral fields exceed the aerodynamic drag force of any plausible mass motions of solar plasma relative to the field to a considerable depth below the photosphere. Thus the magnetic field lines are expected to move essentially as rigid rods anchored to levels at a depth at least 10^4 km below the photosphere. This dominance of $B^2/8\pi$ over ρv^2 helps us to understand how sunspots might move relative to one another and how the different rotation rates of various magnetic structures of different size (see Section 7.1) might act as tracers of plasma motions at different depths in the solar convection zone.

11.1.2 Dynamical Stability

Certain states of equilibrium between two or more forces cannot exist in nature because they are unstable. An example of dynamical instability is the equilibrium of a denser fluid overlying a lighter fluid. If the boundary between the two fluids could be kept absolutely flat, the equilibrium between gravity and the fluid pressure gradient might in principle be maintained. But if an element of the denser fluid is moved infinitesimally below the boundary, it is replaced by the lighter fluid, and the potential energy of the overall system is decreased, which provides kinetic energy to drive further interchange of the two fluids. The result of this Rayleigh-Taylor instability is that the heavy and light fluids rapidly trade places.

Dynamical instability of MHD equilibria seems to play an important role in determining the structure of solar magnetic fields and in how they shed their free energy in flares and filament eruptions. The subject has a large literature, with many basic results taken from plasma fusion research, where the search for stable equilibria is the main goal. We discuss here only a few situations where instability seems to play a key role in determining important solar phenomena.

The long lifetime of sunspots is perhaps the most intensively studied example of MHD stability in solar structures. It is remarkable because in a plasma of uniform density the equilibrium between magnetic field pressure and gas pressure gradient is vulnerable to the so-called interchange instability.

The reason for the instability can be seen in Fig. 11.4. If the interface between field and nonmagnetic plasma is perturbed from a smooth to a fluted shape, the average intensity of the field is decreased. Since this perturbation can occur without changing the total volume of the flux tube or surrounding plasma (so that no net $P\,dV$ work is performed), the lower mean field intensity implies a lower total energy, and the perturbation can grow through this energy release. A net release of energy also accompanies fluting of a more sunspotlike flux tube, whose curvature is concave into the outside plasma (Fig. 11-2), although here the energy decrease is derived from a lowered mean curvature of the field lines and is somewhat more difficult to visualize.

The observation that spots exist despite the concave curvature of their field lines means that somehow the instability is prevented from taking hold. One likely explanation relies on the lower density inside the spot flux tube relative to the photospheric plasma. It can be shown that under these circumstances the fluting of a concave flux tube must do work against gravity, which reduces the net energy released by the perturbation and can stabilize it, provided the field's radius of curvature decreases with height in the atmosphere. In princi-

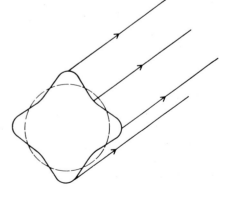

Fig. 11-4 The fluted cross-section of a cylindrical column of plasma that is subject to an interchange instability whose undisturbed cross-section (shown dashed) is circular. Adapted from E. Priest, *Solar Magnetohydrodynamics*, Reidel, 1982.

ple, the equilibrium might also be stabilized by a twist in the field lines or by their rigid tying to an immovable surface. In both these cases, interchange of field lines is made more difficult by their tension, which resists stretching. It then remains to explain how the spot can be stabilized against the gas pressure gradients that must be generated below the umbra by the sunspot's local blocking of heat flow. Present ideas on how spots can form, achieve temporary (metastable) equilibrium, and then decay are described in Section 8.2.3.

Investigation of the equilibrium of horizontal (rather than vertical) magnetic fields within the convection zone confronts us most clearly with the influences of magnetic buoyancy. According to equation (11-1), a region of locally intensified field must have a lower gas pressure if it is to remain in dynamical equilibrium with its surroundings. If the field intensification leaves the flux tube and its surroundings isothermal, the lower gas pressure requires a decrease of density by $\Delta\rho$ within the region of higher field and thus generates a volume force of buoyancy whose magnitude is given by $g\,\Delta\rho$.

This magnetic buoyancy tends to propel the flux tube upward. However, any curvature generated in the initially horizontal field leads to a magnetic tension along the field lines which acts to counteract the buoyancy. The buoyancy wins, and the magnetic field continues to rise, provided

$$g\,\Delta\rho > B^2/4\pi L \qquad (11\text{-}4)$$

or

$$L > 2\frac{kT}{mg} \equiv 2H, \qquad (11\text{-}5)$$

where L is the scale of variation of the field (e.g., the radius of curvature of the curved flux tube) and H is the local gas pressure scale height. We then expect that large-scale magnetic field irregularities would tend to rise to the photosphere, although smaller-scale perturbations might achieve equilibrium between buoyancy and magnetic tension.

The actual importance of magnetic buoyancy in driving the emergence of photospheric magnetic fields depends upon aspects of the model that are difficult to calculate. For one, a very small departure from isothermality between the stronger field region and its surroundings can eliminate the buoyancy effect altogether (see exercises), and it is difficult to predict the energy balance of submerged solar flux tubes to such precision. Also, the rate of rise of the flux tube will be determined largely by aerodynamic drag, and as discussed in the exercises, this uncertain rise time is of considerable importance to studies of solar cycle dynamics.

Analysis of magnetograms indicates that the total photospheric magnetic fluxes measured in the sun's east and west hemispheres are closely similar. If the field were tilted systematically, for instance to either lead or trail solar rotation, an imbalance of the longitudinal flux component would appear between these two hemispheres. Any net tilt seems to be less than 1° over the period 1967–1976. However, the positive and negative polarity flux tubes seem to be tilted toward one another by up to 10° from the vertical near activity maximum, and to become vertical around the time of minimum activity. Such large inclinations measured at the photosphere are difficult to understand if buoyancy is as large as believed since it is difficult to identify any force strong enough to overcome the tendency of such a flux tube to stand vertically, like a tethered balloon.

11.1.3 Thermal Instability

Besides the dynamical instabilities discussed above, thermal instabilities probably also play an important role in determining the solar plasma structure, particularly in the corona and chromosphere. In this case it is the equilibrium of the plasma energy balance [e.g., equation (4-43)] that is inherently unstable, rather than its force balance given, e.g., by the Navier-Stokes equation (4-26). Thermal instabilities are most likely to play a role in the formation of prominences and in the phenomenon of coronal rain.

Two basic effects can act to destabilize the thermal equilibrium of the solar plasma. One is the shape of the plasma radiative loss function $\Phi(T)$ illustrated in Fig. 4-2, which shows that if a plasma of coronal temperature $T \sim 10^6$ K starts to cool, its radiative losses per unit mass increase rapidly. If it is heated, they tend to decrease. This positive feedback caused by the temperature dependence of the plasma radiative losses is one influence that tends to drive the plasma away from a temperature equilibrium between balanced heat inputs and losses.

A second destabilizing factor involves coupling between the plasma's thermal and dynamical stability. If a plasma element initially in temperature and pressure balance with its surroundings begins to cool, its density must increase to maintain the pressure balance. This density rise will sharply increase its radiative losses, which behave roughly as the density squared for an optically thin plasma [equation (4-48)]. The result will be further cooling. Similarly, an initial heating would tend to promote destabilization toward higher temperatures. Why then does the corona not all cool down and collapse or heat up until it escapes from solar gravity?

The main thermal stabilizing factor in the hot coronal plasma is its high thermal conductivity. Whenever a thermal perturbation ΔT is generated locally over some spatial scale L, a conductive heat flux of order

$$F_c \sim K \Delta T / L \tag{11-6}$$

will tend to smooth it out by conducting heat in along magnetic field lines from a hotter plasma if ΔT is negative or away from the perturbation into cooler regions if ΔT is positive. Detailed calculations by G. Field first showed in 1965 that the stabilizing effect of the thermal conductivity results in a critical spatial scale L_c below which the plasma is thermally stable, although larger-scale thermal instabilities could grow.

In an initially isothermal atmosphere where the heat loss per unit mass, given by an expression of the form $C_1 \rho T^N$, is locally balanced by a heating function $C_2 \rho^M$, this critical scale can be expressed as

$$L_c = \left(\frac{2\pi K'}{C_1} \right)^{1/2} T^{(7/4 - N/2)} \rho_0^{-1} (1 - M - N)^{-1/2}. \qquad (11\text{-}7)$$

Here K' is the temperature-independent part of the thermal conductivity K given by equation (4-45) and $C_1 = 4 \times 10^{31}$ cgs. If we take $N = -1$ as the exponent of the power law representing the shape of the radiative loss function around $T = 10^6$ K, we find $L_c \sim 1 \times 10^{10}$ cm at coronal densities of $\rho_0 \sim 3 \times 10^8$ cm^{-3}.

Interestingly, this scale is about an order of magnitude larger than the condensations that are observed to form in coronal loops and prominences. Since we see from (11-7) that the dependence of L_c upon M and thus upon the uncertain coronal heating function is very weak, a more likely explanation of this discrepancy is that other processes (such as compression of the plasma by the magnetic field) locally increases ρ_0 until the instability can take hold. Alternatively, perhaps the picture of a thermal equilibrium is incorrect, and prominences form not as instabilities, but because the heating function is unable to balance plasma energy losses, so no equilibrium is achieved.

11.1.4 Steady Flows

Two examples of apparently steady flows observed in solar flux tubes are the Evershed effect in spot penumbrae and the possibly related phenomenon of coronal rain. An early explanation for the driving force of outflowing penumbral material was the gas pressure excess below the umbra. Such a pressure excess might be expected due to the subphotospheric blocking of heat put forward to explain the umbra's coolness.

The more recent finding that both the Evershed outflow and the penumbral magnetic field are concentrated within the dark penumbral filaments argues against this driving mechanism since such a general pressure gradient might be expected to drive plasma out along the nonmagnetic paths *between* the dark filaments. In an alternative model, the penumbral filaments are interpreted as convective rolls (Section 7.2.2) aligned with the nearly horizontal field of the penumbra. The upward motions in the hot, bright filaments are then expected to carry field lines into the darker and cooler downward-moving filament regions. The Lorentz forces generated in this concentration of field might act to expel the gas horizontally along the field lines as a by-product of the main overturning pattern. This model seems promising, although the forces driving the horizontal flow are difficult to calculate.

An interesting model that has been worked out in more detail is based on siphon flow along field lines between regions of unequal photospheric pressure. The flow along a flux tube arching up through the corona accelerates away from the high-pressure footpoint through a sonic point at the arch apex and becomes supersonic in descending toward the low-pressure footpoint. The downfalling material is braked to subsonic speed at a shock standing above the low-pressure footpoint and continues subsonically to photospheric levels. According to this model, the chromospheric Evershed effect seen in Hα as material flowing into the spot takes place along higher-lying field lines and can also be attributed to siphoning into the regions of very highest field intensity (and thus lowest gas pressure) near the center of the umbra.

The darkness of the photospheric penumbral filaments might be explained simply as radiative cooling if they are, as observations suggest, relatively loosely packed flux tubes running above the photospheric surface adjoining the umbra (Fig. 8-3). In that case, their heating by radiation from the photosphere (subtending 2π sr) would be balanced by radiative cooling into all directions. Calculations indicate that the resulting filament might be cooler than the photosphere by as much as 15–20%.

As mentioned above, the observation that the dark penumbral filaments seem to overlie and partially mask a normal photosphere argues against the convective roll explanation. The observation of the penumbra's sharp outer edge, and of relatively undisturbed granules just outside its perimeter, also seems more difficult to explain in the convective roll model than in the siphon model. In the latter, the filament's opacity might drop rapidly due to a change in field intensity and gas pressure. The evidence for a rapid termination of the Evershed outflow velocity at the penumbral edge is less certain.

The downflows of relatively cool material observed in the corona over active regions at the limb, called coronal rain, are probably related to the chromospheric Evershed inflow seen in spots on the disk and to downflows measured in the magnetic network on the disk in Doppler shifts of transition region lines. Thermal instability of coronal material is likely to play an important role in these downflows, at least after initial compression to densities exceeding roughly $10^9 \, \text{cm}^{-3}$. The criterion for thermal instability given by equation (11-7) is based on a linearized analysis and gives little idea of how the instability develops after its amplitude increases into the nonlinear regime. It appears likely from recent numerical simulations that the plasma cools and condenses to the observed 10^4 K temperature and density of order $10^{10} \, \text{cm}^{-3}$. The time scale is the radiative cooling time scale of the corona, of order 10^{3-4} s.

This relatively dense material is dynamically unstable by the Rayleigh-Taylor mechanism. It will tend to penetrate down into the hotter, more tenuous underlying layers unless it is supported by a stable configuration of horizontal magnetic field lines. For the plasma condensations moving downward along inclined field lines such as in coronal loops and the network, we expect downflows at a substantial fraction of the free-fall velocity from the heights of order 10^4 km. This is in general agreement with the observed downflow velocities of 10–50 km s^{-1}. The free-fall time scale of such plasma is comparable to its cooling time from coronal temperatures, so in such magnetic structures we tend to see radiations from the transition region plasma regime around 10^5 K. In more nearly horizontal field configurations, as in quiescent prominences, the cooling time is much shorter than the free-fall time, so most of the plasma we see is at 10^4 K.

Another process that might play a role in these downflows is the siphoning of material along flux tubes from regions of higher photospheric pressure, as discussed previously in connection with the Evershed effect. Siphoning might be expected to operate not only from the (weaker) penumbral field regions into the (stronger) fields of the network, which might suggest an explanation for both the Evershed outflow and the network downflows, but also from the (now weaker) network fields into the (strongest) fields of the umbral cores. These field lines of the umbra extend to considerable heights and could play a role in the coronal rain phenomenon. On the other hand, the phenomenon of coronal rain is very general in the active region corona, and the thermal instability mechanism seems more likely to play the dominant role. Also, if the siphon effect were the cause of the network downflows it should be absent at times of low activity, when spots are rare. This could be tested observationally.

11.1.5 Oscillations and Waves

The study of oscillatory motions and brightenings in magnetic flux tubes provides information on the magnetic atmosphere, possibly including the subphotospheric layers where the waves might be generated by the buffeting of convection. The wave fluxes observed in sunspot flux tubes are also important in carrying a momentum flux of significance in determining the vertical pressure balance of the umbral atmosphere.

Sunspots present by far the best opportunity to compare observations of magnetic oscillatory phenomena with theory. The clearest examples are the umbral oscillations and flashes (Section 8.1.4), detected particularly easily in chromospheric radiations. These oscillations of rather definite period (in the range 140–200 s for different spots) seem to represent the narrow-band response of two resonant cavities, located in the umbral photosphere and chromosphere, to wider-band forcing by motions below and around the spot. Detailed modeling indicates that the wave mode involved is a hybrid magneto-atmospheric disturbance for which the restoring force arises in part from the magnetic field line tension (as in a pure Alfvén wave) and in part by both the pressure gradients and buoyancy induced by compression of the stratified umbral atmosphere.

One of the two resonant cavities seems to arise from the reflection at the bottom of the umbral atmosphere caused by the rapid inward increase in temperature from the cool umbra to the convection zone, and from reflection at the top by the steep increase in the Alfvén speed in the chromosphere. An overlying chromospheric cavity seems to be bounded above by reflection of essentially acoustic waves propagating along field lines from the steep temperature rise into the corona, and below by the sharp gradient in the acoustic cutoff frequency.

Excitation of the photospheric cavity is thought to be provided by the inhibited convection below the umbra. This excitation may be observed as an umbral flash, followed by oscillation of the whole umbra. Numerical simulations of such an "overstable" equilibrium indicate that, when convective motions attempt to drive the usual overturning cells in a region of strong vertical magnetic field, the resistance to stretching provided by the field line tensions can result in the overshoot of the magnetic field relaxation (hence the term "overstable"). This causes a considerable fraction of the convective energy to propagate as Alfvén waves moving up and down the field lines. The result can be an oscillatory (instead of overturning) convective mode, provided the heat diffusivity by radiation and conduction exceeds the magnetic diffusivity. If the magnetic diffusivity is large, then the freezing-in of plasma and fields ceases to hold, and the convective motions can force plasma past the field lines, tending to weaken the forcing and increasing the ohmic damping.

A model in these terms is able to explain the period and narrow frequency band of the umbral oscillations and is consistent with their observed amplitude, although the relative importance of the photospheric and chromospheric cavities is somewhat unclear. The depth of the photospheric resonant cavity seems to agree with the depth of the layer at which the overstable oscillation is thought to occur. This then would be an estimate of the "depth" of a spot. The high velocities of the oscillations observed in Ca K and Hα relative to photospheric lines can be explained in terms of the steep drop in density into the chromosphere, even though most of the oscillatory energy is confined to the cavity at photospheric layers.

The explanation of the umbral dots also seems to be related to this oscillatory convection, although the dots do not seem to be an oscillatory phenomenon. Different views exist on the explanation of their heating. They may represent the upward penetration of hot convective upflow into small regions of lower magnetic field, or the strong local dissipation of the hybrid wave mode excited by the overstable convection below the umbra. Another interpretation is that they might be identified with the slow, local growth of an oscillatory convection region itself. Observations of these subresolution structures are not yet able to distinguish the relative importance of these mechanisms. Better data on the field strength (is it lower or similar to the umbral value?) and the upflow velocity and its duration are particularly required.

The explanation of the penumbral waves observed running horizontally outward at chromospheric and photospheric levels is also in terms of vertical trapping in a resonant cavity. The same reflection mechanisms that determine the umbral cavity are involved, except that now the magnetic field is close to horizontal. Again, the bulk of the wave energy is trapped in the photospheric cavity, and the more easily observed chromospheric disturbance seems to represent an evanescent wave leaking from this cavity. A difficulty arises in reconciling the much larger (40–90 km s^{-1}) outward phase velocity of the photospheric wave compared to the 15–20 km s^{-1} outward propagation velocity of the chromospheric disturbance. A possible explanation is that the wave fronts are being advected by the Evershed flow outward at about 5 km s^{-1} in the photosphere and inward at about the 20 km s^{-1} chromospheric Evershed velocity when observed in the Hα or Ca K.

The modes that are resonant in umbral and penumbral cavities have periods of around 3 min and thus are significantly shorter than the peak power range of the photospheric 5-min oscillation. The 5-min oscillation can also be detected in spots at reduced amplitude, and it seems to arise from the forced buffeting of the sunspot flux tube by the photospheric oscillation. Recent data indicate that selective filtering of the photospheric oscillations is observed in the spectrum of the umbral 5-min modes. This filtering can be explained in terms of the resonances of the sunspot flux tube. Studies of this kind (sunspot "tomography") could help to determine whether the subphotospheric structure of spots is monolithic or consists of a cluster of separate flux tubes (Section 8.2.3).

The total energy flux carried by the waves discussed above is several orders of magnitude smaller than the very large missing radiative flux in a sunspot, which is roughly 5×10^{10} ergs cm^{-2} s^{-1}. This disparity seems to rule out an appreciable role for these relatively long waves in the spot's energy balance. The vertical gradient of wave pressure near the reflecting boundary, on the other hand, is comparable to the force of gravity, and should be significant in the spot's structure.

A much more significant energy flux might be present in waves of shorter length, given the looser observational upper limits on unresolved umbral motions. Such shorter-period oscillatory motions could have wavelengths much less than the atmospheric depth range over which Fraunhofer lines are typically formed. Thus they act to broaden the line rather than displacing it back and forth as the resolved wave motions of longer wavelength are able to do. However, both the modeling of the waves described above and also direct measurements on the oscillatory power in a coronal line indicate that only a tiny fraction of this wave flux escapes through the upper boundary of the cavity. If wave cooling of spots is important, it would seem to require downward propagation of wave energy into the solar interior. This is made difficult by the lower cavity boundary for such magneto-atmospheric waves, but in real, inhomogeneous spots the actual downward leak might be larger than found in the uniform models of the umbra used so far.

11.1.6 Magnetic Field Dissipation

The appearance of new magnetic fields in the solar atmosphere by emergence from the convection zone, discussed in Chapter 8, is better observed than the process by which such fields are removed from these layers. This removal is effected in part by some combination of subduction back below the photosphere and advection out into the solar wind. But it inevitably also entails some transformation of magnetic energy into kinetic energy of mass motions and into heat. We discuss here the basic mechanisms by which this transformation of magnetic energy seems to take place since they are of considerable importance in understanding energetic phenomena such as flares and filament eruptions and also the more steady processes of coronal heating and of global solar field evolution over the activity cycle.

In terms of the plasma induction equation (4-64), which governs the time rate of change of the magnetic field, we see that if the magnetic Reynolds number is large, the field at a point fixed in space can change only through advection by plasma motions. On the other hand, if the field gradients are sufficiently large and the electrical conductivity is finite, the field lines can leak out of a fixed volume, even in the absence of macroscopic motions, at a rate given by the diffusion equation (4-69). This field line diffusion represents a decay of the current system responsible for the magnetic field over the time scale given in emu as

$$\tau = 4\pi\sigma L^2. \tag{11-8}$$

We thus expect that magnetic field dissipation will be limited in the sun to regions where both the magnetic diffusivity $(4\pi\sigma)^{-1}$ and (mainly) the field gradients, are large. Judging from the temperature dependence of σ [equation (4-99)], the magnetic diffusivity is large around the photospheric temperature minimum and in umbrae. More general calculations of σ that include the important effect of neutral species at the low ionization levels encountered in such relatively cool plasmas indicate that the electrical conductivity of solar plasma around 4000 K is a small fraction of the photospheric value around $\tau_{0.5} = 1$.

In addition, in regions of high magnetic field gradients and pressure, the onset of plasma turbulence can be expected to increase the electrical resistivity. Although the increase expected under solar conditions is uncertain, this anomalous resistivity will increase the rate of field dissipation over the values expected from use of equation (4-99). Spontaneous variations of plasma electrical conductivity within the volume of a current-carrying region can lead to so-called resistive instabilities, which tend to concentrate the current into narrow filaments where the magnetic gradients are larger than average, acting to further increase the dissipation rate.

External pressures on the magnetic fields, such as the emergence of new fields into pre-existing configurations in the chromosphere and corona, can push together two regions where the field has a substantial antiparallel component, as illustrated in Fig. 11-5. The result is a magnetic neutral surface across which the field changes sign. If the magnetic diffusivity of the plasma were zero, such antidirected field lines might coexist in equilibrium. In practice, with finite resistivity, a thin sheath will tend to form within which the

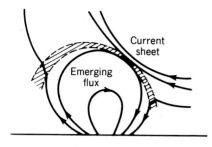

Fig. 11-5 Schematic of reconnection that might occur in a flare between emerging and pre-existing coronal fields. Adapted from Heyvaerts et al. *Astrophys. J.*, **216**, 123 (1977).

magnetic Reynolds number will be sufficiently small to allow field line diffusion, and thus dissipation of magnetic energy, at some appreciable rate. An important result of this dissipation is that antiparallel field lines initially disconnected from one another in the two regions can "reconnect" and form a new, continuous connection pattern across the boundary.

A highly idealized model of this reconnection process is sketched in Fig. 11-6, with a two-dimensional geometry assumed. The main features of this model are the advection of antiparallel field lines and plasma toward the dissipative region, at a velocity $\pm v_x$, and their reconnection to form field lines (which now connect across the initial discontinuity) that are carried out of the region at a velocity $\pm v_y$. A strong electric field $v_x B_y = E_z$ is generated around the region of this so-called X-type neutral point.

The thickness l of the diffusing region must be very small, perhaps of order tens of meters, to achieve the field gradients required to enable the field lines

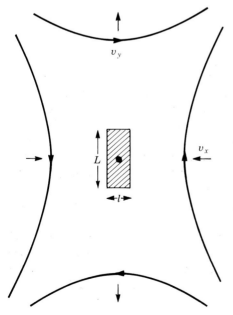

Fig. 11-6 A configuration for steady magnetic reconnection. The curved lines indicate the magnetic field. Adapted from E. Priest, *Solar Magnetohydrodynamics*, Reidel, 1982.

and plasma to slip at the relative velocity v_x at which they are advected into it. Numerical simulations indicate that the reconnection rate, or speed at which the field lines are carried through the diffusion region, is roughly $0.1v_A$, where v_A is the Alfvén speed evaluated just outside the diffusion region.

In practice, both the thickness l and the reconnection speed depend considerably on the boundary conditions of the inflow and outflow at a distance from the neutral point, as well as on the local plasma diffusivity within the diffusion region. For this reason, the quantitative results on how reconnection proceeds are difficult to generalize. However, numerical simulations indicate that dynamically plausible flows can be constructed which might carry field lines into and out of a reconnecting region.

One result of the process can be the restructuring and simplification of the complex field line configurations that must occur in the solar atmosphere given the constant eruptions of new fields and also the ceaseless photospheric motions of the footpoints of flux tubes that extend into the chromosphere and corona. Erupting filaments may represent a particularly spectacular aspect of the process. Another aspect is the ohmic heating of the plasma at a rate which is given by j^2/σ per unit volume. As discussed in Chapter 10, this ohmic heating is considered to heat flares and is also one of the main candidates for a coronal heating mechanism. Finally, the considerable electric fields and turbulence generated at the reconnection region may be important in accelerating energetic particles observed in connection with flares and erupting filaments, as also discussed in Chapter 10.

11.2 ACTIVITY BEHAVIOR OVER THE SOLAR CYCLE

11.2.1 The Sunspot Number and Other Activity Indices

The main record of the sun's magnetic activity has been kept through daily observations of the number of sunspots on the disk. This relatively easily observed index has been shown to correlate reasonably well with many other features of solar activity which require more sophisticated observational techniques.

This spot record exists in most reliable and homogeneous form to the beginning of R. Wolf's tabulations in 1849, in Zurich. Through the efforts of Wolf and others, sunspot behavior has also been reconstructed to the beginning of the 17th century, when telescopic observations first began. Some of the difficulty in the reconstruction of a homogeneous set of data prior to 1849 arises from the nature of Wolf's sunspot index number, defined as

$$R = k(10g + f). \tag{11-9}$$

Here f is the number of individual spots, g the number of recognizable spot groups, and k is a correction factor that is intended to adjust for differences between observers, telescopes, and site conditions. Wolf set $k = 1.0$, but different counting methods instituted in Zurich after 1882 yielded $k = 0.6$, even with the same telescope. Better sites and telescopes used for these observations account for further variations in k, ranging up to several tens of percent, relative to the modern Zurich observations. It is practically impossible to reconstruct the k-values for the sporadic individual observations carried out before the periodicity of sunspot occurrence was first realized during the daily observations by H. Schwabe between 1826 and 1843.

For this reason, and also due to the lack of regular daily observations, the annual mean sunspot numbers assigned before about 1820 are uncertain to approximately a factor of 2, with even larger errors for those before 1750. Modern monthly means are considered accurate to better than 5%. Wolf himself recognized that measurements of the total spot areas on the disk would have been a better index, but he adopted the more arbitrary sunspot number for its convenience. It has been retained for over 135 years for this reason, and also because comparisons with other indices, including the spot areas first tabulated regularly at Greenwich Observatory in 1874, have generally shown good agreement.

The behavior of the annual mean sunspot number is shown in Fig. 11-7 from the present back to the earliest telescopic records. It can be seen that the nominal 11-year cycle varies considerably in amplitude—by at least the factor of 3 seen in the most accurate records between the low maxima of 1884 and 1906 and the largest maximum in 1957. The minima also vary in depth. The one observed around 1913 included extended periods of "immaculate" sun, including a 3-month consecutive stretch when no spots were observed.

The cycle period varies considerably, between roughly 8 and 15 years, with an average of about 11.1 years. The characteristic shape of a sunspot curve, a relatively rapid rise from minimum to maximum followed by a slower fall, is more evident in the cycles of larger amplitude.

The cycle amplitudes seen in Fig. 11-7 seem to be modulated by a longer cycle of roughly 80–100 years length, with recent minima around 1910 and 1810. The reliable spot record is too short to make an entirely convincing case for this cycle identified by W. Gleissberg, but supporting evidence for its existence is found in the much longer auroral and C^{14} records discussed below. There is some evidence for an even-odd alternation in the cycle amplitudes in the relatively reliable post-1850 records, but the significance of this effect is questionable.

The most remarkable feature seen in Fig. 11-7 is the extended period of severely depressed sunspot counts found between about 1640 and 1700. This near-absence of spots was pointed out initially by F. Spörer and discussed by E. Maunder in the late 19th century. It has recently been more extensively documented and discussed by J. Eddy. Good observers equipped with quite

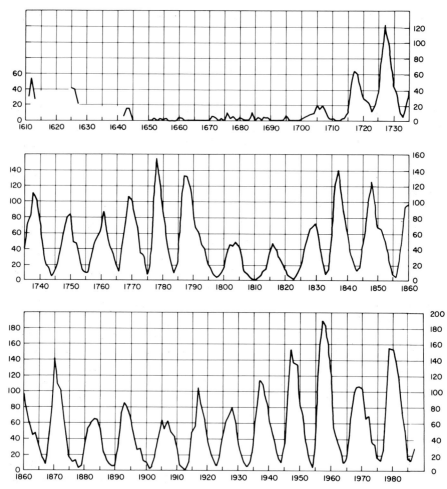

Fig. 11-7 The variation of the sunspot number with time from 1610 to 1987. By permission of J. Eddy.

adequate telescopes specifically reported the very low incidence of spots during this period. The sunspot numbers registered are too uncertain to preclude the continuation of a low-level activity cycle during this period, but the unusually low level of solar activity registered during this Maunder Minimum is corroborated by evidence from several other indices discussed below.

The sunspot record indicates a general rise in the sun's activity level during the 20th century. A key question is when another Maunder Minimum–like depression might be expected. The answer could have important consequences for climate since a significant cooling in Western Europe and North America called the Little Ice Age seems to have coincided broadly with the time of the Maunder Minimum.

11.2.2   Time Behavior of the Sun's Magnetic Field

The remarkably regular 11-year variation of sunspot numbers described above is accompanied by equally regular oscillations in the latitude distribution of spots and in the polarity of their fields. This behavior of the low-latitude fields, together with the 22-year polarity reversal at high latitudes, describes the most fundamental aspects of the sun's magnetic oscillation.

The daily maps of sunspot positions on the disk constructed by Carrington enabled him to point out by 1858 that the latitude band occupied by spots drifted toward the equator as the solar cycle progressed from minimum through maximum to the next minimum. This latitude drift, sometimes known as Spörer's law, is best illustrated by the plot shown in Fig. 11-8, known as the Maunder butterfly diagram. It shows that the zone occupied by the spots, which is typically some 15–20 degrees of latitude wide, moves steadily equatorward over 11 years. The first spots of a new cycle are centered around latitudes 25–30°, while the last spots of that cycle 11 years later are found around 20° closer to the equator.

It can also be seen from Fig. 11-8 that high-latitude spots of a new cycle appear while the low-latitude spots of the preceding cycle are still present. During this overlap period of a few years, the sun actually has four separate spot belts, with short-lived spots seen very occasionally at up to 70° and the lowest-latitude spots sometimes straddling the equator. It is interesting that the periodicity in the latitude migration, determined by fitting curves of mean latitude to the butterfly diagram, is more regular even than the periodicity in spot number.

The solar cycle behavior of sunspot magnetic field polarities was described in 1913 by G. Hale. He and his collaborators at Mt. Wilson found that the magnetic polarity of the bipolar spot groups in a given hemisphere switched from one cycle to the next. During the most recently completed cycle (21), which peaked between 1979 and 1980, the preceding and following spots in the northern hemisphere were predominantly of north and south magnetic polarity respectively, with opposite polarity in the southern hemisphere. The new spots of solar cycle 22 which appeared at high latitudes were distinguished by their opposite magnetic polarity to the last spots of cycle 21 still seen at low latitudes. The so-called reversed polarity groups which violate the Hale-Nicholson polarity rule tend to be short-lived, as are also spots that emerge well outside the usual latitude belts.

This 11-year polarity reversal of the spot fields defines a 22-year magnetic solar oscillation, each full cycle of which includes two 11-year cycles of sunspot number and general activity level. The polarity reversal of the high-latitude (polar) photospheric magnetic fields that occurs near sunspot maximum occurs about a half-cycle out of phase with the polarity switch of the

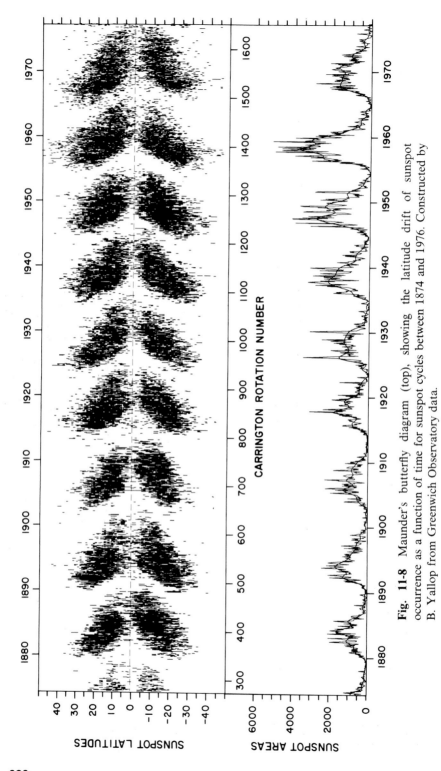

Fig. 11-8 Maunder's butterfly diagram (top), showing the latitude drift of sunspot occurrence as a function of time for sunspot cycles between 1874 and 1976. Constructed by B. Yallop from Greenwich Observatory data.

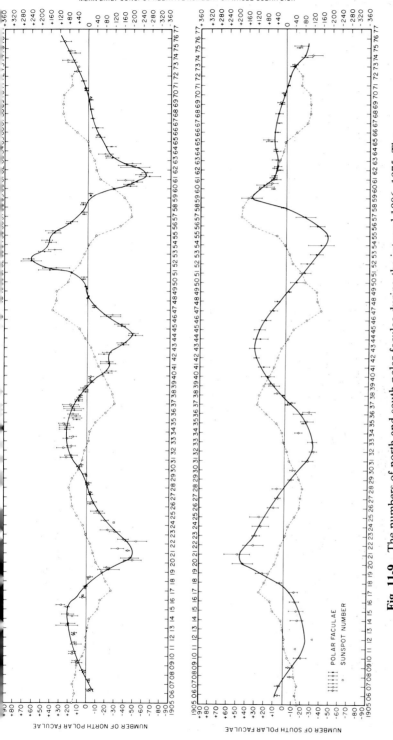

Fig. 11-9 The numbers of north and south polar faculae during the interval 1906–1975. The numbers have been assigned polarities corresponding to the polarities of the associated polar magnetic fields. Also, the numbers represent the magnetic flux normal to the entire surface of each polar cap in units of 0.3×10^{21} Mx ($\pm 50\%$). For comparison, the sunspot number for the whole disk has been plotted with a magnetic polarity assigned that corresponds to that of the following spots in each hemisphere. By permission of N. Sheeley.

spot fields. The switch of the polar fields was first observed at the peak of solar cycle 19 around 1957, with the southern hemisphere field changing first, followed by the northern field about a year later. Similar switches were observed after the peaks of cycles 20 and 21 in data obtained at Mt. Wilson and at Kitt Peak.

The switching of the polar fields can actually be inferred with greater confidence and, going back to much earlier cycles, by using the poleward migration of prominent white-light faculae. The number of these faculae, which are of like polarity in the polar region of a given (north or south) hemisphere, offers a convenient tracer of the total magnetic flux in a polar region. The data back to 1906 are shown in Fig. 11-9. It can be seen that the polar fields estimated in this way seem to have switched polarity roughly every 11 years, always lagging the spot maximum by 1–2 years.

11.2.3 Long-Term Behavior of Solar Activity

Other indices of solar activity supplement the sunspot record in recent times. Figure 11-10 shows the sunspot areas, sunspot numbers, Ca K plage areas, and 10.7 cm microwave fluxes for recent solar cycles. The basic activity cycle is seen equally well in these indices, but the relative amplitudes of the cycles and their shapes vary substantially.

This is not surprising since we know from the discussion of Chapters 8 and 9 that the magnetic structures that give rise to spots and plages differ in their spatial distribution and time behavior. As a more extreme example, plots of the solar wind speed over the two past cycles that are available show no significant cycle variation. It cannot be said that the activity record derived from the spots is more fundamental than the others. Its main significance is that it is the longest historical record with sufficient time resolution and reliability to study the 11-year cycle. Any attempt to deduce solar magnetic field properties from such proxy activity records must ultimately take into account the physical relationship between the index and the underlying magnetic field structure.

The relative longevity of various activity records is shown in Fig. 11-11. The most striking fact is that our longest reliable historical record (daily spot areas) extends for only about 140 years, and many of our most useful records extend only over two solar cycles. The most useful extensions of the record to the pre-telescopic era are based on the counts of naked-eye sunspots and aurorae and on the C^{14} isotope record.

Naked-eye sightings of large spots extend back at least to the 4th century B.C. The record of such sightings averages only a few reports per century, and the influence of sociological biases must be taken seriously given the astrological role of these sightings, particularly in the Orient. Although the data are too sparse and unreliable to yield much information on the early operation of the

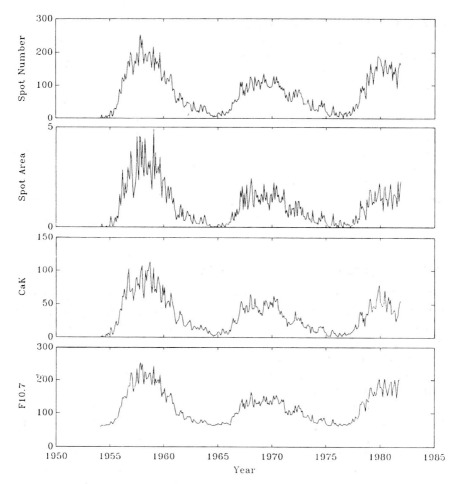

Fig. 11-10 Variation of (from top) the sunspot number, the sunspot area, the Ca K plage area, and the 10.7 cm microwave flux, for recent solar cycles. Data obtained from the NOAA World Data Center.

11-year cycle, they do provide an independent estimate of longer-term activity variations similar to the Maunder Minimum occurring at earlier times.

A better sampling of the activity level is provided by the record of auroral sightings, which extends to the 6th century B.C. The incidence of aurorae seen below the polar circle is known to be positively correlated with the spot cycle, and the number of reports, even in the early part of the millenium, is generally an order of magnitude larger than the number of naked-eye spot sightings, so their record is statistically more convincing.

Measurements of the carbon isotope ratio C^{14}/C^{12} in tree rings of known age provide an estimate of solar activity because C^{14} is formed in the upper

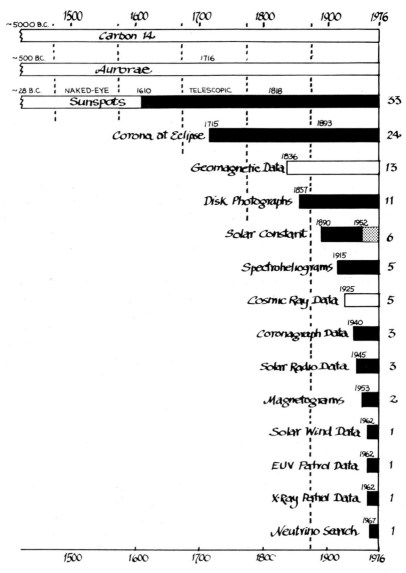

Fig. 11-11 Summary of available data on past solar behavior. Solid bars, direct data; open bars, indirect data. Horizontal scale is years A.D. Vertical lines are 100-year periods from present. Numbers on right margin are numbers of 11-year cycles for which data exists. By permission of J. Eddy.

atmosphere by neutrons produced in the impacts of galactic cosmic rays. These neutrons react with stable N^{14} via the process

$$N^{14} + n \rightarrow C^{14} + p \qquad (11\text{-}10)$$

to produce radioactive C^{14}. When the sunspot number is high, the C^{14} production is observed to be low because the Earth is better shielded from the energetic galactic cosmic rays by the different interplanetary magnetic field (see Chapter 12). The C^{14} formed in this way eventually finds its way into the lower atmosphere and provides a solar activity record going back about 10,000 years, although the natural radiocarbon reservoirs tend to filter out variations as short as the 11-year cycle.

The behavior of solar activity before the telescopic era, as derived from these three main indicators, is shown in Fig. 11-12. The Maunder Minimum period of the 17th century is clearly indicated by a significant rise in C^{14} and by pronounced decreases in telescopic and naked-eye sunspots and auroral sightings. Taken together, these lines of evidence leave little doubt that the sun was unusually magnetically quiet at that time.

These records also agree in indicating a similar activity decrease in the late 15th and early 16th century (referred to as the Spörer Minimum) and an extended period of relatively high activity spanning the 12th and early 13th centuries. The main implication of these records is that the fairly regular behavior of solar activity experienced over the past 250 years cannot be taken as the norm since extended periods of both much lower and somewhat higher activity seem to have alternated over the past millenium. Judging by the spacing of these longer-term modulations, one would not be surprised to see the general activity level begin a drop toward an extended minimum in the near future.

The most important potential implication of these long-term irregularities in the solar activity level is in their possible influence on climate. The evidence for such a connection is less secure than the evidence for the long-term solar activity modulation itself; nevertheless, it warrants close attention. The main finding is that the extended periods of depressed solar activity, such as the Maunder and Spörer Minima, correspond to periods of colder climate, at least judging from weather records kept in Europe and North America. The 12th-century period of high solar activity seems to have been a time of general warming.

The main uncertainty in this correlation arises from difficulties in interpretation of the climate record, rather than of the solar activity record. Some of the difficulties are related to the shortage of reliable records, but others stem from the small number (3) of significant coincidences in the last millenium. Since the climate record is also influenced by nonsolar perturbations, such as large volcanic eruptions, the possibility of chance coincidences cannot be dismissed.

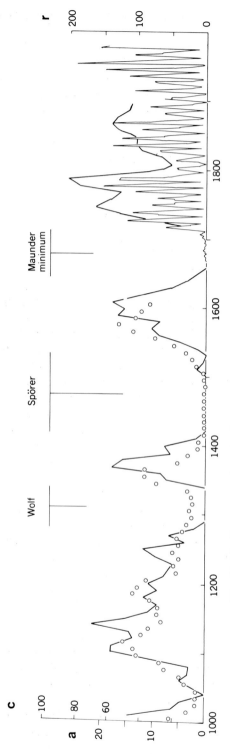

Fig. 11-12 Reconstruction of solar activity over the past millenium from telescopic sunspot observations (light solid line, beginning about 1650 with rapid variations after 1700, corresponding to the index r on the ordinate scale at right), a proxy of the sunspot number, based on C^{14} data (heavy solid line and index c on the left), and northern hemisphere aurorae (circles, in sightings per decade; index a on left). By permission of J. Eddy.

Aside from its obvious practical implications, such a positive correlation between activity level and climate could provide a yet much longer data base to estimate the sun's early magnetic behavior if it were better established over the past few millenia. Measurements of sea temperatures derived from oxygen and beryllium isotope ratios in drill cores in stratified Arctic ice are yielding interesting data. But changes in the insolation of the Earth through cyclic variations of its orbit around the sun (rather than solar luminosity variations) have so far been considered to provide most of the signal.

11.3 DYNAMICS OF THE SOLAR MAGNETIC CYCLE

11.3.1 The Babcock Model of the Solar Cycle

Ideas on how the intense, localized sunspot magnetic fields might be connected to a more general solar magnetic field, and how the sun's magnetic oscillation might be driven, were fragmentary until the development of the magnetograph by C. and H. Babcock in the late 1940s. The much more sensitive measurements of photospheric fields made with this instrument in the 1950s showed that the Hale-Nicholson sunspot polarity rules were also obeyed by fields in the more widespread facular regions. They also showed the behavior of polar fields, in particular their reversal during the 22-year cycle. These new data led H. Babcock to put forward, in 1961, a very important synthesis of how the sun's magnetic field structure at high and low latitudes was related and how the 11-year activity cycle and 22-year magnetic cycle might be driven. This model still provides the best conceptual framework for our current understanding of the main features of the sun's magnetic cycle, namely, the generation of active region magnetic fields, the Hale-Nicholson polarity laws, Spörer's law and the 22-year solar magnetic polarity oscillation.

Babcock's model describes the 22-year cycle in five stages. In the first stage, which coincides with an epoch of low activity about three years before the beginning of a new spot cycle, the solar field is taken to be approximated as a relatively weak (~ 1 G) axisymmetric dipole [Fig. 11-13(a)]. The field lines lie in meridional planes and emerge from the surface only at latitudes above $\pm 55°$. These field lines extend well out into the corona, cross the equatorial plane at a distance of several radii, and connect back to the opposite polar cap. Below the surface, the total flux of about 8×10^{21} Mx is assumed to lie in a thin layer of depth about $0.1 R_\odot$ below the photosphere.

In the second stage, the submerged field is intensified as the field lines are drawn out by the sun's latitudinal differential rotation [Fig. 11-13(b)]. If the surfaces of constant angular rotation rate coincided with the field lines, then isorotation would hold without any intensification. But if these surfaces cut more sharply into the sun than do the submerged field lines at low latitudes, then a given field line joins plasma rotations at different rates, and it will be stretched out, if $R_m \gg 1$. This stretching of the frozen-in field lines transfers

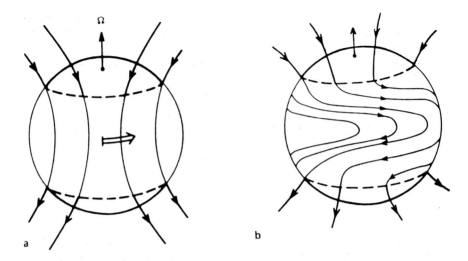

Fig. 11-13 Illustration of the Babcock model. (a) Initial poloidal field before winding up by differential rotation. (b) Predominantly toroidal field generated by differential rotation with equatorial acceleration. Reproduced by permission, from H. Babcock, *Astrophys. J.*, **133** 572 (1961).

energy from the kinetic energy of the sun's differential rotation into the intensification of the field.

Given the sun's higher angular rotation rate at the equator, the initially meridional field lines will be progressively drawn out into an increasingly east-west direction [Fig. 11-13(b)]. The field is then essentially toroidal at these lower latitudes, with the field lines switching direction at the equator. Babcock was able to show from the observed photospheric differential rotation rate that after 3 years, the low-latitude field will have been wrapped about $5\frac{1}{2}$ times around the sun and thus intensified to several hundred gauss.

Babcock also found that the time required to reach a given level of intensification increased toward the equator. He assumed that when a critical intensity is achieved the field erupts to the photosphere in the form of active regions, and the intensification stops at that latitude. The eruption was assumed to be caused by a combination of magnetic buoyancy (Section 11.1.2) and the kinking of a flux tube (like a rubber band) if it is twisted beyond a certain point.

Twisting, and thus further intensification of the field lines, was assumed to occur because of the roller bearing effect of the sun's differential rotation gradient with depth. This twisting acting on nonuniformities in the field due to convection is assumed to result in discrete braided flux ropes of yet an order of magnitude higher intensity than the few hundred gauss toroidal field produced by latitudinal differential rotation alone. These flux ropes of several kilogauss intensity are then taken to define bipolar active regions when they erupt to the photosphere.

The third stage of the model describes the formation of active regions from the flux ropes. The model provides a natural explanation of the Hale-Nicholson polarity law. Each Ω-shaped stitch of flux rope erupting through the surface will produce a bipolar active region with a preceding and following magnetic polarity, and the polarity of the p- and f-portions will switch at the equator since the toroidal field switches sign there also (Fig. 11-14).

Another important success is an explanation of Spörer's law. Given the $\sin^2 \phi$ (ϕ = latitude) term in the sun's differential rotation, the field intensification proceeds most rapidly at around latitude $\pm 30°$, so the first active regions of the cycle are expected to erupt there, and only later at lower latitudes. Babcock showed that the observed time scale of about 8 years between first activity onset at $\pm 30°$, and the last low-latitude spots can be reproduced using the photospheric differential rotation curve. The model is also able to account for the several thousand active regions that erupt during a cycle. The original 8×10^{21} Mx poloidal field (sufficient for about ten active regions) could be amplified by about three orders of magnitude (in total flux) by the differential rotation.

The fourth stage describes the neutralization and reversal of the sun's general poloidal field. The observation that the f-portions of active regions tend to lie at higher latitudes than do the p-portions plays the central role in this process. It is observed further that as an active region evolves, its f-portion migrates toward the nearest pole, while the p-portion migrates toward the

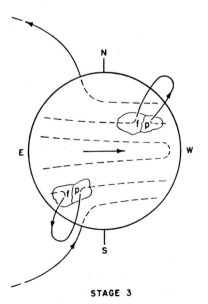

STAGE 3

Fig. 11-14 Bipolar magnetic regions are formed where buoyant flux loops of the submerged toroidal field are brought to the surface. These continue to expand, and the flux loops rise higher into the corona. The letters f and p denote magnetic polarities following and preceding the direction of solar rotation. Reproduced by permission, from H. Babcock, *Astrophys. J.*, **133**, 572 (1961).

equator. The result of this migration is that the f-polarities of the active regions neutralize the existing polar fields, while the p-polarities of the two hemispheres cancel at equatorial latitudes. Further poleward migration leads to replacement of the old polar fields with new ones of opposite polarity.

Figure 11-15 illustrates how the original poloidal field lines are first cancelled and then replaced. As the p and f active region fields separate over a given hemisphere, they also rise into the corona. They are pushed up against the antiparallel field lines of the poloidal field. Reconnection leads first to cancellation and then to replacement by new poloidal field lines connecting the two hemispheres. A by-product is the liberation into the outer corona of large closed loops of flux and the retraction below the surface of flux at equatorial latitudes.

Since the polar fields amount to less than 1% of the total flux emerging in active regions during the solar cycle, this process of cancellation need not be very efficient. About 99% of the active region fields cancel not against the polar fields, but against opposite polarity fields in adjacent active regions.

In stage 5 we find a reversed poloidal field about 11 years after the beginning of stage 1. The submerged fields remaining after the eruption of active regions are expected to be left with an orientation of their poleward ends toward the west and their low-latitude portions lagging rotation. This is

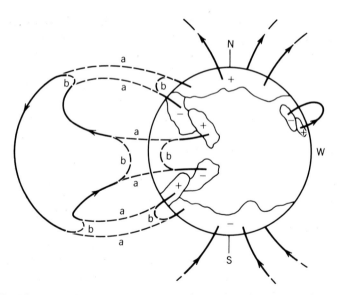

Fig. 11-15   The expanding lines of force above older bipolar magnetic regions move out to approach the lines of force of the main dipole field. Severing and reconnecting gradually occur, and a portion of the main field is neutralized. Also, a large flux loop of low intensity is liberated into to corona. Continuation of the process results in the formation of a new main dipolar field of reversed polarity. Reproduced by permission, From H. Babcock, *Astrophys. J.*, **133**, 572 (1961).

the opposite of the orientation of the active regions, the eruptions of which have carried away a component of the field whose orientation placed the f-portions at higher latitudes than the p-portions. Latitudinal differential rotation will now act first to straighten out these submerged field lines and eventually wind them up again during the second half of the 22-year cycle, but with opposite toroidal polarity.

11.3.2 Dynamics of the Babcock Dynamo Model

The description of the solar cycle given in the Babcock model is mainly kinematical and based on the observed properties of photospheric fields, such as their differential rotation with latitude, their eruption with tilted dipole axes, and the subsequent separation of p- and f-polarities as they evolve. The dynamics behind these phenomena is much less well understood.

Study of the solar-cycle dynamics has for the most part been carried out in the context of dynamo models, in which the intensification of magnetic fields is accomplished by the induction of plasmas trying to move across field lines. The relation governing the growth or decay of fields within a volume fixed in space is the induction equation (4-64)

$$\frac{\partial \mathbf{B}}{\partial t} = \nabla \times (\mathbf{v} \times \mathbf{B}) + \frac{1}{4\pi\sigma} \nabla^2 \mathbf{B}. \tag{11-11}$$

To show that dynamo action can maintain a field at all, the plasma velocity field must be of such magnitude and geometry that the first term on the right outweighs the ohmic decay of **B** represented by the second term. T. Cowling showed in 1934 that purely axisymmetric motions in the meridional plane cannot maintain the poloidal field that was considered at that time to be the geometry of the sun's general field. Since then, many kinematical studies of the dynamo problem have shown that a wide variety of assumed flow fields could in principle maintain both poloidal and toroidal components of **B**.

In recent years dynamo models have moved beyond purely kinematical solutions where **v** is assumed, to construction of fully dynamical solutions of the induction equation (11-11) together with the coupled mass, momentum, and energy relations for the plasma. These models compute a representation of the velocity fields of large-scale solar convection, of the sun's differential rotation profile, and of the magnetic fields that are generated by these motions through the induction equation. The models are based on internally consistent treatments of convection in a rotating, stratified shell, as described in Section 7.3.7. Comparison of the results with recent observations raises important questions about the processes invoked in the Babcock model, although a different model that deals with the phenomena so well, starting from so few postulates, has yet to be proposed. Some of these concerns are outlined below.

In the Babcock dynamo model, the intensification of the field is produced by both latitudinal and radial gradients in the sun's angular rotation rate. The latitudinal gradients produce a toroidal field component from a poloidal field.

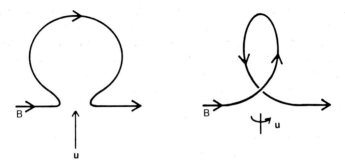

Fig. 11-16 A magnetic field line is bent into a loop and twisted out of the page by a helical velocity field, denoted by **u**.

The radial gradients impart a further twist to the flux tubes through a rollerbearing effect. The physical mechanism most likely to produce a poloidal field component from the toroidal field (i.e., account for the observed tilt of the active region dipole axes that is an important ingredient in the Babcock model) was first put forward by Parker in 1955. In the presence of convection in a rotating fluid, a rising cell of hot gas not only lifts the field lines of a horizontal field upward into an Ω-shaped stitch of flux, but also twists them, due to the Coriolis effect (Fig. 11-16). This twisting into the meridional plane will in general produce a poloidal field component, whose polarity can be opposite to a specified pre-existing poloidal field, provided the amount of twisting achieved by the rising convection is correct.

In the earlier, kinematical, dynamo calculations, this rate of cyclonic swirl, called the α-effect, could be adjusted independently of the rate of winding up of the field by differential rotation, known as the ω-effect. Thus, kinematical $\alpha\omega$ dynamos could claim to reproduce Spörer's law, without confronting the observed sign and magnitude of photospheric differential rotation.

However, as pointed out in Section 7.3.7, dynamical simulations of convection in rotating spherical shells indicate that the observed latitudinal differential rotation of the photosphere is difficult to reproduce unless the sun's angular rotation rate, ω, decreases radially inward. Unfortunately, this sign of $\partial\omega/\partial r$ tends to produce the wrong sign of field migration in Spörer's law. This may not be so much a difficulty of the dynamo model as of the model of differential rotation dynamics. As discussed in Chapters 6 and 7 recent p-mode measurements of $\omega(r)$ in the convection zone do not support the theoretical prediction of $\omega(r)$ being nearly constant on cylinders. This could imply that intensification of the toroidal field is mainly by latitudinal differential rotation (as postulated by Babcock) and that we do not yet understand the dynamics of the latitudinal differential rotation well enough to predict $\partial\omega/\partial r$ from the observed $\partial\omega/\partial\phi$.

A related difficulty indicated by the dynamical MHD models is that the α-effect is several orders of magnitude larger relative to the ω-effect than postulated in the kinematic dynamos. The magnitude of the α-effect is determined by uncertain parameters of the models, such as the eddy viscosity, but in the context of the model's dynamics, these large α-values are associated with the influence of rotation on convection required to produce the observed latitudinal differential rotation. The result is a so-called α^2 dynamo, where the ω-effect is negligible. However, such dynamos exhibit no particular tendency to switch polarities, so they are of little interest in explaining the solar cycle.

The difficulty might be removed if the dynamo models took account of the possible discreteness of the solar field in the convection zone. The flows would then be much less effective in generating both the α- and ω-effects since relative motion of plasma and field could be achieved mainly by relatively easy plasma motion around flux tubes, rather than through them. So far this effect has not been included in the models.

In the Babcock model, the poleward migration of f-polarities and equatorward migration of p-polarities are accepted as an observed fact that is not inconsistent with the uncertain data on the small observed meridional motions of spots. The mechanism that leads to this separation of polarities, and thus to reversal of the polar fields, was elucidated in an important contribution by R. Leighton in 1969. Starting from observed values for the scale and lifetime of the supergranules he was the first to observe properly, Leighton was able to construct a numerical simulation of the relaxation phase of Babcock's magnetic cycle oscillator, given a plausible latitude distribution of erupting fields. Leighton's model incorporated a turbulent diffusion of magnetic field lines caused by the supergranular motions to account for the time scale of the relaxation phase of the oscillator.

The diffusion mechanism put forward by Leighton is based on the idea that in their lateral movements photospheric flux tubes can be considered as atoms that are shuffled about by the fluid motions. When such an atom makes N lateral steps per unit time, each of average length and in random directions, its distance r from its starting point increases with time, on average, approximately as

$$r = LN^{1/2}t^{1/2}. \tag{11-12}$$

If we substitute for L the characteristic scale of a granule or of a super-granule and τ (or N^{-1}) for its lifetime, we find that the area should grow at roughly the same rate due to the shifting of the smaller but shorter-lived granules as due to the larger but longer-lived supergranules. Leighton pointed out that the photospheric fields (e.g., the magnetic network) are seen to organize themselves along the edges of supergranule flows, but do not seem to exhibit any particular disposition relative to the granule flows, so the supergranular contribution to the random walk seems more important. Inserting the supergranular values $L \sim 1.5 \times 10^4$ km and $\tau \sim 7 \times 10^4$ s, we find a time scale of roughly 5 years would be required for active region fields to disperse over an area comparable to the solar disk. More accurate estimates made from the observed dispersal rate of fields on magnetograms now indicate a time scale roughly three times longer.

Leighton studied the evolution of magnetic field distribution on a spherical surface using a diffusion equation

$$\frac{\partial n}{\partial t} = D \nabla^2 n, \tag{11-13}$$

where n is the surface density of the flux tubes and D is the two-dimensional diffusion coefficient

$$D = \tfrac{1}{2} L^2 N. \tag{11-14}$$

He found that if an initially compact magnetic dipole, tilted at a small angle ($\sim 1°$) to the equator, is allowed to diffuse at the supergranular rate in the presence of the observed differential rotation, the f-polarity becomes more drawn out in longitude than the p-polarity. This may help to explain why f-polarity tends to be distributed over a larger area than the p-polarity.

The most important result, however, was that if bipolar fields carrying flux measured in midlatitude active regions are inserted into the model with a small difference in latitude (i.e., tilt) between the p- and f-polarities, the rate of transport of f-polarity to the pole is sufficient to neutralize the polar fields on the observed time scale. Recent numerical simulations by N. Sheeley and co-workers indicate that the longitude and latitude distribution of photospheric fields during cycle 21 can be well reproduced assuming a given (observed) distribution of active region sources, a reasonable value of isotropic diffusivity D, and perhaps a small poleward meridional flow.

Nevertheless, objections have been raised about the field diffusion process. The main point is that a turbulent cascade of energy from the largest spatial scales to the smallest, at which cancellation is supposed to occur by ohmic diffusion, has never been verified. Also, it is not clear that the required high rate of field cancellation between p- and f-portions of active regions at midlatitudes can be achieved. An increased turbulent diffusion rate given by equation (11-13) does not guarantee an increased ohmic dissipation rate, governed by equation (11-11).

Numerical simulations (and observations of, e.g., spot formation) indicate that MHD turbulence can carry energy from small scales to large as well as causing a downward cascade to the small scales where the magnetic Reynolds number is no longer sufficient to avoid significant ohmic dissipation. Also, there is evidence (see Chapter 8) that spot fields are not simply fragmented away by supergranular motions. Rather, their evolution seems to be determined by processes whose fundamental time scale is much longer than the supergranule lifetime.

Although these arguments need to be considered seriously, an observational distinction between a process usefully described as eddy diffusion and one better described as controlled by deeper-seated flows and by flux tube properties such as field line twist has so far remained elusive. Considerable effort has now been invested in detailed time-lapse magnetogram studies of field evolution in network and active regions. Such studies indicate that small, adjacent field elements of opposite polarity are often observed to disappear together. But it is still difficult to differentiate between actual cancellation by reconnection, subduction below the photosphere, and fragmentation into elements smaller than the magnetograph's spatial resolution, which would also lead to disappearance of any net unipolar flux.

The difficulties with dynamo models based on convection zone dynamics have led to increased studies of how dynamos might operate in the overshooting layer below the convection zone, where the constraints on α and ω are less strict, or at least less clear. Unfortunately, such calculations are difficult to test since motions in these layers are essentially unobservable, although the p-mode techniques becoming available are beginning to change this.

ADDITIONAL READING

Dynamics of Solar Flux Tubes and of the Solar Magnetic Cycle

H. Spruit, "Magnetic Flux Tubes," in "The Sun as a Star," NASA Publication SP-450, S. Jordan (Ed.), p. 385 (1981).

P. Gilman, "Global Circulation and the Solar Dynamo," in "The Sun as a Star," NASA Publication SP-450, S. Jordan (Ed.), p. 231 (1981).

T. Cowling, "The Status of Dynamo Theory," *Ann. Rev. Astron. Astrophys.*, **19**, 115 (1981).

E. Priest, *Solar Magnetohydrodynamics*, D. Reidel, 1982.

E. Parker, *Cosmical Magnetic Fields*, Oxford University Press, 1979.

Activity Behavior

K. Kiepenheuer, "Solar Activity," in *The Sun*, G. Kuiper (Ed.), University of Chicago Press, p. 322, 1953.

J. Eddy, "The Historical Record of Solar Activity," in *The Ancient Sun*, R. Pepin, J. Eddy, and R. Merrill (Eds.), Pergamon, p. 119, 1980.

N. Sheeley "The Overall Structure and Evolution of Active Regions," in *Solar Active Regions*, F. Orrall, (Ed.), Colorado Univ. Press, p. 17, 1981.

EXERCISES

1. Use equation (11-1) to calculate the diameter $d(h)$ of a flux tube in magnetohydrostatic equilibrium whose intensity at $h = 0$ ($\tau_{5000} = 1$) is 1 kG. Use the photospheric model given in Chapter 5, and assume the flux tube is fully evacuated. Show that the plasma parameter β is constant with height for a slender flux tube.

2. Calculate the density difference $\Delta\rho$ between a flux tube of 1 kG intensity and a surrounding isothermal plasma in pressure equilibrium with it near the bottom of the convection zone. What percentage deviation from isothermality is required for $\Delta\rho = 0$?

3. Calculate (a) how much farther (in radians) a point on the solar equator travels in a year relative to a point at $60°$ latitude, (b) the sense of Coriolis rotation of rising (and falling) elements of solar gas in the northern hemisphere. How do these values fit with those required in Babcock's dynamo and for an explanation of active region dipole axis tilts?

4. Estimate the order of magnitude of the decay time for a primordial magnetic field that may be present in the sun's radiative core, where field line diffusion through the plasma is likely to be the main cause of decay of such a field. Assume mean conditions for the core using data tabulated in Chapter 6. Could the sun's present field be primordial? Why is a dynamo needed at all?

5. If $R_m \gg 1$, the drag force, F_D, per unit length on a flux tube of diameter D moving at speed v relative to plasma of density ρ can be approximated by the aerodynamic expression $F_D = C_D \rho v^2$ where C_D is a drag coefficient of order unity. Using this expression (a) show that the terminal velocity achieved by a magnetically buoyant flux tube of intensity B rising through the convection zone is approximated by $v_t \sim 10^{-2} B D^{1/2} (T\rho)^{-1/2}$ cm s^{-1} (T is the plasma temperature); (b) estimate the time scale for escape of intense solar fields from the convection zone; (c) comment on implications for operation of a solar dynamo of 11-yr period [see E. Parker, *Astrophys. Journ.*, **198**, 205 (1975)].

12

The Solar Wind and Heliosphere

Measurements from spacecraft moving outside the Earth's magnetic shield, or magnetosphere, over the past 25 years have demonstrated that a continuous flux of charged particles streams outward from the sun past the planets and into interstellar space. Convincing arguments for such a continuous plasma outflow from the sun were put forward by L. Biermann in the early 1950s, based on his study of the acceleration of plasma structures within comet tails. His work suggested that both these accelerations and the deviation of the plasma comet-tail orientation from the sun-comet vector could best be explained by the plasma pressure of a continuous radial outflow of charged particles from the sun.

Much earlier evidence for clouds and streams of charged solar particles emitted intermittently by flares and other solar phenomena had been found from research on aurorae and geomagnetic storms dating back to the 19th-century studies of Carrington and Sabine mentioned in Chapter 1. Work on the penetration of solar charged particles into the geomagnetic field by K. Birkeland and C. Stoermer was influential in furthering these ideas around the turn of the century. The demonstration by A. Schuster that such streams must necessarily disperse rapidly if they consisted of either positive or negative charges alone eventually led to recognition that they must consist of plasma streams.

An understanding of the dynamics and solar sources of the plasma outflow has been much more recently acquired. S. Chapman's 1957 model of the corona was influential in showing that the high thermal conductivity of this hot plasma could result in relatively high temperatures and appreciable densities very far from the photosphere. Thus Chapman's model gave the first good explanation of how a relatively high density of matter at 1 AU and beyond might be a direct consequence of typical conditions in the solar atmosphere. Other sources considered were primordial dust or plasma associated with the formation of the solar system or due to intermittent solar eruptions such as flares.

Close examination of the evidence put forward by Biermann, and of Chapman's model, led E. Parker in 1958 to put forward the hydrodynamical model of a continuously expanding corona, or solar wind, that has been most influential in guiding thinking in the past 30 years. Abundant spacecraft measurements since the first systematic studies from the *Mariner 2* Venus probe in 1962 have confirmed the basic features of Parker's once controversial arguments for a supersonic wind. Good coronal imaging on the disk in the EUV and X-rays showed in the early 1970s that the high-speed solar wind streams that give rise to the well-known 27-day recurrence of many geomagnetic disturbances were traceable to areas of low coronal density, called coronal holes.

Much has been learned from in situ spacecraft measurement ranging in as close to the sun as 0.3 AU (the *Helios* mission) and as far out as the *Pioneers* and *Voyagers*, which have moved to beyond 40 AU both in the direction of the sun's motion through the Galaxy and in the opposite direction. These measurements have shown that the gross structure of the plasmas and magnetic fields near the ecliptic plane agrees well with the dynamical model. Important insights have also been obtained about how disturbances to the geomagnetic field are generated in coronal holes, flares, and filament eruptions and transmitted through the complex environment of interplanetary space to the Earth orbiting 150,000,000 km away.

Still, important gaps remain in our knowledge, in part because the orbits of the spacecraft used so far have not departed much from the ecliptic plane. Consequently, the three-dimensional structure of the heliosphere, the volume of interstellar space defined by the excess pressure of the sun's expanding atmosphere, is much less well measured than are conditions near the Earth. Interesting techniques using radio emission and transmission characteristics of the heliospheric plasmas, and also the behavior of comet tails, have provided some clues to solar wind behavior near the sun's poles. The instruments on board the *Ulysses* spacecraft should tell us much more when this mission is finally launched into an orbit that swings out first toward Jupiter and then is hurled back by that planet's gravity, first over the south pole of the sun and then over its north pole roughly a year later.

12.1   STRUCTURE OF THE SOLAR WIND

12.1.1   In Situ Measurements of Particles and Fields

Spacecraft whose trajectories take them beyond the Earth's magnetospheric cavity are able to directly sample the charged particles and fields flowing out from the sun. Such in situ measurements account for most of our understanding of the solar wind near the plane of the Earth's orbit.

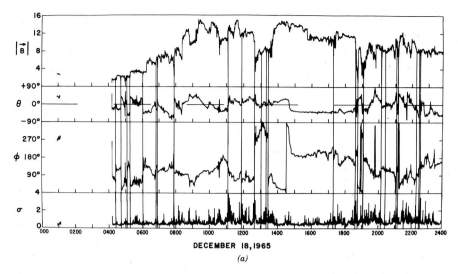

DECEMBER 18, 1965

(a)

Fig. 12-1(a)   30-s averages of the magnetic field intensity (top panel) and direction (2nd and 3rd panels) observed by *Pioneer 6* on December 18, 1965. The quantity in the bottom panel is the standard deviation of the observed fields for each averaging interval. Reproduced with permission, from A. Hundhausen, *Coronal Expansion and Solar Wind*, copyright 1972 by Springer-Verlag, N.Y.

The density, temperature, velocity, and chemical composition of the wind are derived from measurements of the energy per unit charge of electrons and ions and of their direction. In a few of the measurements made so far, the particle energy and charge can be measured separately. Mostly, the signatures of various ionic species must be deduced from the shape of the measured distributions of energy per unit charge. The vector magnetic field measurements are obtained with an accuracy in field intensity and direction of a few percent. A sample plot of plasma density, temperature, velocity, and field over a time interval of a few hours is shown in Fig. 12-1.

A synopsis of conditions in the solar wind as measured over many months is given in Table 12-1. The plasma consists mainly (about 95% by number) of protons and electrons, with the rest consisting almost entirely of doubly ionized helium nuclei (α-particles). As discussed later, the helium abundance is highly variable during energetic transient events. The total plasma density measured 90% of the time lies between 3 and 20 particles per cm^3, with a mean of about 10 cm^{-3}. The plasma temperature is found from the particle velocity dispersion in the frame of reference of the plasma bulk motion. The mean values of the proton and electron components are both between 100,000 and 150,000 K, while the α-particles are four to five times hotter. Judging from comparison of instruments on separate spacecraft, the absolute calibration uncertainties of the density are about 30% and about 15% for the proton and electron temperatures.

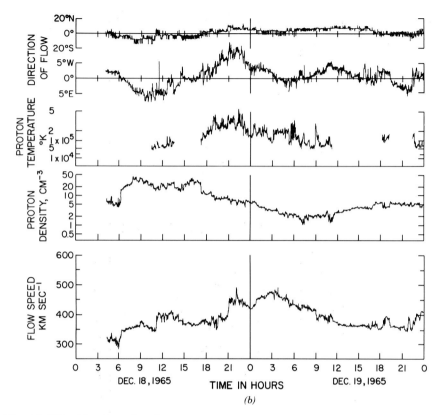

Fig. 12-1(b) The flow direction, proton density, temperature, and flow velocity observed by *Pioneer 6* on December 18–19, 1965. Reproduced with permission from A. Hundhausen, *Coronal Expansion and Solar Wind*, copyright 1972 by Springer-Verlag, N.Y.

The solar wind velocity ranges between about 300 and 700 km s^{-1} 90% of the time, with negligible absolute error. This speed varies little with radius between 1–20 AU. To within the measurement error of about 1°, the velocity vector is radial and lies in the ecliptic plane. For reasons discussed later, any systematic deviation of the plasma velocity vector is of great importance to estimates of the sun's angular momentum loss rate. Present evidence suggests a small deviation of less than 1° trailing solar rotation.

The magnetic field intensity ranges between 2 and 10 γ (1 $\gamma = 10^{-5}$ G). The average direction of the field vector in the ecliptic plane is consistent with the roughly 45° lag from the radial direction predicted by the dynamical model discussed in Section 12.3, but variations of many tens of degrees are common. The radial variation in the main solar wind parameters has been studied, and the density, mass flux, and bulk kinetic energy have been found to agree with an inverse square dependence, at least in the range $0.81 < r <$

TABLE 12-1 Plasma Characteristics of Various Types of Solar Wind Flows

Parameter	Average			Low Speed			High Speed		
	Mean	σ	% Variation	Mean	σ	% Variation	Mean	σ	% Variation
N (cm^{-3})	8.7	6.6	76	11.9	4.5	38	3.9	0.6	15
V (km s^{-1})	468	116	25[a]	327	15	5[a]	702	32	5[a]
NV (cm^{-2} s^{-1})	3.8×10^8	2.4×10^8	63	3.9×10^8	1.5×10^8	38	2.7×10^8	0.4×10^8	15
ϕ_v (degrees)	−0.6	2.6	430	+1.6	1.5	94	−1.3	0.4	31
T_p (K)	1.2×10^5	0.9×10^5	75	0.34×10^5	0.15×10^5	44	2.3×10^5	0.3×10^5	13
T_e (K)	1.4×10^5	0.4×10^5	29	1.3×10^5	0.3×10^5	20	1.0×10^5	0.1×10^5	8
T_α (K)	5.8×10^5	5.0×10^5	86	1.1×10^5	0.8×10^5	68	14.2×10^5	3.0×10^5	21
T_e/T_p	1.9	1.6	84	4.4	1.9	43	0.45	0.07	16
T_α/T_p	4.9	1.8	37	3.2	0.9	28	6.2	1.3	21
$\langle \delta V^2 \rangle^{1/2}$ (km s^{-1})	20.5	12.1	59	9.6	2.9	31	34.9	6.2	18
N_α/N_p	0.047	0.019	40	0.038	0.018	47	0.048	0.005	10
Average			98			45			17

Source: From W. Feldman et al., in *The Solar Output and Its Variation*, O. White (ed.), Colorado University Press, 1977.
[a] Not included in the average percentage variation.

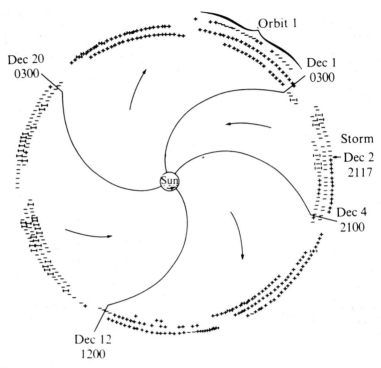

Fig. 12-2 The sector structure of the interplanetary magnetic field observed by the *IMP 1* satellite. The plus or minus polarities correspond to field directed out of and into the sun. By permission of N. Ness.

1.5 AU. The magnetic field dependence was found to follow a $r^{-1.25}$ relation. In both cases this behaviour is consistent with the model discussed in Section 12.3.

The most prominent feature of the solar wind's magnetic variability near 1 AU is the well-ordered alternation of polarity seen in the radial component B_r of the magnetic field. As solar rotation sweeps the interplanetary field past the Earth in about 27 days, the sign of B_r switches abruptly several times, indicating the presence of several sectors of uniform and alternating B_r in the field measured near the ecliptic plane (Fig. 12-2). This sector structure, discovered in 1965 by N. Ness and J. Wilcox in data from the *IMP 1* satellite, has been shown to be quite stable for years. But it is also capable of undergoing slow changes between a four-sector to a two-sector configuration. At the time of its discovery, there were four sectors, three of equal size and one half as wide.

The fundamental reason for this sector structure observed near the ecliptic plane can be seen in Fig. 12-3. The predominant north and south polarities of the large-scale solar magnetic field extend far outward into the interplanetary medium, and near the ecliptic plane the oppositely directed field lines are

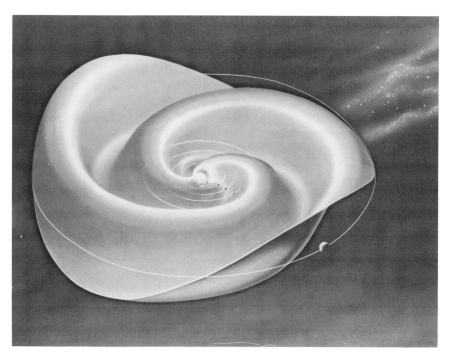

Fig. 12-3   An illustration of the warped heliospheric current sheet, drawn to scale with some planetary orbits. By permission of T. Hoeksema.

separated by a thin current sheet (defined by $\nabla \times \mathbf{B} = 4\pi\mathbf{j}$). This sheet is not strictly planar since it ultimately maps into the neutral line separating photospheric magnetic structures of the northern and southern hemispheres, shown in Fig. 8-16. Since this neutral line does not exactly follow the solar equator, its curves introduce warps into the interplanetary current sheet. As this sheet rotates past the Earth every 27 days, we observe alternatively northern or southern hemisphere polarities, i.e., "sectors." Thus, the sectors are not expected to be a real feature of the heliosphere and would not be observed at higher latitudes. The heliospheric current sheet that gives rise to the sectors is an important feature whose detailed structure is the focus of much current study. One important question is how the current sheet changes every 11 years, when the solar field that produces it changes polarity.

Solar wind variations observed on time scales longer than about a day are found to be caused by the large shocks and high-speed streams associated with coronal holes, both discussed in Section 12.2. In addition, considerable microstructure is observed in the solar wind on shorter time scales. Given the range of solar wind speeds, this implies structures of spatial scales between roughly 10^6 and 10^7 km.

The microstructure consists in part of transverse, large-amplitude variations of velocity and magnetic field identified as Alfvén waves. The variations measured in these two quantities can approach the magnitude of the Alfvén velocity and interplanetary field intensity. Another kind of variation consists of large-amplitude density and temperature fluctuations out of phase, so that the total pressure is unchanged as the disturbance passes, in distinction to a sound wave. On the shortest time scales, below 10 min, rapid discontinuities in the field direction are the most common form of microstructure, often rotating the field by tens of degrees with no significant changes in other plasma properties. Some of these discontinuities can be ascribed to sharply crested Alfvén waves and others to sharp gradients in plasma pressure oriented transverse to the bulk flow, referred to as tangential discontinuities.

12.1.2 Observations Out of the Ecliptic Plane

The in situ measurements described above have so far been made only close to the ecliptic plane. Since this plane is inclined to the solar equator by about 7.5°, the Earth's annual motion enables us to sample only a small range in heliographic latitude.

So far, our knowledge of the solar wind properties away from the ecliptic plane is based mainly on remote sensing using observations of the propagation of signals from cosmic radio sources and on the behavior of comet tails. Unlike the orbital planes of planets, those of comets can be inclined at any angle to the solar equator. Their plasma tails act as wind socks whose orientation can thus be used to infer the solar wind velocity throughout the heliosphere. Further information on the wind's flux and magnetic structure can be gleaned from the extent of a comet's hydrogen cloud as observed in its Lyman alpha radiations, and perhaps also from the occurrence of disconnection events during which the entire tail becomes detached from a comet's head.

Other information on heliospheric structure both in and out of the ecliptic plane has also been gathered from analyses of variations in the galactic cosmic ray flux and of geomagnetic variations, from radar probing, and from phase delays in radio signals transmitted from interplanetary spacecraft. Most recently, the study of long-wave (kilometric) plasma radio emissions from the interplanetary disturbances themselves is providing important information on how such transients propagate from the sun to the Earth's orbit and beyond.

Probably the most important information on solar wind behavior at high latitudes has been obtained from interferometric studies of radio wave propagation from cosmic sources. Information on magnitude and orientation of plasma density fluctuations over scales comparable to the interferometer baseline of tens of kilometers can be derived from observations of the apparent size of sources such as the Crab nebula as our line of sight to such a source moves through the outer corona. This increase in angular size is associated with scattering of the radio signals off refractive index variations caused by the plasma density fluctuations. An approximate formula for the apparent diameter ϕ in radians at a radio frequency of f megacycles s^{-1} is given by

$$\phi \sim 1.5 \times 10^{-4} \frac{\delta N_e n}{f^2}, \qquad (12\text{-}1)$$

where ΔN_e is the rms electron density variation (cm^{-3}). The parameter n is the number of density irregularities along the line of sight.

Observations of small-diameter cosmic radio sources also exhibit flux variations, or scintillations, on time scales of 0.1–10 Hz attributed to interference between the scattering patterns caused by plasma density variations. Movement of the irregularities with the solar wind leads to a drift of the interference pattern across an observer on Earth, causing the flux variations. Spatial coherence of the pattern can be verified by comparison of the scintillations observed at antennas separated by several tens of kilometers.

The phenomena of radio source scattering and of scintillation described above are similar to the optical phenomena of "seeing" and scintillation. The former causes the familiar blurring of objects seen along a line of sight crossing hot pavement and is due to variations in refractive index between convecting elements of air relatively nearby. The latter is due to atmospheric refractive index variations far from the observer and causes the twinkling of stars.

The radio observations integrate along a line of sight extending through the heliosphere, so in general a model is required to localize the region responsible for most of the scattering. In some cases, the interplanetary scintillation (IPS) studies of the solar wind velocity in the ecliptic plane have been compared to in situ spacecraft results. Good agreement is found, provided the solar wind flow is dominated by steady velocity structures rather than by relatively short-lived transients.

The results of the IPS work indicate that the solar wind velocity increases with latitude at an average rate of roughly 2 km s^{-1} deg^{-1} up to the highest observable latitude of approximately $65°$. Plots of the wind velocity mapped back to the solar surface indicate also that high-velocity wind streams tend to emanate from regions of low coronal brightness—in particular from the coronal holes always present at the poles.

To the extent that the heliospheric magnetic structure can be sampled away from the ecliptic during the extreme excursions of the Earth to $\pm 7.5°$ heliographic latitude, spacecraft measurements indicate that the radial component of the heliospheric field tends to be that of the north polar solar field in the northern heliospheric hemisphere and that of the south pole in the southern hemisphere. However, evidence from comet tail disconnections at high ($\sim 45°$) latitudes suggest that the warp in the heliosphere current sheet might be quite large, so that the sector structure and polarity mixing might extend to much higher latitudes.

12.1.3 Cosmic Rays

In addition to the continuous outflow of solar wind plasmas of relatively low energy (< 0.1 keV) particles, frequent bursts of energetic (occasionally relativistic) ions and electrons pass outward through the heliosphere. The ions range in energy between roughly 10 MeV and 10 GeV in the most easily detected bursts (Fig. 12-4), while the discrete electron events detectable with existing instrumentation contain particles of about 45 KeV or more. Besides these solar energetic particles whose acceleration is discussed in Chapter 10, the heliosphere is also permeated by cosmic rays originating elsewhere in the Galaxy. Their ion energy spectra peak broadly around 200–300 MeV per nucleon and range up to 10 GeV or more.

Although the continuous flux of galactic cosmic rays has been measured since their first detection in 1912, the discovery of solar cosmic rays is much more recent, dating back only to 1942. They were detected first through increased count rates in ionization chambers at ground level (the presence of galactic cosmic rays was first inferred from the increase in atmospheric ionization with height measured with balloons). Since about 1954 the use of ground-level monitors that detect secondary neutrons produced by the impact of energetic ions in the upper atmosphere has become more common. However, such neutron monitors are sensitive only to primaries having energies in excess of about 500 MeV, even at high latitudes.

An indirect measure of the flux of much less energetic protons in the few MeV range is provided by absorption of background cosmic radio noise measured with instruments called riometers available since 1952. Increased absorption, referred to as a polar cap absorption or PCA, is produced by an increase in the flux of solar protons into the magnetosphere and down to a level around 80 km.

Continuous spacecraft measurements of energetic particles having less than 10 MeV per nucleon have only been available since the mid-1960s, thus for the last two solar cycles. They show that although only a few dozen relativistic proton events have been detected since 1942, discrete events consisting of the lower-energy particles occur more or less all the time, and it is difficult to relate all these low-energy events to discrete solar events.

Our basic understanding of a large flare-induced solar particle event is illustrated in Fig. 12-5. Typically, in such events, the most energetic relativistic particles hit the upper atmosphere within roughly the light transit time to the Earth of 8 min and produce a sharp increase in the neutron counts. The flux of these most energetic nuclei usually decays within a few hours. The flux detected at the Earth may actually exhibit a short-term drop a few days later if the flare-induced shock wave creates conditions in the interplanetary medium that exclude the flux of galactic cosmic rays. This subsequent reduction is called a Forbush decrease.

The time profile of the particle fluxes detected at 1 AU exhibits a progressively slower rise and fall for protons of energy decreasing from about 100

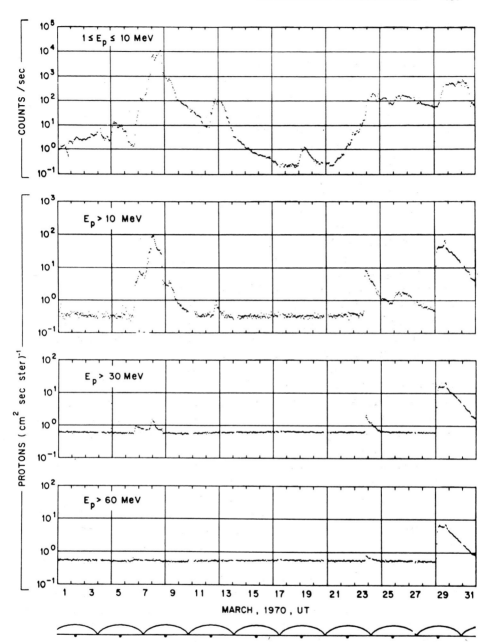

Fig. 12-4 Solar protons in four energy intervals measured in interplanetary space by the solar proton monitoring instrument on *Explorer 41* during March 1970. Reprinted by permission, from L. Lanzerotti in *The Solar Output and Its Variation*, O. White (Ed.), Colorado University Press, 1977.

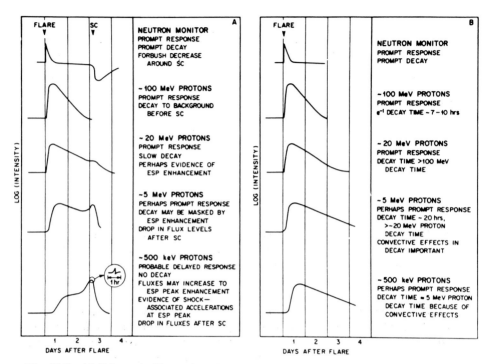

Fig. 12-5 (a) Schematic illustration of proton fluxes that might be measured at 1 AU after a western hemisphere flare when a flare-produced shock wave is observed at the Earth. (b) Same as in (a), but if no flare-produced shock wave were observed at the Earth. Reprinted by permission, from L. Lanzerotti, in *The Solar Output and Its Variation*, O. White (Ed.), Colorado University Press (1977).

MeV down to below 1 MeV. For the least energetic protons, a secondary increase a few days after the flare is associated with particle acceleration within the flare-induced interplanetary shock wave itself.

The outward propagation of solar energetic particles is complicated by their interaction with the complex magnetic structure of the corona and solar wind. For this reason, the sample of solar events observed at the Earth must be interpreted carefully. For instance, the locus of origin of relativistic proton events is centered at about 60° west of solar disk center. This is understandable if these particles propagate along the pre-existing spirally wound solar wind field lines, which map from the Earth to roughly 60° west solar longitude. The particles responsible for both the PCA events and the geomagnetic storms cluster more nearly around the sub-Earth point at 0° solar longitude. This seems to agree with our understanding of these phenomena as caused by the bulldozing out of a direct path between the sun and Earth by a flare-induced shock wave, thus allowing the lower-energy particles to propagate more directly to the Earth.

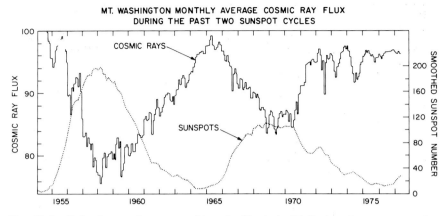

Fig. 12-6 Galactic cosmic rays reaching the Earth (solid line) and sunspot number (dots) plotted for two solar cycles. Drawing by courtesy of A. J. Hundhausen from cosmic-ray data supplied by J. A. Lockwood.

The chemical composition of the higher-energy solar particles is roughly consistent with photospheric abundances, at least in regard to the few heavier elements such as iron that can be identified with some confidence. The situation is much more variable from flare to flare for the lower-energy particles of a few MeV per nucleon. In some flares the abundance of elements heavier than helium is considerably enriched relative to photospheric values. Isotopic composition is also found to be variable, with some events exhibiting more He^3 than He^4. The reasons for these striking compositional variations require further study. They may provide important insights into the basic acceleration processes acting in flares and in interplanetary shocks.

In addition to the complex behavior of solar particle fluxes outlined above, the galactic cosmic ray background (due to sources well outside the solar system) also exhibits a clear solar-cycle modulation illustrated in Fig. 12-6. Galactic cosmic ray fluxes decline when the solar activity level rises and increase toward activity minimum. The amplitude of 11-year modulation is about a factor of 5 for protons around 200 MeV and drops to only a few percent for the most energetic ~ 10 GeV cosmic ray muons. The reasons for this modulation are only partially understood, but they seem to be associated with restructuring of the heliospheric magnetic field (including its polarity) over the solar cycle, resulting in changes in the efficiency of its shielding against galactic cosmic rays.

12.1.4   Interplanetary Gas and Dust

When the partially ionized plasma of the interstellar medium encounters the heliospheric boundary, or heliopause, the ions and electrons are excluded by the heliospheric magnetic field, but neutrals can pass right through the heliosphere with little dynamical interaction, except for those that are ionized by solar EUV radiation. Photometric observations of the neutral hydrogen using scattered solar radiation in Lyman α and in the neutral helium resonance line at 584 Å have been carried out from the Soviet *Prognoz* and *Venera* spacecraft. These observations provide some insight into the properties of the neutral component of the interstellar medium encountered by the sun as it moves through galactic gas clouds.

The velocity of the interstellar neutral wind is measured to be roughly 20 km s^{-1}, and its direction is very approximately that expected from the sun's motion relative to other stars, as discussed in Chapter 13. The temperatures of the neutral hydrogen and helium are roughly 8000 and 16,000 K, and their densities are about 0.06 and 0.01 atoms cm^{-3} respectively. The fundamental difficulty of interpreting such photometric observations (which integrate along the line of sight) in terms of a three-dimensional distribution of neutral gas in

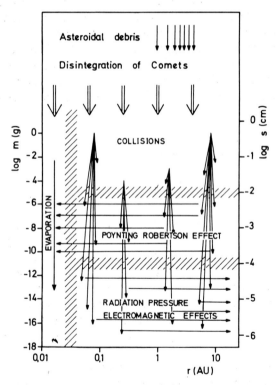

Fig. 12-7 Schematic diagram of dynamical effects which change mass and heliocentric distance of interplanetary meteoroids. Sources for meteoroids are comets and asteroids; loss processes are ejection from the solar system and evaporation near the sun (this material is ionized and carried away with the solar wind). Reproduced, with permission, from G. Morfill et al., in *The Sun and Heliosphere in 3-D*, R. Marsden (Ed.), Kluwer, 1986.

the heliosphere should be considerably alleviated when instruments designed to directly sample the neutral helium in the heliosphere are flown on NASA's *Ulysses* mission.

The interplanetary medium also contains solid particles ranging in size from submicron grains up into the range of small, boulder-sized asteroids. The presence of a continuous spectrum of such solid "dust" particles was demonstrated by studies of cratering in the lunar surface. The sources of these particles are mainly the disintegration of comets, asteroidal debris, and interstellar grains. Their mass density near 1 AU is actually about an order of magnitude higher than that of the solar wind plasma.

Fig. 12-7 summarizes the processes that determine this mass density distribution with size. Particles of intermediate size (between about 1 μm and 0.1 mm) spiral toward the sun due to the drag of solar photons and ions (the Poynting-Robertson effect). Smaller particles get blown out of the heliosphere by radiation pressure.

The concentration of solid particles toward the ecliptic plane accounts for the appearance of the zodiacal light and for the F-corona seen at eclipses. It is also responsible for the gegenschein, a faint glow seen in the antisolar direction under conditions of exceptionally dark sky. Photometric studies of the white-light brightness distribution and polarization, and of their IR emission, provide interesting information on the type of particles responsible for the zodiacal light, their size distribution, and on their spatial distribution in the heliosphere.

12.1.5   Structure of the Heliosphere

The basic structure of the heliosphere is illustrated in Fig. 12-8. The features that have been determined with some confidence from comparison of observations with theoretical models are its approximate size and shape and the magnetic field structure near the ecliptic plane. Beyond these gross features, the representation in Fig. 12-8 is based only on untested extrapolation.

Probably the most firmly established aspect of the heliosphere is the spiral geometry of the field near the ecliptic plane. The pitch angle is about 45° to the sun-Earth direction (i.e., trailing rotation) around 1 AU and increases outward. This spiral pattern has been verified by magnetometer measurements between 1 AU and 8.5 AU by the *Pioneer* spacecraft. No systematic north-south component of the magnetic vector relative to the ecliptic plane has been determined.

The existence of a vast current sheet in the ecliptic plane separating magnetic fields whose radial component is oppositely directed in the north and south heliospheric hemispheres is also well established by spacecraft magnetometer observations ranging in to about 0.3 AU and out to 25 AU. *Pioneer 11* measurements up to 16° helio-latitude indicate also that the current sheet is confined to low latitudes, at least near activity minimum. Mapping of the current sheet deformations in towards the sun produces reasonable agree-

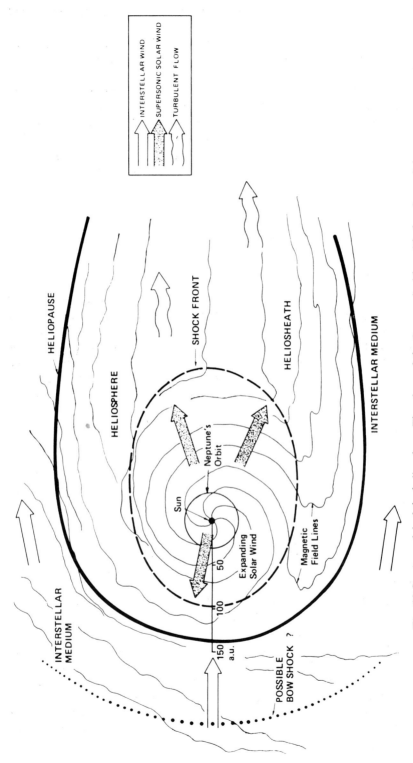

Fig. 12-8 A model of the heliosphere. The plane of the figure coincides with the plane of the sun's equator, which is approximately the general plane of planetary orbits.

ment with large-scale neutral lines of the photospheric magnetic field, when the 4–5 day transit time of such fields from the sun to Earth is taken into account.

The source of the warps in the neutral sheet seems to be mainly the shape of the neutral line observed at the photosphere, although dynamical effects in the solar wind may also play a role. Observations of the spatially smoothed photospheric field discussed in Chapter 8 indicate that warps in the north-south neutral line are confined to relatively low latitudes during activity minimum, but become accentuated with increasing activity and reach the polar regions during activity maximum.

12.2 TRANSIENT FEATURES IN THE SOLAR WIND

12.2.1 High-Speed Streams

Studies of solar wind velocity structure show prominent high-speed regions whose mean plasma velocity of about 700 km s^{-1} is 50% higher than the mean solar wind value and more than twice the lowest speed encountered. Studies of the time and space coherence of these high-speed conditions have demonstrated that they represent long-lived streams of high-speed wind issuing generally from the coronal holes discussed in Chapter 9, over periods of as long as several solar rotations. The streams corotate with the spiral magnetic field geometry of the solar wind, and the successive rotations past the Earth of the longest-lived streams at roughly a 27-day interval account for the well-known tendency to recurrence of geomagnetic storms.

The most obvious feature of streams observed near 1 AU is the characteristic rapid rise (over about 1 day) and slow decay of the plasma bulk flow speed and of the proton temperature. The plasma density and magnetic intensity are also enhanced along the stream's leading edge, associated with a high-pressure ridge which is produced by the dynamical interaction of the stream with its surroundings as it propagates out from the sun. Over most of the stream, however, the density is low relative to the ambient plasma. Many other properties, such as the hydrogen-to-helium ratio, have been noted to vary systematically across the streams.

The high-speed streams are closely associated with the sector structure seen in the ecliptic at 1 AU, although individual streams are less stable than the sectors in which they are found. Several streams may coexist within a given sector, and this correspondence can be understood by recalling that a sector corresponds to a single warp in the interplanetary current sheet, which corresponds at the photosphere to either a northward or southward excursion of the photospheric neutral line. The coexistence of several streams within a given sector simply means then that more than one of the coronal holes that give rise to the streams may exist within the area included in a warp of the neutral line.

12.2.2   Interplanetary Shock Waves

Fig. 12-9 illustrates a class of sharp discontinuity, transverse to the flow direction, that is observed in solar wind speed, density, and temperature and lasts for a few minutes in the reference frame of the moving spacecraft. Such rapid upward jumps of about 10% in solar wind speed and by about a factor of 2 in proton density and temperature are caused by shock waves.

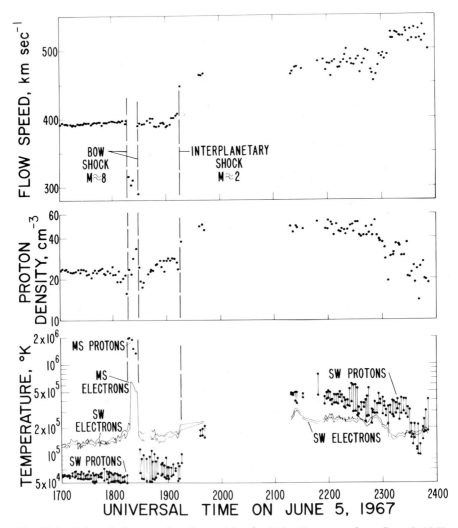

Fig. 12-9   Solar wind properties observed by the *Vela 4B* spacecraft on June 5, 1967. The abrupt change at 1915 UT signals the passage of an interplanetary shock. Reproduced by permission, from A. Hundhausen, *Coronal Expansion and Solar Wind*, Springer-Verlag, N.Y., 1972.

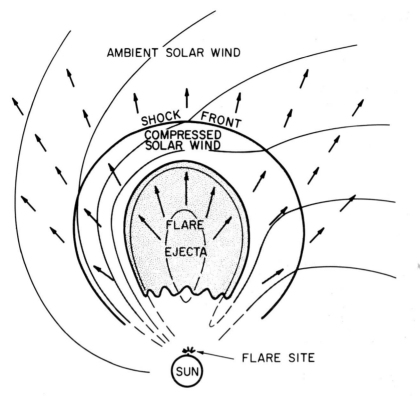

Fig. 12-10   A qualitative sketch, in equatorial cross-section, of a flare-produced shock wave propagating into an ambient solar wind. The arrows indicate the plasma flow velocity and the light lines indicate the magnetic field. Reproduced by permission, from A. Hundhausen, *Coronal Expansion and Solar Wind*, Springer-Verlag, N.Y., 1972.

Shocks of the magnitude illustrated in Fig. 12-9 are observed typically 10–15 times per year, and in most cases can be identified with solar flares occurring some 2–3 days before. Figure 12-10 illustrates how the plasma ejected from a flare plows outward, pushing and compressing the interplanetary medium ahead of it. A shock is formed by the gradual steepening of this pressure jump during its outward propagation at about 500 km s^{-1}. Behind the shock lies a region of compressed plasma, typically 0.1 AU thick, followed by the flare-ejected coronal plasma itself. Since the shock is traveling supersonically relative to the solar wind ahead of it, by 1 AU it has far outdistanced the slower-moving coronal material whose ejection originally generated it.

Measurements of the chemical composition show that the helium content of the plasma is sometimes enriched by about a factor of 2 in a shell about 1–10×10^6 km thick, detected some 5–12 h after the shock passage. The helium enrichment may signal the arrival near 1 AU of the coronal plasma whose ejection during the flare also caused the shock wave, and most of the shocks associated with helium enrichment have been identified with flares.

The detailed time profiles and geometrical shapes of observed shocks vary greatly. An interesting distinction is made between those in which the disturbance peaks and then subsides rapidly (within hours) and those in which the speed, density, and temperature continue to rise gradually over several hours after the initial sharp increase. The former condition is characteristic of a blast wave in which the shock is created by a short-lived impulse. The second corresponds to a driven shock in which the "piston" that accelerates the material continues to push on it for several hours.

In general, coronal flares and filament eruptions can be identified as the drivers of most interplanetary shocks, although the correspondence between size of flare and shock magnitude is not straightforward. In other cases, shocks seem to be generated within the solar wind by steepening of the compressed region at the leading edge of a high-speed stream. Determination of the shock geometry and orientation, which is required to make such distinctions, requires simultaneous measurements from well-separated spacecraft. Such measurements have been made, but they require careful coordination. A promising new technique for studying these shocks relies on observations of their Type II radio emission from spacecraft. The extended range of low frequency radio waves observable above the ionosphere makes it possible to track these shocks all the way out to 1 AU.

12.2.3 Coronal Transients

Observations of the outer corona from space-borne coronagraphs have revealed huge eruptions of coronal plasma moving outward into interplanetary space. An example of such a coronal mass eruption or "coronal transient" is shown in Fig. 12-11. When located a few solar radii above the photosphere, their diameters often exceed that of the whole sun. Their morphology is varied: some look like loops, some more like bubbles, and some are irregular.

Coronal transients seem to occur at a rate of roughly one per day, although fewer than half of these can be observed, given our limited coverage from Earth. With masses in the range 10^{15}–10^{16} g and propagation speeds typically between 200 and 500 km s^{-1} (with some much higher velocities observed to exceed 10^3 km s^{-1}), they make a small but appreciable contribution of perhaps as much as 10% of the total solar wind mass flux measured near the ecliptic plane.

The space coronagraphs used to observe the transients in visible white light achieve a very low level of scattered light (below 10^{-6} of the sun's disk brightness) by using an additional external occulting disk placed outside the objective. Their design typically occults the central portion of the image out to about 1.5 $R_{\odot}$, or $0.5R_{\odot}$ above the limb. This makes it more difficult to identify the structures of the lower corona that gave rise to the transient event observed further out.

Fig. 12-11 A coronal mass ejection photographed on August 10, 1973. The High Altitude Observatory's white-light coronagraph aboard *Skylab* had an annular field of view ranging from the occulting disk at the center ($1.5R_\odot$ in radius) to approximately $6R_\odot$ at the outer edge. A strong gradient of coronal radiance has been suppressed in this image by instrumental vignetting and by dodging the print. By courtesy of D. Sime.

Statistical studies indicate that only about half the transients observed can be associated with flares or filament eruptions on the visible disk. It is likely that the other half of the transient events are produced by such activity on the invisible hemisphere. Certainly, all flares observed to eject H_α material produce transients. Nevertheless, transients were found to be produced more often by filament eruptions than by flares.

Neither the mechanism responsible for triggering the transient nor the forces that propel it outward have been conclusively identified. The trajectory is certainly not ballistic since most transients are observed to be continuously accelerating on their way across the coronagraph field of view, thus at least out to $6R_\odot$ from sun center. In some cases, a transient is launched before its associated flare, so the trigger mechanism is not the flare itself. The driving force is most likely to be the gradual release of magnetic stresses in looplike coronal structures. Neither plasma pressure gradients nor the momentum likely to be imparted by coronal waves seem to suffice by many orders of magnitude to provide the kinetic energies of order 10^{30}–10^{31} ergs measured for typical transients.

The association of coronal transients with interplanetary phenomena is not straightforward, although over 1000 transients have been studied by coronagraphs on four satellites since their discovery in 1971. It appears that at least for the faster-moving eruptions the pressure wave pushed ahead of the ejected plasma steepens into an interplanetary shock detectable at heliocentric distances beyond about 0.3 AU and occasionally near 1 AU. The close association of the more energetic, flare-associated white-light transients with Type II meter wave bursts, which are known to be emitted by shocks (Section 10.2.1) supports this interpretation.

The ejected plasma itself, associated with moving Type IV meter wave bursts nearer the sun, may also have been identified with magnetic bubbles in the interplanetary medium. These are volumes found to exhibit relatively low plasma temperatures and pressures. It would appear from such analyses that coronal transients are the outer coronal signature of flares and filament eruptions, so their study provides the link we seek between these low coronal eruptive phenomena and transient events in the interplanetary medium.

12.3 DYNAMICS OF THE SOLAR WIND

12.3.1 Thermal Conductivity of the Corona

With acceptance of the high coronal temperatures of order a million degrees in the 1940s came recognition that the thermal conductivity of the coronal plasma, which is proportional to $T^{5/2}$ (see Section 4.2.4), implies very efficient outward (as well as inward) heat transport from the inner corona. The influence of this high thermal conductivity on the radial extent of the corona was first investigated in 1957 by S. Chapman.

His study showed that if steady outward thermal conduction is taken to be the only heat source to a static corona in which all other heating or cooling such as radiation are neglected, the coronal energy balance obeys the relation

$$F_c = 4\pi r^2 K \, dT/dr. \tag{12-2}$$

Here F_c is the (constant) conductive flux, r is the heliocentric distance, and K is the plasma thermal conductivity given by

$$K \simeq \text{const} \times T^{5/2}. \tag{12-3}$$

Equation (12-2) can be integrated to yield the radial temperature profile of such a corona, which is

$$T(r) = T_0 (r/R_0)^{-2/7} \tag{12-4}$$

where $T_0 = T(R_0)$ and T is assumed to vanish at $r = \infty$. The high thermal conductivity thus ensures a relatively slow decrease of $T(r)$, so that equation (12-4) predicts temperatures around 1 AU still in excess of 10^5 K.

It can be shown that the corresponding hydrostatic density at 1 AU produced by this temperature dependence would be in excess of 100 cm^{-3}. The agreement between this density and the electron density that seemed to be required to explain the brightness of the zodiacal light led, in the late 1950s, to some early confidence in this result. It soon turned out that the interpretation of the zodiacal-light measurements in terms of primarily electron scattering was incorrect—the scattering is now known to be caused mainly by interplanetary dust. The further difficulty with matching the high hydrostatic plasma pressure to the low pressure expected in the interstellar medium at large distances from the sun was pointed out by E. Parker, who went on to formulate a dynamical theory of the coronal extension into interplanetary space.

12.3.2 Expansion of the Corona

If the corona is a dynamic, rather than hydrostatic plasma, its behavior will be governed by the equations of mass, momentum, and energy balance discussed in Section 4.2. For a steady flow, the mass balance condition of equation (4-34) becomes

$$\nabla \cdot \rho v = 0 \tag{12-5}$$

or, for a purely radial flow,

$$\frac{1}{r^2} \frac{d}{dr} (r^2 \rho v) = 0, \tag{12-6}$$

which can be integrated to yield

$$\rho v A = \text{const.} \tag{12-7}$$

The momentum balance is given by equation (4-25), which for a steady radial flow becomes

$$\rho v \frac{dv}{dr} = -\frac{dP}{dr} - \frac{GM_\odot}{r^2}\rho. \tag{12-8}$$

Here the solar gravitational acceleration g has been expressed in terms of the sun's mass and heliocentric distance r. We see that the forces acting on the plasma are the pressure gradient and gravity. In addition, we will later consider the possibility that a unit mass of the fluid might gain momentum at a rate D through the direct action of other forces such as waves. In this case an additional term ρD would be added to the right-hand side of equation (12-8).

The total energy balance of this steady flow, described by equation (4-54), includes fluxes of kinetic and potential energy as well as heat. In general, (4-54) would need to be solved simultaneously with (12-7) and (12-8). However, the basic features of the solution can be illustrated by assuming that the corona is isothermal. This is the approximation made by Parker in his first approach to the problem in 1958.

For an isothermal corona, the density may be eliminated between equations (12-7) and (12-8), and we find that the flow is governed by

$$\left(v - \frac{v_s^2}{v}\right)\frac{dv}{dr} = \frac{2v_s^2}{r} - \frac{GM_\odot}{r^2}, \tag{12-9}$$

where v_s is the isothermal sound speed (which is lower by a factor of $\gamma^{1/2}$ than the usual adiabatic sound speed introduced in Section 4.4.2).

Equation (12-9) can be integrated analytically (see exercises), and some typical solutions are plotted in Fig. 12-12. These solutions can be divided into five basic types. Since we are interested in single-valued continuous solutions that describe plasma motions from near the sun to large r-values, Types I and II, which are double-valued and confined to small r and large r respectively, can be excluded. Solutions of Type III do not satisfy the observation that general coronal plasma velocities near the sun are subsonic.

Solutions of Type V are subsonic everywhere, and they predict speeds of approximately 10 km s^{-1} at 1 AU. These are the so-called solar breeze solutions advocated in the late 1950s by J. Chamberlain. The solution favored by Parker is of the Type IV, starting subsonically near the sun and achieving supersonic speeds. Such a solution passes through a critical point, where dv/dr is undefined, at $v = v_c$, $r = r_c$. At this point the coefficient of dv/dr and the right-hand side of equation (12-9) vanish simultaneously.

Many factors figured in the controversy between advocates of the solar breeze and solar wind solutions. It was shown that the solar wind solutions had the desirable property of predicting vanishing pressure at infinity, while

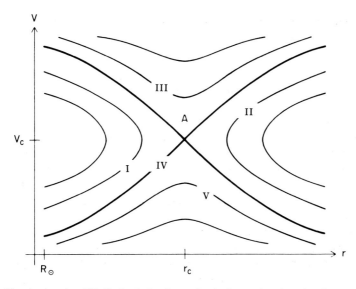

Fig. 12-12 A sketch of E. Parker's isothermal solutions, showing the different classes I, II, III, IV, V. Type IV (solar wind) passes through the critical point (A). Type V represents the subsonic, solar breeze solutions. Adapted from E. Priest, *Solar Magneto hydrodynamics*, Reidel, 1982.

the solar breeze solutions produced finite pressure at infinity and thus did not avoid the problem posed by Chapman's hydrostatic coronal model. By relaxing the requirement that the corona be isothermal, it was shown also that the solar wind speed remained finite at infinity.

Despite these attractive features, the requirement that the sun choose that particular solution passing through the critical point was one consideration that postponed acceptance of the solar wind solution as the correct type until after the *Mariner 2* measurements showed the high plasma velocities of the order predicted by Parker. As discussed further below, the solar wind model successfully predicts the rather complex observed interplanetary magnetic field geometry. It also predicts the measured radial dependences of plasma density, velocity, and to a lesser extent, temperature. The most important difficulty is that dynamical models starting with the temperatures and densities observed in the coronal regions expected to give rise to the wind predict mass fluxes at 1 AU that are roughly a factor of 2 below those observed. Also, they cannot match the wind speeds observed, particularly in high-speed streams.

This discrepancy has been the focus of considerable theoretical effort, mainly to determine whether the acceleration of the solar wind by simply heating the corona and allowing it to expand under its pressure gradient is sufficient. Since the pressure gradients developed will depend in part upon the efficiency of heat conduction from the region where heat is deposited, some of this work has investigated the role of the magnetic field, and also the low plasma density, in limiting the outward conductive flux. Other work has focussed on the influence of heat deposition (e.g., by waves) at various levels. Yet another line of inquiry has worked out the influence of having the wind flow outward along a "nozzle" defined by magnetic field lines that diverge faster than radially.

The main result of this work has been to show that an additional outward force besides the pressure gradient must be acting directly on the flow to reproduce the observed mass fluxes at 1 AU. Such a force, represented by an additional term ρD in equation (12-8), is most likely caused by the wave pressure given by equation (4-77). The basic form of this equation holds even for MHD waves, although the sound speed would be replaced by the phase speed of the relevant mode, such as an Alfvén wave. It also appears most likely that this wave pressure is already present quite low in the corona, below the sonic point of the flow. Only in that way can the relatively high mass fluxes be produced. Momentum addition to the wind above the critical point accelerates it, but cannot increase its mass flux since information about the additional acceleration cannot travel inward (against the supersonic flow) to the mass reservoir in the low corona.

12.3.3   Geometry of the Interplanetary Magnetic Field

The outflowing solar wind interacts dynamically with the solar magnetic field through the Lorentz force described in Section 4.3.1. In the low corona, the plasma thermal and kinetic energy densities are generally well below the magnetic energy density, so the magnetic field determines the loop-like shape of low coronal structures. Higher in the corona, and in the interplanetary medium, the kinetic energy of the supersonic flow dominates the field and determines its geometry. In both regions, the scales of motion and electrical conductivities are sufficiently large that the freezing-in approximation of ideal MHD is assumed to hold at least throughout most of the plasma volume.

The radial outflow of solar wind plasma would simply carry the magnetic field lines radially outward from their photospherically anchored footpoints if the sun did not rotate. For a solar angular rotation rate ω rad s^{-1}, the radial and tangential components of plasma motion relative to a stationary frame of reference (i.e., an observer looking down on the ecliptic plane) are v_r and ωr respectively (see Fig. 12-13). The resultant streamlines, and thus also the magnetic field lines, are given by the relation

$$\frac{1}{r}\frac{dr}{d\phi} = \frac{v_r}{v_\phi} = \frac{v_r}{-\omega r} \tag{12-10}$$

where ϕ is the azimuthal angle in spherical polar coordinates. For constant v_r,

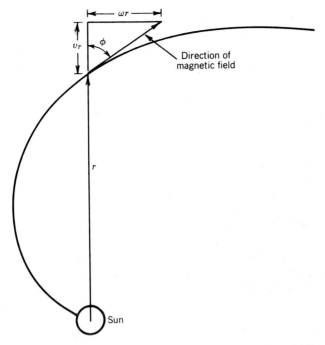

Fig. 12-13   The local orientation of a flow streamline or magnetic field line in a frame of reference rotating with the sun. Reproduced with permission, from A. Hundhausen, *Coronal Expansion and Solar Wind*, Springer-Verlag, N.Y., 1972.

which is a good approximation at $r \gg r_c$ in the solar wind solution illustrated in Fig. 12-12, this equation can be integrated (see exercises) to yield the form of the streamlines given by

$$\phi - \phi_0 = -\frac{\omega}{v_r}(r - r_0). \qquad (12\text{-}11)$$

Here ϕ_0 denotes the azimuthal angle at a heliocentric reference distance r_0.

The spiral magnetic field line geometry defined by equation (12-11) is illustrated in Fig. 12-14. The value of $\phi \sim 135°$ observed near 1 AU is found for $v_r \sim 430$ km s^{-1}, which is in the typical range of solar wind outflow speeds. More generally, if the latitude variation of the spiral geometry is required, the constant angular velocity ω can be replaced by $\omega \sin \theta$, where θ is the spherical polar colatitude.

It is important to keep in mind that the solar wind plasma always moves radially outward, following a trajectory that has been compared to that of a phonograph needle following the spirally wound grooves of a record rotating backward.

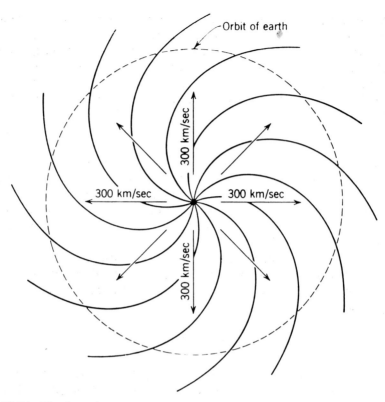

Fig. 12-14 The interplanetary magnetic field configuration in the solar equatorial plane for a constant solar wind speed. Reproduced with permission, from A. Hundhausen, *Coronal Expansion and Solar Wind*, Springer-Verlag, N.Y., 1972.

12.3.4 Energy and Angular Momentum Fluxes

The total energy of the solar wind can be estimated from the sum of its individual terms evaluated from measurements near 1 AU. The expression (4-54) for the total energy of a moving gas can be integrated, for a steady flow, to yield the relation for the constant energy flux F through a sphere of radius R, namely

$$4\pi R^2 \left(\frac{1}{2}\rho v^3 + \frac{5}{2}nkTv - \rho\frac{GM_\odot v}{R} - K\frac{dT}{dr} \right) = F. \qquad (12\text{-}12)$$

The measurements near 1 AU indicate that most (about 90%) of the total energy flux is found in the kinetic energy flux term $\frac{1}{2}\rho v^3$ and only a few percent is carried by thermal conduction. The total energy flux in the wind at 1 AU is estimated to be about 0.2 ergs cm^{-2} s^{-1}. This flux is only about 10%

of the total coronal energy budget, most of which goes into radiative losses and into conductive losses back to the photosphere.

The sun's magnetic field plays an important role in determining the solar angular momentum loss rate. In the earliest stage of solar evolution, the sun's rotation may have been braked significantly by the drag on magnetic field lines connecting like spokes from its interior out into the much larger and more nearly stationary cloud of partially ionized plasma from which it was born.

However, a magnetic star like the sun can continue to lose angular momentum throughout its lifetime even in the absence of a dense surrounding cloud to create an external drag, if it is losing mass into interstellar space. The solar wind plasma outflow along magnetic field lines will tend to conserve angular momentum, so that its angular velocity ω decreases with distance as r^{-2}. Since the field lines themselves tend to rotate as rigid spokes with constant ω, the deceleration in their frame of reference of the plasma frozen to them creates a torque acting to curve them backward.

The total rate of angular momentum transport outward by this magnetized plasma can be estimated from observable parameters of the plasma flow and magnetic geometry. The angular momentum loss rate per unit volume due to the plasma motion is

$$L_{\mathrm{p}} = \rho v_r v_\phi r, \tag{12-13}$$

where v_r and v_ϕ are the radial and azimuthal plasma velocity components.

The corresponding angular momentum loss rate due to magnetic stresses is

$$L_m = \frac{1}{4\pi} B_r B_\phi r. \tag{12-14}$$

Observations indicate that solar wind plasma outflow might be slightly nonradial by about $1°$, so that v_ϕ is nonzero. When expressions (12-13) and (12-14) are evaluated from observations, it is found that $L_{\mathrm{p}}/L_m \sim 3$ and $L_{\mathrm{p}} + L_m \sim 6 \times 10^3$ g cm^2 s^{-1} cm^{-2} s^{-1}. Assuming uniform conditions over the low solar latitudes, this angular momentum flux implies a torque sufficient to decelerate the sun substantially from its present rotation rate within its remaining main sequence lifetime.

MHD models of the solar wind outflow in the presence of a magnetic field exhibit similar properties to the gas-dynamic model described above in Section 12.3.2. One difference is their multiple critical points where the wind speed accelerates through, e.g., the Alfvén speed, as well as the sound speed that figures alone in the gas dynamic case.

Such models indicate that the total angular momentum flow per unit volume can be expressed as

$$L = \rho \omega R_{\mathrm{A}}^2, \tag{12-15}$$

where R_A is the heliocentric radius at which the flow becomes super-Alfvénic. This may be interpreted simply to mean that the total torque is approximately that expected if the plasma were to rotate with the solar angular velocity out to $r = R_A$ and then lose its ability to exert a torque on the sun thereafter. This is not unreasonable since for $r = R_A$ the energy density of the accelerating plasma comes to dominate the energy density of the field.

The value of R_A is uncertain because the observations required to evaluate the solar wind speed and Alfvén speed in the critical region between about $5R_\odot$ and $20R_\odot$ are particularly difficult. This region lies inward of any measurements made by the near-sun probe *Helios* yet on the outer edge of coronograph access. New data on velocities in this interesting region are now being obtained through Doppler shift measurements of the Lyman α line scattered from the neutral hydrogen component of the solar wind.

12.3.5 Sources of the Wind

The solar wind clearly arises from a variety of sources, and the identification of a clear link between phenomena in the low solar corona and solar wind features observed in interplanetary space has proved relatively difficult. Considerable progress has been made through careful comparison of solar wind data and photospheric and coronal phenomena. Nevertheless, the answer to the obvious question "where on the sun does most of the solar wind plasma come from?" is still not easy to answer.

A chain of events has now been firmly established between large flare or filament eruptions in the low corona, the subsequent generation of a coronal transient propagating through the outer corona, and the eventual development of an interplanetary disturbance, often of moderate shock strength, moving out to 1 AU and beyond. However, our understanding of the factors influencing the strength and propagation direction of individual interplanetary disturbances are still insufficient to produce reliable predictions of geomagnetic effects, even when good observations of solar eruptions are available. Nevertheless some useful insights into the dynamics of such shocks can be obtained from comparison of idealized theoretical models with the basic features of the observations. The fundamentally transient nature of the phenomenon requires that the time derivatives in the equations of momentum, mass, and energy conservation (4-25), (4-34), and (4-54) be retained. These time-dependent equations can be solved only for greatly simplified assumptions on the geometry and energy balance (purely adiabatic) during the expansion.

Perhaps the most useful feature of the solutions is the distinction between shocks caused by an instantaneous "blast" and "driven" shocks, in which energy is supplied over a considerable time by an outward-moving piston. In both cases the density (and also the temperature and flow speed) rise instantly at the arrival of the shock. For the driven shock, they continue to rise after the initial impulse passes, whereas they decrease after a blast-type shock.

Comparison of this theoretically predicted behavior with measurements indicates that shocks exhibiting driving (see, e.g., Fig. 12-9) seem to be most common, although shocks that more closely resemble blast-wave-generated disturbances also occur. Analysis of the numerical solutions indicates that driven shocks might be expected from coronal events that last more than about 10% of the shock wave transit time to the point of observation, while the blast type is expected from coronal events of duration shorter than about 1% of that transit time. Given the typical sun-Earth shock transit time of about 55 hr, this suggests that most shocks are produced by enhanced coronal expansion over much longer than the few tens of minutes associated with the flare explosion itself.

Interplanetary shock waves can lead to transient mass and energy fluxes from the corona that significantly exceed the global losses in the "background" solar wind over the few-hour duration of such an event. But averaged over time, only about 10% of the total solar wind mass and energy transport occurs during such events. The high-speed solar wind streams discussed earlier in this chapter also can produce considerable enhancements in both density and velocity relative to the time-averaged background wind. But as found for the shock waves, their time-averaged contribution to the total solar wind energy flux is an order of magnitude below that of the globally averaged wind of some 5×10^{26} ergs s^{-1}. Their time-averaged contribution to the mass flux seems even lower since the density increase in the shock is often followed by a rarefaction.

A simple explanation of the high velocities measured in the wind streams has not yet been provided. One promising explanation seemed to lie in the observation that the coronal hole magnetic fields diverge much more rapidly than radially over the first few solar radii. To understand the influence of such divergence, we can eliminate the density from the equation (12-8) for the solar wind expansion [using equation (12-7)] to obtain

$$\left(v - \frac{v_s^2}{v} \right) \frac{dv}{dr} = \frac{v_s^2}{A} \frac{dA}{dr} - \frac{dv_s^2}{dr} - \frac{GM_\odot}{r^2}. \qquad (12\text{-}16)$$

This form of the solar wind equation shows that the plasma acceleration dv/dr is determined in part by the outward increase of the cross-sectional area A of the wind flow as well as by the gradient of sound speed (thus temperature) and by the gravitational force. The acceleration is increased as A grows more rapidly with r, so the diverging magnetic field line geometry of coronal holes should tend to accelerate the wind more rapidly than a spherically symmetric expansion. More detailed study of this effect showed, however, that the diverging field lines also increase thermal conduction out into the wind, thus bleeding away heat that otherwise would be available for conversion to kinetic energy of the flow.

The considerable work that has been carried out on progressively more complete solar wind models indicates that the high speeds measured in the streams require at least that heating of the wind take place over an extended range in radius, perhaps beyond even r_c. Simple expansion from a point of localized heat input in the low corona seems to be insufficient. The additional acceleration achievable by direct wave pressure may also prove necessary. For instance, dissipation of the currents associated with Alfvén waves might heat the plasma, thus accelerating it through an increase in the gas pressure gradient. Direct acceleration might also be produced by the Alfvén wave pressure although the results of recent calculations argue against this.

The question still remains as to the sources of the "garden variety" background wind measured near the ecliptic plane. Some contribution probably comes from the polar cap coronal holes, which seem to be a permanent feature of the high-latitude corona. The field lines from these high-latitude holes eventually return toward the ecliptic to join the heliospheric current sheet. Also, although regions of closed coronal magnetic fields certainly inhibit solar wind flow, it is likely that some outflow is able to reach the interplanetary medium along locally generated current sheets that might be expected to exist at all latitudes. The coronal streamers prominent in eclipse photos may represent the high density flows predicted by models of how coronal wind expansion is expected to proceed over regions of closed field in the low corona (Fig. 12-15), and these may contribute significantly to the quiet wind also.

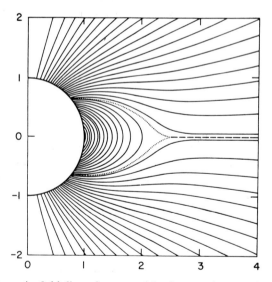

Fig. 12-15 Magnetic field lines for a model of coronal expansion from a dipolar surface field. The dotted curve indicates the current sheet extending away from the neutral point. (After G. Pneuman and R. Kopp, *Solar Phys.*, **18**, 258 (1971).

The present discrepancy of about a factor of 2 between wind speed measurements in streams and simple models may prove possible to explain by adjusting existing models through consideration of wave pressure, heat transport in the collisionless plasma, and other detailed effects. But real understanding of solar wind sources and acceleration processes may require detailed analysis of their links to the (also unsolved) problems of the acceleration of spicules and high-speed jets in the chromosphere and transition region and their contributions to coronal heating and mass balance.

ADDITIONAL READING

Observations of the Solar Wind and Heliosphere

W. Feldman, J. Asbridge, S. Bame, and J. Gosling, "Plasma and Magnetic Fields from the Sun," in *The Solar Output and Its Variation*, O. White (Ed.), Colorado University Press, p. 351, 1977.

A. Hundhausen, *Coronal Expansion and Solar Wind*, Springer-Verlag, 1972.

R. Marsden (Ed.), *The Sun and the Heliosphere in Three Dimensions*, D. Reidel, 1985.

Dynamics of the Solar Wind and Cosmic Rays

L. Fisk, "Solar Modulation of Galactic Cosmic Rays," in *The Ancient Sun*, R. Pepin, J. Eddy, and R. Merrill (Eds.), Pergamon, p. 103, 1980.

A. Hundhausen, *Coronal Expansion and Solar Wind*, Springer-Verlag, 1972.

R. Kopp, in *The Sun as a Star*, NASA Publication SP-450, S. Jordan (Ed.), p. 373 (1981).

E. Parker, *Interplanetary Dynamical Processes*, Interscience, 1963.

EXERCISES

1. Describe and draw the geometry of the heliosphere including the extension of solar magnetic fields, the current sheet, the direction of solar wind outflow, and the interplanetary magnetic field configuration in the ecliptic plane. Plot the latest positions of the *Pioneer* and *Voyager* spacecraft on this drawing, together with a few of the planets.

2. Explain the reasons for (a) the spiral geometry of the interplanetary magnetic field and (b) the slightly nonradial outflow direction of the solar wind. Integrate equation (12-10), and use the result [equation (12-11)] to evaluate ϕ at 1 AU, for reasonable v_r. Explain the interpretation of the angle ϕ and its relation to the angle between the interplanetary field vector component in the ecliptic plane and the sun–Earth radius vector.

3. Draw a picture of the warped heliospheric current sheet, and explain how it maps back onto photospheric structures. Explain also the reason for the magnetic sector structure observed by spacecraft. How many sectors would be observed if the solar field were simply a dipole with axis tilted relative to the solar rotation axis?

4. Describe the solar sources and phenomenology of (a) a high-speed wind stream and (b) an interplanetary shock.

5. Integrate the solar wind equation (12-9) and discuss the form of the solutions, including interpretation of the critical point.

6. Use expression (12-15) to estimate the sun's angular momentum flux density with a plasma density appropriate for the given range of R_A. Compare this value of L with that given in the text of Section 12.3.4. Estimate the sun's total angular momentum loss rate, and give the time scale for halving its total angular momentum.

13

The Sun,
Our Variable Star

The sun is by far the most practically important star in the sky, for its huge output of heat and light supports life on our planet. Its ultraviolet and X-ray emissions determine the structure of the Earth's upper atmosphere. Its outputs of plasma and magnetic fields mold the geomagnetic field and cause perturbations detected in RF transmissions and navigation. Solar energetic particle outputs, and also the modulation of galactic cosmic ray fluxes by the heliospheric fields over the solar cycle, determine the Earth's energetic particle environment.

This direct coupling between the sun and Earth has given rise to a well-developed discipline called solar-terrestrial research, in which the transmission of the sun's variable outputs is traced through the interplanetary medium, and their influence on the Earth's magnetosphere and upper atmosphere is studied in detail. As more objective techniques are brought to bear on the links that may exist between solar variability and the troposphere, the range of solar-terrestrial research will undoubtedly be expanded to include the possibly variable solar effects on climate.

Study of the sun has produced a large body of physical understanding that has been transferred to research on the properties of other stars. Stellar research has repaid this investment with unique opportunities to test models of nuclear energy generation in the solar interior by comparing the model predictions on solar evolution with observational results on large samples of stars. In more recent years observations of sunlike stars have revealed variations similar to the sun's magnetic activity cycle. Stellar observations from space in X-rays and in the ultraviolet have helped to constrain the processes that heat and accelerate plasmas in the sun's nonthermal atmospheric layers and in its outflowing wind. These developments have given rise to the fruitful new topic of solar-stellar studies, in which time-varying properties of solarlike stars are scrutinized to add to our understanding of our own variable star and its effects on the past and future of our planet.

13.1 THE SUN COMPARED TO OTHER STARS

13.1.1 The Sun's Location and Proper Motion in the Galaxy

The sun is located about 10 kiloparsec (1 parsec = 3.086×10^{13} km) from the center of the Milky Way galaxy, about halfway out in the galactic plane. As seen in Fig. 13-1, it is positioned at present on the inner edge of the Orion spiral arm of our galaxy. Such spiral arms are defined by local enhancements in density of gas, dust and young stars. The next closest is the Sagittarius arm, which contains the star clusters and nebulae so prominent in that constellation of the summer sky in the northern hemisphere.

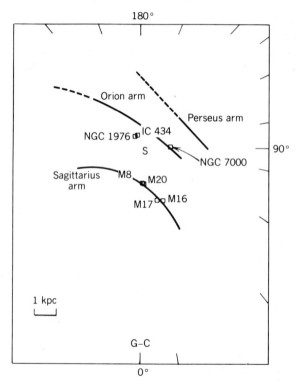

Fig. 13-1 Plot of the sun's position s in the galactic plane. The principal nearby spiral arms are shown, along with the locations of some well-known star clusters and galactic nebulae, such as the Orion nebula (NGC 1976) and the Lagoon nebula (M8) in Sagittarius.

The sun takes part in the general galactic rotation at about 250 km s^{-1} around the center of the Milky Way, and it completes a revolution in approximately 250 million years. Its motion relative to nearby stars is at about 20 km s^{-1}. This motion is directed in galactic coordinates (see Fig. 13-1) toward longitude 56° and latitude +23°, roughly along the axis of the Orion arm and somewhat northward out of the plane of the galaxy.

Forty-five stars are known to be located within 5 parsecs (16.3 light years) of the sun. The closest is the α-Centauri system, located 4.3 light years away, whose most luminous component is a star very similar to the sun. By far the most luminous stars in the solar neighborhood are Sirius, the brightest star in the night sky; Altair; and Procyon. Both Sirius and Altar are very hot, relatively young stars, while Procyon is closer to the solar temperature.

13.1.2 Mass, Radius, and Chemical Composition

The sun's basic global properties, such as its mass, radius, chemical composition, and luminosity, are unexceptional. This average quality increases the likelihood that comparative studies of the detailed physics of the sun and most other stars might be fruitful.

The sun's mass, $M_\odot = (1.9889 \pm 0.0003) \times 10^{33}$ g, places it near the lower end of the range, which extends from about $0.1 M_\odot$ for the coolest dwarf stars to over $100 M_\odot$ for the hottest and most massive supergiants. Its value can be obtained from Kepler's third law

$$\frac{P^2}{a^3} = \frac{4\pi^2}{G(m + M_\odot)}, \tag{13-1}$$

where a and P are respectively the semimajor axis of the Earth's orbit and its orbital period. The quantity m is the Earth's mass, which has been determined to relatively high accuracy from the motions of artificial satellites. The uncertainty in $M_\odot$ is determined by the error in G, the gravitational constant, which is known only to four significant figures.

The chemical composition of the sun is discussed in Section 5.6. To within measurement errors of typically a factor of 2, the solar abundances agree with those found for the approximately 90% of all stars that are considered "normal" on the grounds that they fit within the standard scheme of spectral classification discussed below. A pronounced deficiency of elements heavier than helium (called metals by astronomers) in some stars is linked to their much greater age than the normal stars of the galactic disk, such as the sun. Other factors are thought to affect the surface abundances rather than the star's overall chemical composition. The best known are the strong magnetic fields of so-called peculiar A stars (see below), which appear to inhibit convective mixing of elements in these stars. The finding that the solar abundances agree well with those of the interstellar medium, and of most stars, further supports the view that the physical processes of stellar physics that are sensitive to the plasma composition, such as nuclear burning, can be usefully generalized.

The sun's absorption line spectrum discussed in Chapter 3 is classified as G2V according to the criteria of the Morgan-Kennan (MK) scheme. In this scheme stars are categorized according to the two parameters temperature and luminosity, the former determined from the relative strengths of certain spectral lines. The main temperature classes are denoted (in decreasing temperature) by the letters O, B, A, F, G, K, and M, with the numbers serving to interpolate between the letters, so G2 is roughly 20% of the way between a G0 star and a K0 star. For a given effective temperature, a star's luminosity is determined by its radius. As discussed later, the luminosity classes are denoted by the Roman numerals I–VI. They characterize stars called respectively supergiants, bright giants, giants, subgiants, main sequence and dwarfs together, and finally subdwarfs and white dwarfs together.

The spectral criteria used in the classification scheme center on the relative strengths of the hydrogen and helium lines to those of the metals. In the hottest stars of types O and B, lines of ionized and neutral helium, both of which require large excitation energies, are strong. In A stars the hydrogen lines are at their maximum strength. In F stars metallic lines such as the H and K of Ca II are getting stronger, and in K stars they are at maximum strength. The coolest M stars have strong molecular bands such as TiO. A G-type spectrum such as that of the sun is defined by the relative intensity of certain metallic and hydrogen lines, such as Ca λ 4227 relative to Hδ, and Fe λ 4325 relative to Hγ.

13.1.3 Luminosity, Effective Temperature, and Spectrum

The sun's luminosity, $L_\odot = (3.84 \pm 0.04) \times 10^{33}$ ergs s^{-1}, is more than a thousand times greater than that of the dimmest dwarf stars, but almost a million times lower than the prodigious output of the hottest supergiants. The value of $L_\odot$ is obtained from the total solar irradiance S measured above the Earth's atmosphere at 1 AU. Assuming that the sun's global radiation field is isotropic, the relation is

$$L_\odot = 4\pi R_\oplus^2 S, \tag{13-2}$$

where $R_\oplus$ is the mean radius (or equivalently, the semimajor axis) of the Earth's orbit. The main uncertainties in $L_\odot$ arise from error in the absolute radiometry of S, which is accurate to no better than 0.2%, and the dependence of S upon solar activity (see Section 13.3.3).

The sun's radius, $R_\odot = (6.959 \pm 0.0007) \times 10^{10}$ cm, places it at the lower end of the stellar size range, nearer to the smallest dwarf stars of radius about $0.1R_\odot$ than to the largest supergiants, whose radii of about $10^3 R_\odot$ would extend almost to the orbit of Jupiter. This value of $R_\odot$ is obtained from the sun's angular size, which is $959\overset{''}{.}63$ at the mean Earth–sun distance.

The Earth–sun distance can be obtained from the solar parallax defined as

$$\pi_\odot = r_\oplus/a, \qquad (13\text{-}3)$$

where $r_\oplus = 6.378160 \times 10^8$ cm is the Earth's radius and a is the semimajor diameter of its orbit, the so-called astronomical unit (AU). The best determination of a is derived from radar measurements of distances to other planets and asteroids, which provide the absolute scale of the solar system. The value of $a = 1.49597892 \times 10^{13}$ cm then follows from Kepler's third law [equation (13-1)]. The uncertainty in $R_\odot$ is thus chiefly due to error in measuring the location of the solar limb to determine the sun's angular size.

The sun's effective temperature $T_{\text{eff}} = 5776$ K $\pm$ 15 K is obtained from values of $R_\odot$ and $L_\odot$ using the defining relation (2-38)

$$\sigma T_{\text{eff}}^4 = L_\odot/4\pi R_\odot^2. \qquad (13\text{-}4)$$

The solar effective temperature is about twice the value determined for the coolest dwarfs, but much lower than the values of $T_{\text{eff}} = 35,000$ K found for the bluest and hottest supergiants or the even higher temperatures of certain highly evolved white dwarf stars.

13.1.4 Chromospheric and Coronal Radiations

The solar outputs in the X-ray, far ultraviolet, and radio regions of the spectrum originate mainly in the sun's nonthermally heated outer atmosphere. In the soft X-ray region between about 3 Å and 40 Å, the total flux of the quiet sun is approximately 5×10^{27} ergs s^{-1}. This is within the range of 10^{27}–10^{30} ergs s^{-1} found for other F and G dwarf stars on the main sequence. It is far below the output of the most powerful steady stellar X-ray emitters, the hottest O and B stars, whose X-ray luminosities arc between 10^{32} and 10^{34} ergs s^{-1}.

The solar continuum output drops rapidly from the blue into the ultraviolet (Fig. 3-1), and it varies greatly from line to line in the EUV (Fig. 3-3), where quiet sun surface fluxes are of order 100 ergs cm^{-2} s^{-1} around 1000 Å. The sun's UV emissions are stronger by about a factor of 100 than those of the weakest detectable late-type UV-emitting stars, the M giants. They are comparable to those of other G stars, but up to two orders of magnitude weaker in the hottest coronal lines than the strongest emitters, which are close binary stars. The spectra of the stronger UV emitters resemble the spectra of solar plage regions. As discussed in Section 13.3.2, total luminosity and effective temperature seem to play little role in determining a cool star's UV emissions.

Microwave observations of nearby main sequence stars of late spectral type have also revealed slowly varying emissions from M dwarf flare stars and certain binary stars. The emission at 6 cm is similar to the microwave emission from flaring solar active regions, although the time scales of the flaring tend to be longer and the dimensions of the flaring regions must be much larger. The mechanism seems to be gyrosynchrotron radiation, as in the sun's microwave bursts.

13.1.5 Stellar Winds and Mass Loss

The measurements of the solar wind discussed in Chapter 12 describe a flow at an average speed of roughly 470 km s^{-1} with an average particle density of about 10 protons cm^{-3}, yielding a mean proton flux of about 4×10^8 cm^{-2} s^{-1}. If this flow is assumed to be isotropic (solar wind data have so far been obtained almost entirely from the ecliptic plane, so we do not know how good an assumption this is), we can estimate a solar mass loss rate of roughly 2×10^{12} g s^{-1}, or 2×10^{-14} $M_\odot$ per year. At this rate, the sun would have lost less than 0.01% of its mass in its lifetime—an insignificant amount by evolutionary standards. As discussed later, it is likely that this rate was much higher for the young sun, so mass loss may have played a more significant role in its evolution than the present value implies.

Observations of absorption lines, such as Ca H and K and Na I D, originating in circumstellar shells, indicate evidence for mass outflows from cool giant and supergiant stars at speeds between 5 and 100 km s^{-1}. Wind speeds as high as 600–3500 km s^{-1} have been observed on the much more luminous and hot stars of early spectral type.

The mass loss rates for cool stars are highly uncertain since they depend not only on the outflow speed (which can be estimated relatively well) but also on the geometry and density of the flow. Major uncertainties arise in relating the outflow speeds measured in a few species to the total mass flux present in all atomic ionization states and also in the abundant molecules and dust observed in infrared and microwave studies of these same circumstellar shells. Mass loss rates for cool supergiants, such as Antares, range between 10^{-6} and 10^{-7} $M_\odot$ year^{-1}, at least 10 million times greater than the sun's present mass loss rate.

13.1.6 Angular Momentum and Magnetism

The Doppler measurement techniques used to measure the sun's photospheric rotation rate to an accuracy of a few percent have marginal sensitivity for study of stellar rotation (by line broadening, rather than line shifting observed on the sun's resolved disk) much below 5 km s^{-1}. However, the rotation of plages across the disks of magnetically active stars similar to the sun (see Section 13.3) modulates their chromospheric light output in lines such as Ca K and thus enables accurate tracer measurements of rotation rates as slow as that of the sun. Such studies indicate that the sun's rotation is about typical, compared to other stars of similar age and mass.

As discussed in Section 6.5.5, the stellar angular momentum per unit mass, J, can be estimated on the assumption of rigid rotation at the mean surface rate. Figure 6-6 illustrates that the sun's J-value lies about a factor of 5 below that of more massive stars of similar age. The reason for this markedly lower J-value is not known. It is not likely to be found in the (large) angular momentum of the sun's planetary system, unless we are to believe that the solar system is unique.

When the sun's kilogauss photospheric magnetic fields are observed with a Babcock magnetograph at progressively lower spatial resolution, the mean field measured declines rapidly. When the sun's disk-integrated light is placed on the magnetograph slit, the net field measured is about 0.2 G around activity minimum and about a factor of 10 higher around maximum activity. This mean field represents the resultant of the active regions and large-scale quiet-sun polarity structures, weighted to those near the solar disk center. The net field is observed to change from day to day as different large-scale structures dominate the signal (Section 8.4.3).

This low value of net magnetic field is mainly the result of the fragmented structure of solar field patterns that reverse polarity on small scales and occupy in total less than a few percent of the disk area. The peak fields measured at the photosphere appear to be of respectable magnitude compared with the strongest fields measured directly in any other stars. Even in the "peculiar" A stars, whose photospheres seem to be entirely permeated by an organized poloidal field, typical intensities are a few thousand gauss, with a highest detected value of 34 kG.

The net fields below 1 G expected in stars similar to the sun are too weak to be detected in reasonable integration times with the Babcock magnetograph, given the much lower light level. A more sensitive technique for this application is to measure the total Zeeman broadening of magnetically sensitive lines. This avoids the cancellation of the Zeeman signals produced in a Babcock magnetograph by opposite magnetic polarities within the field of view (Section 1.5). The most convincing measurements of this kind have been carried out in the infrared, where the Zeeman effect is much larger than in the visible region and the small Zeeman broadening can be distinguished with some confidence from other influences on the line width.

Measurements on some late-type stars using this technique with lines formed in plages lead to field estimates of between 1900 and 2500 G, thus comparable to solar plage fields. The relative equivalent width of the shifted and unshifted components yields the fractional area covered by magnetic and nonmagnetic photosphere. This area coverage appears to be as high as 10–45% in the stars studied, thus much higher than the solar value of a few percent (Section 8.4.2). As noted above, this higher area coverage in many late-type stars is supported by the stronger chromospheric emissions detected in lines such as Ca K.

13.2   EVOLUTION OF THE SUN

13.2.1   The H-R Diagram and Stellar Evolution

Of the sun's global properties described in Section 13.1, only its mass and initial chemical composition seem to determine its evolution. This remarkable result is based on observations of how stars are distributed in a plot of luminosity against surface temperature. The underlying physical principles that govern this evolution are provided by the theory of nuclear energy generation in stars, whose application to the sun is discussed in Chapter 6.

Observationally, it has been known since the studies of H. Russell in 1913 that in a plot of luminosity against temperature most stars fall along a relatively narrow strip called the main sequence. Such a plot, known as a Hertzsprung-Russell diagram, is shown in Fig. 13-2 for stars in the sun's neighborhood.

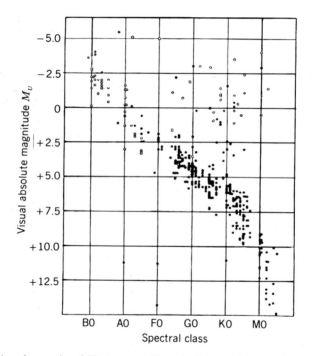

Fig. 13-2   An observational Hertzsprung-Russell diagram for stars in the solar neighborhood. The ordinate is the visual absolute magnitude, a logarithmic measure of a star's luminosity, with the stars at the top about a factor of 10^8 more luminous than those at the bottom. The abscissa is the spectral class, a measure of the stellar temperature, with the hottest stars at the left. The sun, of spectral class G2, has a visual absolute magnitude $M_v = 4.8$. Reproduced with permission, from A. Unsold, *The New Cosmos*, Springer-Verlag, N.Y., 1967.

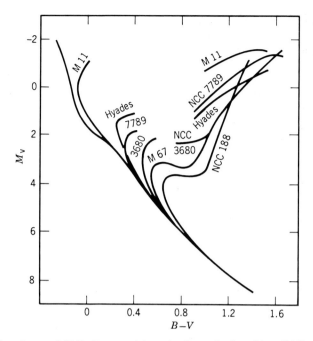

Fig. 13-3 An observed H-R diagram (plotted schematically with solid lines instead of the individual stars seen in Fig. 13-2) for open clusters of various ages. The abscissa plotted here is the *B-V* color index, a measure of the stellar temperature which decreases from left to right over about the same temperature range as covered by the spectral classes plotted in Fig. 13-2. By permission of A. Sandage.

We see that the main sequence is a curved strip running diagonally across the H-R diagram from upper left (the hottest and most luminous O, B, and A stars) to lower right (the coolest and dimmest M stars). The relatively few stars found in the upper right of the diagram must be much larger than main sequence stars since their luminosity is so much higher for a given temperature. These are called giants; the hottest and very largest are the supergiants found along the upper boundary of the diagram. The stars of more common size found along the main sequence, such as the sun, are called dwarfs. Even smaller stars, called subdwarfs, fall to the lower left of the H-R diagram. The smallest stars, known as white dwarfs, occur some ten magnitudes below the main sequence.

Observational studies of the H-R diagrams of star clusters have been successfully interpreted using calculations of a star's evolution on this diagram as it burns its nuclear fuel. From such studies have emerged our ideas on the sun's past and future evolution. Figure 13-3 shows the H-R diagrams of galactic clusters, gravitationally bound groupings of stars of similar age and distance from the sun. This plot shows that stars of spectral type later (i.e., cooler and dimmer) than a certain turnoff value follow the main sequence

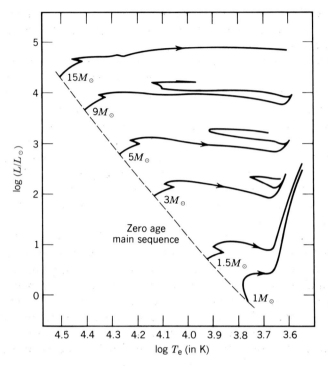

Fig. 13-4   Theoretical evolutionary tracks calculated for stars of different masses. By permission of I. Iben.

closely. Stars brighter than this turnoff are no longer found on the main sequence, but are mainly located to the upper right of it, in the red giant region.

The reason for this turn off the main sequence has been made clear by theoretical studies of the evolution of stars of different masses. Figure 13-4 shows the development of stars of masses between $1M_\odot$ and $15M_\odot$. Their evolutionary tracks after the onset of nuclear burning take them to the right of the main sequence into the red giant region, as the star expands and becomes more luminous due to changes in its internal energy generation from hydrogen to helium.

The time elapsed between onset of hydrogen burning and substantial movement off the main sequence decreases dramatically for stars of increasing mass. These stars lie higher on the main sequence, which defines a progression toward more massive stars from right to left. The more massive stars evolve off the main sequence more rapidly because their luminosity per unit mass is much higher than for less massive stars like the sun [see equation (6-37)]. Therefore, they consume their nuclear fuel much faster. For the most massive and luminous O stars the time scales associated with the evolutionary tracks illustrated in Fig. 13-4 are almost 1000 times shorter than for the sun, whose track is shown at the extreme lower right.

The decreasing evolutionary time scales upward along the main sequence suggest that the location of the main sequence turnoff found is determined by the age of the cluster. Thus the galactic clusters shown in Fig. 13-3 increase in age from about 2×10^8 years for M11 to over 5×19^9 years for M67 and 10^{10} years for NGC 188. Taken together, the stellar evolution computations and the observational data indicate that a star of known chemical composition and mass, such as the sun, can be dated to fair accuracy by comparing its position on the H-R diagram with the predicted evolutionary path that it should follow. Moreover, its past and future behavior in luminosity, radius, and temperature can also be determined reasonably well, at least over time scales of many hundreds of millions of years.

Several tests can be made of this theory of stellar evolution. For one, the observed dependences of stellar luminosity on effective temperature and on mass, and also the relation between stellar radius and mass, can be explained using straightforward dimensional arguments and values of the nuclear energy generation rates and opacities calculated from detailed stellar models (see Chapter 6 exercises). Furthermore, the good agreement between the fairly complex observed distribution of cluster stars of similar age in the H-R diagram and their theoretical distribution predicted from only their mass and initial chemical composition is also encouraging.

An independent test is provided by the theoretical prediction that massive stars which have evolved to the point of helium burning in their core should modify their internal structure in a way that makes them unstable to internal pulsations observable as light-output variations. The theory makes quite definite predictions on the location of such Cepheid variables in the H-R diagram. It even predicts the relation between their pulsational period and mass, and thus their detailed evolutionary track. Observations confirm that the Cepheids lie in the relatively narrow strip of the H-R diagram between $L/L_\odot$ of about 100 and 10^4, and at an effective temperature only slightly higher than that of the sun.

The theory of stellar evolution also successfully predicts the behavior of the stellar abundance of the light element lithium in the H-R diagram. According to the arguments outlined in Section 6.5.2, stars in which convection is absent should have a relatively higher surface abundance of this element than stars with convection zones that circulate photospheric material to depths at which temperatures exceeding $2-3 \times 10^6$ K are expected to occur. Stellar structure theory predicts that main sequence stars somewhat more massive than the sun should be in radiative equilibrium. Such stars only develop significant convection zones after they evolve off the main sequence. Observations confirm that stellar lithium abundances are relatively high in the more massive main sequence stars and decrease off the main sequence for these stars.

The only important discrepancy discovered so far between stellar evolution theory and observations is in the low measured solar neutrino flux as compared to solar interior model predictions. The nature of that discrepancy and some possible explanations have been given in Section 6.5.1.

13.2.2 The Sun's Future Evolution

The theory outlined above makes relatively well tested predictions for the behavior of stars (at least for those more massive than the sun) after they begin to burn hydrogen on the main sequence. The rate of evolution for stars less massive than the sun is so slow that in order for such stars to have moved significantly off the main sequence, the cluster age must be greater than 10^{10} years. Very few nearby clusters are so old.

We begin with a discussion of our future, since the picture of where the sun is headed in the next two to three billion years seems to be better understood presently than how it was born and arrived on the main sequence. As the sun gradually burns the hydrogen in its energy-generating core into helium, its mean molecular weight increases and its luminosity rises along the path for a star of $1M_\odot$ in Fig. 13-4. When the sun's core hydrogen has been exhausted about 9×10^9 years after its arrival on the main sequence, the core will contract to maintain the pressure required to support the outer layers by drawing on its gravitational potential energy. The sun's estimated age of 4.6×10^9 years places it roughly halfway along its main sequence lifespan since its pre–main sequence evolution was very rapid. Contraction of the core will eventually be halted when the density reaches such high values that the matter becomes degenerate and further condensation is prevented by the quantum exclusion principle for electron energy states.

The core will then gradually grow while hydrogen is burned in a shell that moves outward. During this outward movement, which is expected to last about 10^9 years, the sun will slowly expand while maintaining roughly constant luminosity at decreasing temperature, thus moving to the right in the H-R diagram. Eventually, a deep convection zone will develop, encompassing half of the stellar mass. In this stage, the sun is expected to expand enormously (by about a factor of 30 or more from its present radius) within some 10^9 years and become a highly luminous red giant.

Predictions of the sun's final state rely on uncertain ideas regarding mass loss in the red giant phase and its effects on further evolution. It appears most likely that a star like the sun will eventually use up its hydrogen fuel and contract back until the internal temperatures ignite helium burning. A phase similar to that which occurs during hydrogen shell burning would follow, but the accompanying expansion would now be sufficient to entirely eject the sun's outer envelopes into a huge circumstellar shell observable as a planetary nebula.

The sun's core left exposed by this ejection would slowly cool and become a white dwarf, a tiny star of radius comparable to that of the Earth, but containing a substantial part of $1M_\odot$. The densities of the degenerate matter in such stars are between 100,000 and 1,000,000 times higher than the 150 g cm^{-3} value presently encountered in the sun's core.

Needless to say, the scenario for future solar evolution outlined above is not hospitable to life on Earth as we know it. Even the steady increase of $L_\odot$ by about the factor of 2 expected over the next $4-5 \times 10^9$ years as the sun moves off the main sequence will lead to a very significant warming of climate. This scenario would put the Earth in a radiation environment similar to that now encountered on Venus, whose surface temperature is about 700 K with a pressure of around 100 atmospheres! On the other hand, the same model predicts that the solar luminosity should have already increased by about 40% in the past 4.6 billion years, but the Earth's climate seems to have changed relatively little (Section 6.5.4).

In any case, the enormous expansion and brightening of the sun toward the red giant phase would incinerate the Earth, which would then find itself orbiting a star covering a substantial fraction of the sky and emitting a radiative flux almost 1000 times more intense than the present value.

Solar changes on much more imminent time scales are likely to challenge our ability to survive on this planet. A considerable amount has been learned about the likely future behavior of the sun's magnetic activity level. Discussion of this material is left until Section 13.3, where factors governing the magnetism and activity of sunlike stars are laid out in more detail, along with ideas on the sun's past and future outputs of X-rays, ultraviolet radiation, charged particles, and fields.

13.2.3 The Early Sun

The circumstances surrounding the sun's birth and relatively rapid evolution onto the main sequence are less clear than the later, hydrogen-burning phases that occupy most of its life described above. The main difficulty is that although the sun's present state provides good initial conditions to study its future, our ideas on how it reached the main sequence are quite sensitive to assumed initial conditions at its birth. The most convincing description relies on a combination of theoretical modeling and observations of stars of roughly solar mass that are thought to be very young.

The sun probably originated in the gravitational collapse of a huge, tenuous interstellar gas cloud. This collapse may have been triggered by the compression of a nearby supernova explosion, by a collision between two clouds, by a shock wave propagating with the Galaxy's spiral arm, or by thermal instability of the interstellar medium.

Whatever the trigger was, if a cloud of gas of given mass is compressed below a critical radius, its negative gravitational potential energy outweighs its internal thermal energy, and the cloud will continue to contract inexorably. This critical radius can be estimated as follows. The thermal energy of the cloud of density ρ, temperature T, radius R_s, and molecular weight μ is

$$E_t = 2\pi R_s^3 \left(\frac{\rho RT}{\mu} \right),$$
(13-5)

whereas the gravitational potential energy is

$$E_g = -\frac{3}{5} \frac{GM^2}{R_s} = -16G\rho^2 R_s^5.$$
(13-6)

Comparison of equations (13-5) and (13-6) shows that the ratio E_g/E_t increases as R_s^2, so that for given values of ρ and T found in interstellar clouds a cloud larger than a critical radius must contain enough mass to have negative total energy. Alternatively, if we specify the mass M of the cloud, then the critical radius is determined by

$$R_c = \frac{\mu GM}{RT} \sim \frac{0.2}{T} \frac{M}{M_\odot} \text{pc},$$
(13-7)

and the cloud will only contract if its radius is below R_c.

These conditions for contraction are based on a simple model which assumes that additional contributions to the internal energy from magnetic fields, rotation, and internal turbulence can be neglected. Nevertheless, even without these factors, which all act to impede contraction, we find from equation (13-7) that only very large dust clouds containing thousands of $M_\odot$ within a radius of tens of parsecs will contract (see exercises). It seems that the initial collapse may have occurred in several stages. First, a large gas and dust complex could begin to contract, triggering contraction within much smaller clouds with mass comparable to the sun's after a much higher density was reached.

Initial collapse of the cloud's inner core would have been rapid, almost a free fall that might have required no more than 1000 years to reduce its diameter from a light year or more to a few tens of present solar radii. Eventually the internal pressure builds up sufficiently to halt the free fall, as half the released gravitational energy goes into heating the star (see Chapter 6 exercises). The star is then in hydrostatic equilibrium and continues to contract much more slowly as it radiates away its gravitational energy.

The details of the sun's hydrostatic contraction from somewhere in the upper right of the main sequence down onto it, where hydrogen is ignited and the contraction is halted, are still controversial. In the early stages of this contraction phase, the proto-star is expected to have been relatively cool because, with its large surface area, the effective temperature required to radiate away one half the gravitational energy released is relatively low. It is also likely that during this early contraction the sun was fully convective since

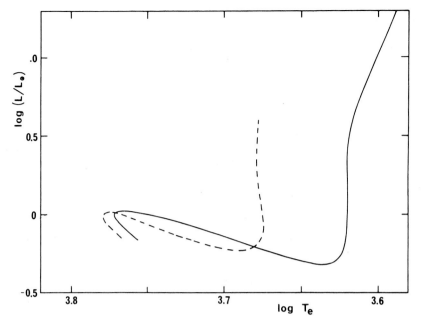

Fig. 13-5 Pre–main sequence evolution of two models of the early sun. The dashed and solid curves show the effects of different assumed chemical composition. Both curves terminate on the main sequence. By permission of I. Iben.

the opacity of the cool gas was too high to enable radiation to carry the heat efficiently.

Fully convective models of the proto-sun's time behavior were first calculated by C. Hayashi in 1961. A more recent computation of H-R diagram paths for such a proto-sun is shown in Fig. 13-5. The rate of evolution is determined by the rate at which radiation can remove energy from the proto-stellar photosphere since the efficiency of convection is extremely high. The sun would then have decreased its luminosity by roughly a factor of 10 to about its present value by contracting at roughly constant temperature over about 10^7 years. This so-called Hayashi phase would have taken the sun almost vertically downward in the H-R diagram.

Eventually the heating of the sun's interior would have decreased the opacity to the point where energy transport by radiation takes over. Now the proto-sun should have evolved almost horizontally to the left in the H-R diagram, as the photosphere heats up to compensate for its shrinking surface area. The rate of evolution over a few tens of millions of years in this radiative cooling phase is determined mainly by the efficiency of the internal radiative transport.

Finally, nuclear burning is initiated. First, for a brief period C^{12} burns, then the main hydrogen fuel ignites. The final readjustment of the star's internal temperature structure from heating by gravitational energy release throughout the interior to nuclear energy release (only in the hottest core) accounts for the last loop in Fig. 13-5. This lowers the luminosity and effective temperature before the sun settles onto the main sequence.

The main questions about this picture of solar pre–main sequence evolution arise from uncertainties in the initial conditions of chemical composition, rotation, geometry, and magnetic fields in the original gas cloud. These matters cannot be resolved by modeling alone, and comparison with observations of stars in the expected region of the H-R diagram has been very helpful.

Identification of the objects in the sky that might correspond to the proto-stellar contracting clouds and to the pre–main sequence stars fueled by gravitational collapse, is less straightforward than the identification of the much older objects that correspond to hydrogen burning on the main sequence and to evolution off it. The clear, relatively dramatic signposts predicted by the theory of post–main sequence evolution such as the red giant phase, the Cepheid pulsations, the expulsion of planetary nebulae, white dwarfs, and supernovas are less clear for the pre–main sequence objects. In part this is due to uncertainties in the predictions made by the difficulty theory of highly dynamical objects. But it is also due to the complex phenomenology of stars often associated with obscuring atmospheres and envelopes of gas and dust, probably associated with their evolution.

The most likely class of stars to be associated with the sun's early evolution are those similar to the prototype star T-Tauri. One reason for making this association is the location of T-Tauri stars in large and dense gas clouds rich in molecules and dust, which are sometimes illuminated by very bright young supergiant stars giving rise to diffuse nebulae such as the object called the Lagoon shown in Fig. 13-6. In part it is also based on their location in the H-R diagram, where they are found to the right of the main sequence, in the region expected for stars of mass between $0.1 M_\odot$ and $1 M_\odot$ that have spent between 10^5 years and 10^6 years in the Hayashi phase of hydrostatic contraction.

A relatively high lithium abundance is one parameter that differentiates the T-Tauri stars from much older post–main sequence red giants that lie in the same region of the H-R diagram. This would be expected in very young stars whose internal nuclear reactions have not yet had a chance to deplete lithium. Also, observations in UV, X-rays, and microwave radiations confirm optical and IR observations that the T-Tauri stars have strong chromospheric emission and highly varying and anisotropic extended outer envelopes which appear to be associated with strong outflowing winds.

T-Tauri stars are also found to be in rapid rotation at up to ten times the sun's rate. For reasons discussed in more detail in Section 13.3, chromospheric emission, rotational speed, and magnetic activity are found to decay with time over much of a star's life. This also suggests that T-Tauri stars are young objects. If we accept this evidence, the detailed observations of T-Tauri stars

Fig. 13-6 Photograph of the Lagoon nebula, M8, located in the constellation Sagittarius. The intricate dark structures are caused mainly by opaque dust clouds associated with the nebula. These often intervene along the line of sight between us and the glowing interstellar gas. The nebula is roughly 35 pc across, and the mass of the ionized gas alone is approximately $3000 M_\odot$. Lick Observatory photograph.

and their associated phenomena are at least consistent with the following general scenario for early evolution of the sun and its planetary system.

We can visualize the proto-sun as collapsing first in essentially free fall from a dense gas cloud within about 1000 years. This applies to at least the densest central parts—the rest of the sun's mass may well have accumulated more slowly by accretion onto this core. Since angular momentum will tend to be conserved, a disklike extension of the central condensation may well have formed in the plane of rotation, and material accreting onto the disk might then have slowly found its way inward as angular momentum was redistributed in the disk by turbulence and by the braking effect of magnetic field lines dragging through surrounding gas.

Meanwhile, the most condensed central proto-sun within this solar nebula would have entered the Hayashi phase of hydrostatic contraction. Its radius would be some $2–5 R_\odot$, and its rotation speed at the photosphere could be at least ten times higher than the sun's present rate. Given its high rotation rate and fully convective interior, dynamo activity should be strong, and the proto-sun's magnetism and activity levels would be much higher than present values. Both the total light output and its spectrum would vary on all time scales due to large spots and plages.

Every few thousand years, huge bursts of light lasting several years and increasing the star's luminosity by a factor of 100 or more might occur, as observed in the T-Tauri star FU Orionis. These could be caused by instabilities in the accretion disk whirling around the proto-sun and causing powerful and rapid heating of the photosphere. This proto-sun would be a powerful emitter of UV and X-rays from a deep chromosphere, with surface fluxes a thousand times higher than at present. When observed from a great distance, it would also exhibit strong IR emission since the partial absorption of this harder radiation in the surrounding gas and dust remaining after the free fall should lead to powerful re-emission of IR, particularly in the continuum from heated dust.

The strong magnetic activity level and the heating of the photosphere by frictional interaction with the more rapidly rotating accretion disk would tend to produce a hot corona, as observed in X-rays and EUV. This corona could not be gravitationally contained, and its outward pressure would eventually overcome the inflow of material, at least over some parts of the proto-sun's photosphere. The result might be a complicated inflow-outflow pattern which could be highly variable. Typical outflow speeds might be well below 100 km s^{-1}, but they could reach 300 km s^{-1} and mass loss rates as high as $10^{-5} M_\odot \text{ year}^{-1}$ in short-lived events.

As the sun aged for some 10^7 years on the Hayashi track and its interior came into radiative equilibrium, its rotation would have been progressively braked by the wind. The planets would have formed progressively from the accretion disk, and both the mass loss rate and also the X-ray and UV outputs would have gradually decreased as the star evolved onto the main sequence.

13.3 SOLAR AND STELLAR VARIABILITY

13.3.1 Observations of Stellar Activity

In the past 15 years the main features of solar activity have been detected in other stars and have been calibrated against observations of the sun as a star, in disk-integrated solar light. Probably the most striking illustration of activity in other stars came from the detection of cycles in chromospheric emission similar to the sun's 11-year cycle. Figure 13-7 illustrates the three basic types of behavior seen after 20 years of monitoring about 100 F, G, and K stars. Roughly 40% of the selected stars exhibit cycles whose lengths vary between about 7 years and the maximum of about 15 years that might be discovered with a data base of this length. The cycle amplitudes range between about 10% (roughly the amplitude of the sun's cycle observed using the same measure of

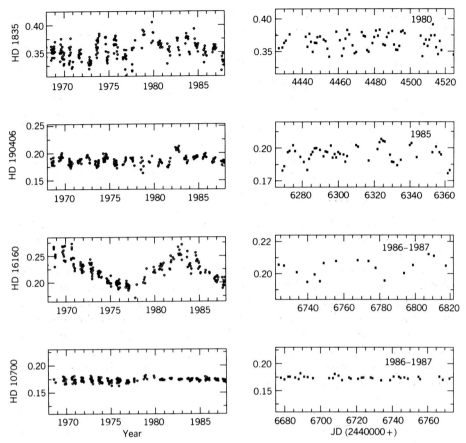

Fig. 13-7   Time variations in Ca II H- and K-line emissions for four dwarf stars. The left panels show variations observed between about 1968 and 1988. The first panel from the top shows the behavior of a very variable star with no clear period that rotates in 7.8 days. It represents the behavior expected of the young sun of age 10^9 years. The next two panels show stars with activity cycles of 2.6 years and 13 years. The right panels show the variations of the same stars on an expanded time scale with the abscissa now in days. The second and third stars from the top exhibit rotation periods of 14 and 45 days. The last star at the bottom exhibits no chromospheric variations on either rotational or cyclic time scales above the measurement errors. Data from the Mt. Wilson HK Project. By permission of S. Baliunas.

disk-integrated Ca K) and about 35%. About 20% of the stars show no variation, while the remaining 40% show relatively short-period variations of large enough amplitude that any underlying cycle would be obscured. The irregularly varying stars tend to be those with large Ca K fluxes.

When the Ca K fluxes of all the stars studied are plotted against an index of their temperature (Fig. 13-8), we see that the sun is among the weakest Ca K

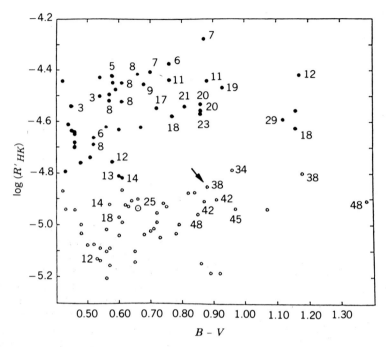

Fig. 13-8   A plot of Ca II H and K emission (expressed as a logarithmic fraction of total luminosity) for stars of various spectral types with temperature increasing from left to right. Numbers next to each point give the stellar rotation period in days. Filled and open circles represent stars above and below the Vaughan-Preston gap discussed in the text. From R. Noyes et al., *Astrophys. J.*, **279**, 763 (1984).

emitters of the sample. As discussed below, the strength of Ca K emission in these late-type stars is correlated with age, so the tendency for the weaker Ca K emitters such as the sun to have cycles also suggests that young, highly active stars tend to exhibit more irregular activity variations, while more regular cycles appear with increasing stellar age.

When the Ca K variability of these same stars is studied with better time resolution, a rotational modulation of typically 5–10% can be discerned, with periods of characteristically 20–40 days (right-hand side of Fig. 13-7). This modulation is produced by rotation of stellar plages across the star's disk. As mentioned above, it provides a more sensitive determination of the rotation rate for stars rotating at roughly the solar rate, which is barely detectable spectroscopically in disk-integrated light.

The physical properties of the plages that presumably give rise to most of the Ca K variations on both rotational and activity-cycle time scales are best studied in their ultraviolet spectra. Relatively little can be derived about the plage atmospheres in main sequence dwarfs similar to the sun, except that their chromospheric, transition, and coronal emissions exhibit similar relative

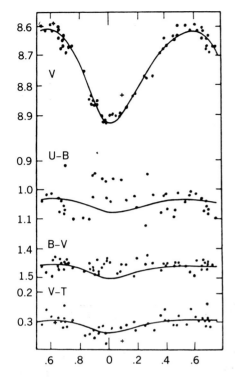

Fig. 13-9 Plot of time changes in visible light (top plot) and color (lower plots) of the spotted star AU Mic. The ordinate is given in magnitudes, a logarithmic intensity scale used by astronomers. A decrease of 0.1 magnitudes represents an intensity increase of about 10%. Variations in the three photometric color indices in the lower plots show that the star gets significantly redder at its minimum brightness. Reproduced by permission, from C. Torres and S. Ferraz Mello, *Astron. Astrophys.*, **27**, 231 (1973).

emission measures to those seen in solar plages, but are generally much brighter.

However, for a class of rapidly rotating and strongly convecting late-type stars, very interesting results have been obtained on the temperatures, dimensions, surface locations, and lifetimes of both plages and huge, dark starspots. These stars of the BY Draconis and RS Canum Venaticorum types exhibit much more intense chromospheric emissions than the sun. Their total light output is also modulated at the level of tens of percent by dark spots (Fig. 13-9). Study of these systems yields information on extremes of magnetic activity that could be helpful in understanding solar activity phenomena.

The spots on the RS CVn and BY Dra stars are found to be approximately 1000–1200 K cooler than the star's photospheric temperature of about 4600 K. Direct spectra of these starspots show the strong molecular bands of TiO expected at the spot temperature of about 3500 K. To produce the observed dips of the star's T_{eff} as seen from the Earth, such spots must then cover several tens of percent of the visible hemisphere.

Longer-term photometric variations in wide-band continuum have also been detected on spotted stars such as BY Draconis from studies of photographic plates dating back to the early 1900s. These variations occur over 40–60 years and have amplitudes similar to the starspot rotational modulation quoted above.

Study of at least certain RS CVn stars indicates that the plage regions may occupy a substantially smaller fractional area than the 25–30% that the starspot seems to cover. This is the opposite of the solar situation, where the plages typically cover several times more area than any associated spots. Also, if the emission lines of ions such as Mg II, C IV, C II, and He II originate in such a small area, they imply surface fluxes thousands of times higher than those in solar plages. X-ray studies of eclipses in certain binary systems also indicate that the coronal emission can be concentrated in an active region around the spot. The coronal emissions in such stars seem to originate in plasma loops of temperatures between 4 and 200×10^6 K and pressures orders of magnitude higher than measured in the solar corona.

Direct Zeeman broadening measurements of the field intensity in stellar active regions has been carried out for several dozen late-type stars, along with estimates of the fractional area f that they cover. The field intensities range from below 1 kG to 3–4 kG, with f between 0.1 and 0.8. These values can be compared with the solar figures of 1.5 kG and 0.02 obtained by the same technique. These results probably refer to photospheric faculae on such stars since their spot emissions contribute relatively little to the star's total light. The dimness of the spots makes a direct proof of their magnetic character elusive, although a satisfactory explanation of the rotational modulation of the stellar continuum without invoking a magnetic starspot on the surface has yet to be advanced.

Flares on late-type stars have been observed and carefully monitored in optical radiations for about 40 years, thus well before stellar activity cycles were detected. Optically detectable flares are found on peculiar dwarf stars of spectral types K and M whose spectrum exhibits unusually strong hydrogen line emission. These dKe and dMe star flares typically last for less than a minute, a much shorter time than their solar counterpart. In the largest events, an energy output of about 10^{34} ergs seems to be indicated. This is about a thousand times greater than the largest solar flares, and flare output seems to account for as much as 0.1% of the time-averaged total luminosity of certain stars such as UV Ceti. Flares are harder to observe on stars hotter than the M dwarfs because the contrast of the photospheric and chromospheric brightening decreases.

Stellar and solar flares share many characteristics. Their light curves indicate a rapid rise and slow fall. They seem to be localized on the surface and concentrated in active regions since they occur more frequently near the minimum light output of spotted stars. They also exhibit similar intensification of radio, UV, and X-ray emissions, including Type II emission indicating particle and shock wave propagation. The inherent variability of chromospheric and coronal emissions observed in Ca K and X-rays down to time scales of minutes suggests that large individual flares are only the extreme of a continuum of transient energy release events that accounts for up to a few percent of the total luminosity of highly active late-type stars.

13.3.2   Mechanisms of Stellar Activity

Studies of other late-type stars have increased our understanding of how stellar activity depends upon basic parameters, most notably on age and rotation. The observed connection between magnetic activity level and rotation can be understood from the role of (differential) rotation velocity in the dynamo mechanism outlined in Chapter 11. A further connection between rotation and age is expected from the ideas on angular momentum loss through the solar wind presented in Chapter 12.

The basic evidence is shown in Fig. 13-10, which shows plots of the rotation velocity, Ca K emission strength, and lithium abundance as a function of age for late-type main sequence stars. One of the points is derived from observations of the sun at its present age of 4.6×10^9 years. The others are based on data on younger stars in the Hyades, Pleiades, and Ursa Major clusters ranging in age down to some 5×10^7 years. The curves show the interesting result that both rotational velocity and Ca K emission seem to decrease with age t, roughly as $t^{-1/2}$. The lithium abundance also follows this decrease initially, but then falls off more rapidly.

The decrease of chromospheric emission and rotation speed is consistent with the idea that, in convective stars, the magnetic fields, chromospheres, and coronae are generated by turbulent gas motions whose magnitude is related to

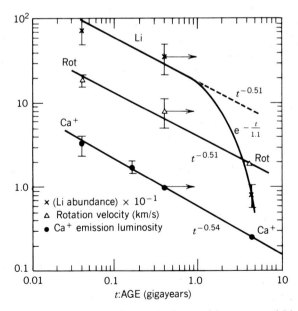

Fig. 13-10   Age dependences of the Ca II emission, rotation rate, and Li abundance for groups of stars of mass similar to the sun. Reproduced by permission from A. Skumanich, *Astrophys. J.*, **171**, 565 (1972).

the star's rotation speed, although the exact processes are not yet well understood. The hot corona also determines the rate of outflow of mass in the accompanying stellar wind, although it turns out to be the lever arm determined by the mean photospheric field intensity which sets the angular momentum loss rate from the star (Chapter 12). As the star ages and spins down (at least in its outer layers), its magnetism as measured indirectly by the Ca K chromospheric emission gradually decays. Studies of activity in binary stars have demonstrated very neatly that it is rotation, and not age per se, that determines activity level. In the binaries studied, rotation is tidally locked to orbital revolution, so the stars rotate as fast as younger single stars and exhibit similar activity levels.

The mechanism that determines a star's level of magnetic activity from its rotation velocity can be studied using recently acquired observations of Ca K rotation periods for slowly rotating stars similar to the sun. These show that the flux of chromospheric emission per unit area of the star's surface decreases with increasing period P. An even closer relation is defined in Fig. 13-11 between the same Ca K emission index normalized to the star's overall luminosity and the Rossby number $R_0 = P/\tau_c$, where τ_c is the turnover time of convective motions in the star's interior. While P is measured directly, τ_c is estimated from models of stellar convection zones. The Rossby number defines the relative importance of Coriolis and inertial forces in a flow. It can be considered a measure of the efficiency of dynamo activity in the sense that smaller R_0 values indicate greater influence of rotation upon convection, thus

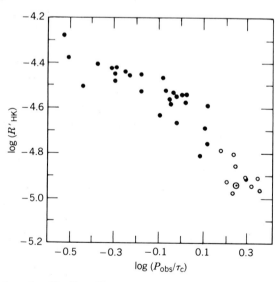

Fig. 13-11 A plot of stellar Ca II H- and K-line emission (expressed as a logarithmic fraction of the total luminosity) against the logarithm of the Rossby number. Reproduced by permission, from Noyes et al., *Astrophys. J.*, **279**, 763 (1984).

increasing the α-effect and dynamo field generation. Figure 13-11 indicates that τ_c can be chosen so that a surprisingly small dispersion is seen about a curve that indicates a decreasing level of activity with increasing Rossby number.

The closeness of the relation over an order of magnitude range in R_0 is remarkable since the stars in Fig. 13-11 represent a wide range of spectral type, age, and activity level in late-type dwarfs. It implies that for all these stars, the mechanism that determines the level of magnetic activity is similar. It also indicates that the separate mechanism that determines chromospheric emission does not change much either.

This is particularly surprising in view of a finding that is evident in Fig. 13-8, where a Ca K index is plotted against an index of the star's temperature, for a group of stars similar to those used in Fig. 13-11. A gap running from left to right is evident near the midrange of the Ca K index. This gap, named after its co-discoverers, A. Vaughn and G. Preston of Mt. Wilson Observatory, indicates that for all the late spectral types represented here there is a marked lack of stars in the intermediate range of chromospheric activity.

The gap does not seem to be a statistical fluctuation caused by a lack of stars of intermediate rotation speed in the solar neighborhood. It might indicate an abrupt change in dynamo activity between stars rotating rapidly and the slower rotators. This explanation is favored by the finding that the more rapidly rotating stars found above the gap tend to exhibit irregular time behavior of chromospheric emission, while those below exhibit more regular cycles of the kind seen on the sun. It might also indicate a change in the chromospheric heating produced in these two classes of stars. Alternatively, it could imply rapid passage of stars through the intermediate range of rotation velocities, due to increased spin-down at this stage in the evolution of their stellar winds.

The evidence for decreasing stellar activity with rotation is also supported by studies of X-ray coronal emissions. Figure 13-12 shows a plot of X-ray luminosity L_x against the observable projection of rotation velocity, $V \sin i$, about a rotation axis in the sky plane. X-ray output clearly increases with rotation speed, although the real slope of the curve is made uncertain by selection effects in both L_x and $V \sin i$. A decreasing X-ray output with increasing age has also been noted for late-type stars, particularly for the G-type stars most like the sun. For these G stars, X-ray luminosity is highest in the Pleiades, a young cluster, and the distribution of luminosities shifts to lower values for the older Hyades and for the oldest field stars such as the sun.

Rotation rate, determined by age (at least for single stars), thus seems to be the main determinant of activity levels in late-type stars. The spectral type and luminosity class seem to play a role mainly in determining the limits of the region in the H-R diagram where activity is found. Thus activity in main sequence stars is found only in spectral types later than about F5 and earlier than M5. The F5 limit is thought to correspond to the onset of convection in the main sequence. The M5 limit corresponds roughly to the onset of fully

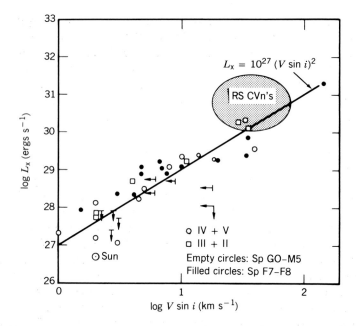

Fig. 13-12   A plot of the logarithm of soft X-ray luminosity against the logarithm of the stellar rotation rate. Symbols denote various spectral and luminosity classes of stars. The straight line indicates the relationship $L_x = 10^{27}(V \sin i)^2$. Reproduced by permission, from G. Vaiana and R. Rosner, *Ann. Rev. Astron. Astrophys.*, **23**, 428 (1985).

convective interiors. Since it is believed that dynamos in stars like the sun might operate most efficiently in the boundary layer between the convection zone and the radiative core (see Section 11.3.2), a fully convective star might have little or no dynamo activity.

The evidence outlined above supports the view that magnetic activity in stars similar to the sun is the product of the convection found in these stars together with their rotation. Acting together, these two intrinsic characteristics of the star produce magnetic fields, possibly through some form of dynamo activity as described in Chapter 11. The heating of chromospheres and coronae by a combination of fluid turbulence and magnetic dissipation produces a stellar wind which progressively slows the star's rotation.

The activity correlations with rotation and age, and the limits set by spectral type, define this scenario. More information on a larger sample of stars should help to study the many questions that remain on the nature of the magnetic field generation process. For instance, much interesting information is being compiled on the relations among stellar rotation periods and activity cycle lengths and amplitudes.

The stellar observations could also prove helpful in constraining the mechanisms responsible for heating the chromospheres, coronae, and winds of late-type stars. Solar observations show that the chromospheric and coronal plasmas are denser and radiate more in regions of higher average field strength, such as the network spicules and active region plages. However, they do not really discriminate between heating caused by dissipation of magnetic energy and heating by essentially acoustic shocks channeled along magnetic field lines.

The stellar observations that show strong chromospheric and coronal emissions over a wide spectral range on the main sequence also do not clearly discriminate between ohmic and shock heating. However, magnetic energy dissipation provides the most reasonable explanation of the large differences in chromospheric and coronal activity levels observed between stars of similar surface temperature, in correlation with their different rotation rates.

Photometric investigation of starspots and stellar plages enables us to study how such structures influence a star's total radiative outputs on longer time scales than are accessible in the record of precise solar radiometry. For instance, it has been found that younger sunlike stars become dimmer as their activity level increases over the star's quasi 11-year cycle. As discussed below this is the opposite of behavior observed in the present sun. The difference may indicate that when the sun was a few billion years younger, rotating faster, and more magnetically active, its dynamo favored the formation of spots rather than faculae.

13.3.3 The Sun's Variable Outputs

The variability of the sun's total luminosity is measured by radiometers whose receivers are blackened cones with closely similar absorptivity of solar radiations from at least 0.2 μm in the UV to beyond 10 μm in the IR. Since this spectral range includes more than 99.99% of the solar power output, variations in the radiative heating of such a blackened cone flown above the Earth's atmosphere can be taken to indicate changes in the total solar irradiance, S. Some variations in S due mainly to passage of dark spots across the sun's disk are shown in Fig. 13-13. Peak-to-peak fluctuations recorded over time scales of about a week can be as large as 0.2% during times of high solar activity. The missing heat flux is most likely stored in the convection zone, causing it to heat up and expand slightly over long time scales (see Section 8.2.2). Thus the irradiance changes can be interpreted as solar luminosity variations, not compensated by solar radiation variations in other directions away from the sun–Earth vector.

However, the sun does not get dimmer at solar activity maximum as might be expected if spots were the only influence on the total irradiance. The variation of S caused by photospheric activity on time scales of several years (Fig. 13-14) has now been reliably detected by simultaneous measurements

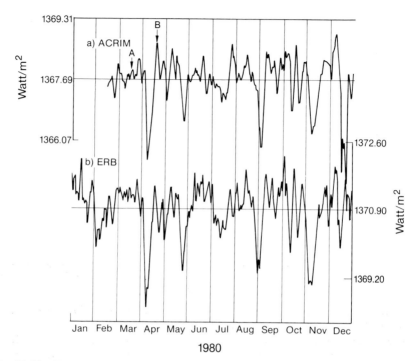

Fig. 13-13 Two plots of total solar irradiance variations measured simultaneously by radiometers on separate satellites during 1980. The large dips are caused by sunspots. The higher absolute value measured by the ERB radiometer (lower panel) relative to the ACRIM instrument (upper panel) indicates the absolute calibration uncertainty in total irradiance measurements of roughly 0.2%. From P. Foukal and J. Lean, *Astrophys. J.*, **328**, 347 (1988).

from radiometers on separate spacecraft. Over months and years the level of total solar irradiance seems to be determined by the continuum radiations from the bright faculae, whose total area greatly exceeds that of spots, although their photometric contrast is lower. The increasing facular area with increasing activity level appears to more than compensate for the sunspot blocking of heat flow to the photosphere, which probably explains why the sun is typically about 0.05% brighter at activity maximum than at minimum. These irradiance variations also seem to represent slow changes in total solar luminosity, caused by the influences of magnetic activity on photospheric heat flow.

It is quite likely that even slower variations of S occur (not necessarily in synchronism with photospheric activity), but radiometry with the required long-term repeatability has not been available for long enough to properly investigate their existence. Possible causes for such variations might be slow changes in solar convective efficiency due to variations in deep-seated convec-

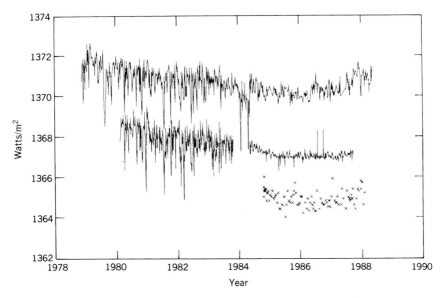

Fig. 13-14 Plots of total solar irradiance variation over the period 1980–1988, as measured by the ERB radiometer on the *Nimbus* 7 satellite (above), the ACRIM radiometer on the Solar Maximum Mission spacecraft (middle), and by the ERBS satellite radiometer (below). Note the slow decrease from the peak of cycle 21 around 1981 to mid-1986 (when activity minimum recurred) and then the gradual increase. Data by courtesy of J. Hickey, R. Willson, and R. Lee, respectively.

tive patterns. Changing magnetic field patterns deep within the convection zone may play a role, but relatively straightforward calculations indicate that they should be less important than the photospheric influences of spots, faculae, and network.

The spectral dependence of the total irradiance changes discussed above is below the 1% accuracy so far achieved in absolute measurements of the solar spectral irradiance in the infrared and visible. But we expect them to behave roughly as black-body variations in T_{eff}, i.e., proportionately smaller in the IR and increasing into the blue. The sun's output variability in the UV and EUV regions between about 3500 Å and 100 Å has been studied in more detail. Figure 13-15 shows the short-term (solar rotational variability) in several UV passbands. Variations of about 1% in the continuum near 2800 Å increasing to more than 25% in the EUV lines below 1500 Å are observed, caused mainly by rotation across the disk of bright plages. Changes in intensity of these plages produce higher-frequency variations on time scales down to minutes. The most prominent of these are due to flares, which can increase the solar UV output by up to 20% at Lyman α and by up to a factor of 10 in coronal EUV lines.

The more interesting issue of the sun's UV variability over the solar cycle remains largely unanswered because of the difficulties in maintaining the

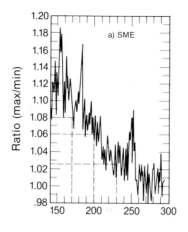

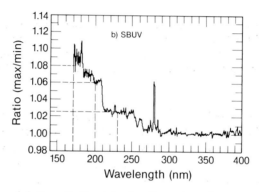

Fig. 13-15 The modulation of ultraviolet irradiance by solar activity caused by solar rotation over the wavelength range 150–400 nm (1 nanometer = 10 angstroms). The amplitude of the modulation is given as a ratio of maximum to minimum intensity. The top panel is based on data from the SME spacecraft, the lower panel from the SBUV experiment on the *Nimbus* 7 satellite. Both data sets refer to observations in July and August 1982. Reproduced by permission, from J. Lean, *J. Geophys. Res.*, **92**, 839 (1987).

photometric calibrations of UV sensors in space for years. The best measurements are shown in Fig. 13-16. They indicate perhaps a factor-of-2 modulation over the cycle in the EUV, decreasing rapidly with increasing wavelength, to less than 10% at 2000 Å.

The same plage area variations that seem to account for much of the solar UV variation also produce changes in the sun's radio emissions. These are conveniently described in terms of three components. The minimum level of RF flux at any wavelength is determined by the so-called quiet-sun component, whose flux density decreases from millimeter to decameter wavelengths

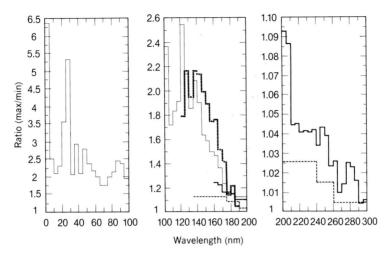

Fig. 13-16 Observational estimates of the ratio of maximum-to-minimum ultraviolet irradiance below 3000 Å during solar cycle 21, which peaked around 1979. The curves correspond to measurements from three spacecraft and also from a set of rocket experiments. Reproduced by permission, from J. Lean, *J. Geophys. Res.*, **92**, 839 (1987).

(Fig. 13-17). Superposed on this constant signal is the so-called gradual or slowly varying (S-component) emission from plages, which changes on time scales of days and weeks. Its flux is generally below that of the quiet component and peaks at centimeter wavelengths, as seen also in Fig. 13-17.

Noise storms (often referred to as Type I storms) consist of many short-duration bursts (about 1-s) occurring over hours or days. Most complex of the radio emissions are the transient bursts associated with flares and filament eruptions. Their flux can far exceed the quiet and S-component, but they are relatively rare and their time-integrated contribution is small. The mechanisms of solar radio bursts are discussed in Chapter 10.

In the X-ray region, the variability on all time scales is even much more pronounced than in the UV and EUV. At around 10 Å, the modulation on time scales of days due to active region plages is about a factor of 10. It reaches a factor of 100 or more during moderate-sized flares. The change of the X-ray flux over the solar cycle is easy to measure since the flux is between a factor of 500 and 1,000 times higher at solar maximum than at minimum, integrated over the 1–8 Å wavelength range.

The sun's fluxes of plasma and fields as sensed by spacecraft at the Earth's orbit vary rapidly during the passage of interplanetary shocks and their associated sudden commencements of geomagnetic activity. However, time variations in the sun's global outputs of plasmas and fields are difficult to reconstruct from observations in the Earth's vicinity alone, given the complex latitudinal, longitudinal, and radial structure of the heliosphere (Chapter 12).

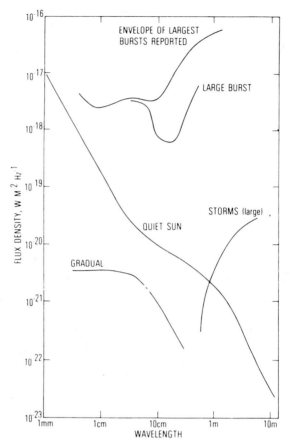

Fig. 13-17 Logarithmic plot of the flux density spectra of solar radio bursts. Reproduced by permission, from F. Shimabakuro, in *The Solar Output and Its Variation*, O. White (Ed.), Colorado University Press, p. 133, 1977.

Slow variations in solar wind structure over the activity cycle can be detected, although the data mainly pertain to the last two 11-year cycles and need to be checked over a longer time base. The best-established 11-year variations seem to appear in the increased incidence of interplanetary shocks with increasing sunspot number and the increasing incidence of high-speed wind streams with declining activity after spot maximum. How these changes relate to global changes in mass and momentum fluxes over the cycle is still uncertain.

Variation of energetic particle fluxes from the sun is closely related to the incidence of flares. The output of protons below about 1 MeV is relatively continuous. But more energetic, relativistic protons of greater than 10-MeV

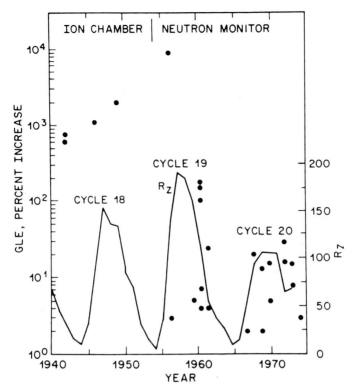

Fig. 13-18 Magnitudes of ground-level solar cosmic ray events (filled circles) between their discovery in 1942 and 1975. The solid line gives the annual mean sunspot number. From L. Lanzerotti, in *The Solar Output and Its Variation*, O. White (Ed.), Colorado University Press, 1977.

energy are correlated with flares, although the mechanism of their acceleration is insufficiently understood (Chapter 10) to predict which flares will produce them.

Given this association with flares, it is not surprising to find that the sun's energetic particle outputs are well correlated with the sunspot number (Fig. 13-18). In addition to this positive correlation between the occurrence of solar protons and spot number, increasing activity levels produce a decrease in the flux of galactic cosmic rays within the heliosphere (Chapter 12). The sudden decrease in galactic cosmic rays observed to follow a large flare, known as the Forbush effect, seems also to be caused by restructuring of the interplanetary fields. Over the solar cycle such a change in the heliospheric shielding against cosmic rays manifests itself as an anticorrelation between spot number and galactic cosmic ray primaries. The magnitude of this modulation is about a factor of 2 for 500-MeV protons, but decreases to a few percent for the most energetic cosmic rays of 10 GeV or more.

13.3.4 Prediction of Solar Activity

A practical need exists for prediction of solar activity on time scales ranging from a few hours to an 11-year cycle or more. Since 1928, when the first broadcasts of geomagnetic data were made from Paris, warning centers have been developed by many countries, and an intentionally coordinated service, the URSIGRAMS, has been provided to assist in telecommunications and other areas where solar activity causes disruptions.

One application of these predictions lies in the choice of radio communication frequencies, both in traditional ionosphere-reflected communications and also in the trans-ionospheric transmission employed in satellite links, where ionospheric turbulence plays a role. Another important application exists in satellite operations where changes in solar EUV outputs determine the upper atmospheric pressure scale height and thus the orbital drag. Estimates of satellite lifetime thus rely on predictions of solar activity levels many years in advance. Predictions of geomagnetic activity are also of importance in operation of large-scale electric power systems at high latitudes since such systems are susceptible to power surges induced by geomagnetic storms. Such currents are also induced in oil and gas pipeline systems at high latitudes, where they cause corrosion and interfere with attempts to develop anticorrosion safeguards. A separate area of application lies in the prediction of large flares whose associated energetic protons pose a hazard to astronauts and even to airline passengers on flights over the polar regions.

Solar activity can be usefully predicted on several distinct time scales. The shortest is a few hours to about two weeks, which covers the flaring and gradual evolution of individual active regions present on the visible disk. Activity levels can also be predicted one or more 27-day solar rotation periods in advance given the recurrence of the long-lived large active regions, active complexes, and coronal holes. Significant enhancements of flare activity are also noted on time scales of several (three to five or more) rotations. These enhancements have been related to specific longitude strips where activity seems to cluster (see Section 8.4.3). The reason for the periodic enhancements of flaring within these active complexes is quite unknown and deserves close attention.

On even longer time scales, the 11-year and 22-year cycles offer relatively stable predictions of the levels and polarities of activity. The envelope of solar-cycle amplitudes provided by the Gleissberg cycle period even offers some basis for prediction of gross activity levels several cycles ahead. However, the likely recurrence of a Maunder Minimum type of depression in solar activity (which is overdue, see Chapter 11) makes prediction on such longer time scales quite uncertain.

The prediction of flares and filament eruptions is the main aim of the short-term predictions on active region evolution since both types of events can produce the geomagnetic variations and energetic protons that are of the greatest practical concern. The empirical and physical bases of such predictions rest on observations of the geometry of the active region, such as its magnetic complexity (and thus its magnetic field gradients), and its twisted-ness as measured by the rotation of its magnetic neutral line out of the east-west direction. Its previous record for producing flares is of over-riding importance, since the success rate in predicting the first flare from a given region is markedly lower than for subsequent flares.

The evolution of the active region's plage area and intensity is also of interest as an index of potential flare activity. In addition, it provides the main basis for predictions of X-ray, EUV, and UV radiations, which are dominated by the sum of the active region contributions.

The need for prediction of the level of solar activity several years in advance has led to a wide range of methods for determining the amplitude and epoch of the next solar activity maximum and, to a lesser extent, the shape of the cycle. The basis for these predictions lies mainly in empirical relationships. The dynamo theory of the solar cycle described in Chapter 11 is far from being well enough specified to have much usefulness in predicting from one cycle to the next.

Two basic approaches are used most often. In one, the past sunspot record is Fourier analyzed. The power spectrum is then used to form a Fourier series to estimate future solar cycle behavior. An alternative approach is to use specific properties of one cycle to predict the next. Statistical relations have been found, for instance, between increasing skewness of the nth spot cycle and an increasing ratio of its amplitude to that of the next ($n + 1$), cycle.

One relation of this kind is based on a correlation between the minimum value of geomagnetic activity in a given 11-year cycle and the height of the next cycle. The regression found was

$$R_M = 1.845M_{min} + 66.3, \qquad (13\text{-}8)$$

where R_M is the spot number at maximum of cycle $n + 1$ and M_{min} is the minimum average annual value of index $M = 10 \Sigma(K_p - 10)$, where K_p is the daily geomagnetic activity index. Using this index, the maximum of cycle 21 was predicted (in 1979) to be about 185, based on data obtained for the previous cycle.

Empirical correlations of this latter type seem to have a physical basis in that the strength of a cycle is expected, from the Babcock model, to depend on features of the previous cycle's declining phase. From the discussion in Chapter 11 one might expect, for instance, that the strength and extent of the sun's polar fields measured in the preceding cycle might play a role in determining the amplitude of the next maximum of solar activity. As another example, connections found between the product of the number of active region eruptions in the declining phase and their mean lifetime on one hand and the amplitude of activity found in the next cycle, might be understood at least qualitatively in the context of the Babcock model.

ADDITIONAL READING

Background in Stellar Astrophysics and Star Formation

A. Unsold, *The New Cosmos*, Springer-Verlag, 1967.

F. Shu, F. Adams, S. Lizano, "Star Formation in Molecular Clouds," *Ann. Rev. Astron. Astrophys.*, **25**, 23 (1987).

Solar-Stellar Connections

R. Bonnet and A. Dupree (Eds.), *Solar Phenomena in Stars and Stellar Systems*, D. Reidel, 1981.

L. Hartmann and R. Noyes, "Rotation and Activity in Main Sequence Stars," *Ann. Rev. Astron. Astrophys.*, **25**, 271 (1987).

S. Baliunas and A. Vaughan, "Stellar Activity Cycles," *Ann. Rev. Astron. Astrophys.*, **23**, 379 (1985).

R. Rosner, L. Golub, and G. Vaiana, "On Stellar X-ray Emission," *Ann. Rev. Astron. Astrophys.*, **25**, 413 (1987).

Solar Variability

O. White (Ed.), *The Solar Output and Its Variation*, Colorado University Press, 1977.

J. Eddy, "The Historical Record of Solar Activity," in *The Ancient Sun*, R. Pepin, J. Eddy, and R. Merrill (Eds.), Pergamon, p. 119, 1980.

G. Newkirk, "Solar Variability on Time Scales of 10^5 to $10^{9.6}$ yrs," in *The Ancient Sun*, R. Pepin, J. Eddy, and R. Merrill (Eds.), Pergamon, p. 293, 1980.

P. Foukal, "Physical Interpretation of Variations in Total Solar Irradiance," *J. Geophys. Res.*, **92**, 801 (1987).

J. Lean, "Solar Ultraviolet Variations: A Review," *J. Geophys. Res.*, **92**, 839 (1987).

Prediction of Solar Activity

P. Simon, G. Heckman, and M. Shea (Eds.), "Solar-Terrestrial Predictions," U.S. Department of Commerce Publication (1986).

EXERCISES

1. Calculate the seasonal variation in total irradiance S caused by variation in the changing Earth–sun distance due to the eccentricity value of 0.0016 of the Earth's orbit. Compare this to the changes in S caused by photospheric activity, over time scales between weeks and decades.

2. Derive equation (13-7), and show that the sun is unlikely to have contracted alone from a single gas cloud. Also, estimate the magnitude of the sun's internal magnetic field if the energy density of an original interstellar field

of $B \sim 10^{-6}$ G were increased in the same proportion as the thermal energy density of the plasmas that formed the sun's interior.

3. Assuming that the cloud which contracted to form the sun shared the mean angular velocity of stars presently in the sun's neighborhood (take the spin velocity of this neighborhood as $v_s \sim 10^{-2} R$ km s^{-1} pc^{-1}, where R is the radial distance from the center of this spinning system), calculate the sun's main sequence rotation rate assuming no loss of angular momentum during its contraction from a cloud of radius 10 pc and mass $10,000 M_{\odot}$.

4. Construct a time line (on a logarithmic time scale) of the sun's evolution from a gas cloud to helium burning, labeling the key phases, and give the fraction of the sun's lifetime spent in each phase.

5. Calculate the decrease in total luminosity of a star after a spot forms, assuming the spot is about 1000 K cooler than the star's 4500 K photosphere and covers 10% of the total photospheric area. Where might the missing energy be stored?

6. Derive the expression used here for the Rossby number by taking the ratio of the inertial term $\rho v \cdot \nabla v$ [e.g., in equations (4-24) or (7-20)] to the Coriolis force $2\rho\omega \times v$ on a convective cell of dimension L, velocity v in a star of radius R, rotating at angular velocity ω (see Section 7.3.7). Also estimate the Rossby numbers for solar granulation, supergranulation, and for a possible deep solar convection of characteristic speed less than 10 m s^{-1}. Comment on the relative effectiveness of these convective scales in contributing to a solar dynamo.

7. Use equation (4-94) to show that the gyro radius of galactic cosmic ray protons of energy 200 MeV in the heliospheric magnetic field of magnetic intensity $B \sim 10^{-5}$ G is of order 6×10^5 km. Compare this radius to the heliospheric dimensions, and comment on the efficiency of the calculated shielding. The observed modulation at these energies is a few tens of percent between solar activity maximum and minimum.

Index

Page numbers appearing in *italics* indicate the primary references. When there are no italics, the first page listed will be the primary reference.